Handbook of Chitin and Chitosan: Properties and Preparation

Handbook of Chitin and Chitosan: Properties and Preparation

Editor: Jute Cassie

STATES
ACADEMIC PRESS
www.statesacademicpress.com

States Academic Press,
109 South 5th Street,
Brooklyn, NY 11249, USA

Visit us on the World Wide Web at:
www.statesacademicpress.com

ISBN: 978-1-63989-253-2 (Hardback)

Cataloging-in-Publication Data

Handbook of chitin and chitosan : properties and preparation / edited by Jute Cassie.
p. cm.
Includes bibliographical references and index.
ISBN 978-1-63989-253-2
1. Chitin. 2. Chitosan. 3. Polysaccharides. 4. Carbohydrates. 5. Chemistry. I. Cassie, Jute.
QD321 .H36 2022
547.78--dc23

Table of Contents

Preface

Chitin is a polysaccharide which is a long chain polymer of N-acetylglucosamine. It is found in the cell walls of fungi and exoskeletons of arthropods such as insects, crustaceans, cephalopod beaks and scales of fish. It has a structure comparable to cellulose. Chitosan is formed by the treatment of chitin cells of crustaceans with an alkaline substance like sodium hydroxide. It is a linear polysaccharide which consists of randomly distributed β-(1→4) linked D-glucosamine and N-acetyl-D-glucosamine. Chitin is used to create edible films in food industry. It is also used as an additive to stabilize and thicken foods and food emulsions. Chitosan is used as a pesticide to fight plant fungal infections. It also finds applications in beverage industry as fining agent and as an antibacterial agent in bandages to help reduce bleeding. This book outlines the properties and preparation of chitin and chitosan in detail. The topics included herein on chitin and chitosan are of utmost significance and bound to provide incredible insights to readers. This book is a vital tool for all researching or studying chitin and chitosan as it gives incredible insights into emerging trends and concepts.

This book is the end result of constructive efforts and intensive research done by experts in this field. The aim of this book is to enlighten the readers with recent information in this area of research. The information provided in this profound book would serve as a valuable reference to students and researchers in this field.

At the end, I would like to thank all the authors for devoting their precious time and providing their valuable contribution to this book. I would also like to express my gratitude to my fellow colleagues who encouraged me throughout the process.

Editor

1

A Review of Chitosan-Based Materials for the Removal of Organic Pollution from Water and Bioaugmentation

Carlos Escudero-Oñate and Elena Martínez-Francés

Abstract

Chitin is a natural polymer extracted mostly from shrimp or crab shells and is the Earth's second most abundant polysaccharide. After a simple deacetylation procedure, chitin is converted into chitosan that consists in a polysaccharide structure of deacetylated-β-glucosamine. Chitosan has been largely employed in wastewater treatment the removal of colloids through coagulation-flocculation processes. Different chitosan based materials have been produced and tested in the removal of inorganic pollutants such as toxic metals and metalloids, nutrients, dyes, micropollutants and hydrocarbons. Sorbents such as magnetic-activated carbon chitosan have been successfully tested in the removal of antibiotics (ciprofloxacin, erythromycin and amoxicillin) from water. Raw chitosan and ZnO nanoparticles entrapped in chitosan have demonstrated an excellent potential for the removal of the insecticide permethrin from aqueous effluents. Chitin and chitosan in flake and powder form have also demonstrated a promising effectiveness in the removal of oil spilled in seawater. Superhydrophobic and superoleophilic sponges modified by thioles have been also prepared from chitosan and used for the removal of oil spills. Chitosan hydrogels have been tested as well as entrapment matrices for the immobilization of hydrocarbon-degrading biomass for oil spills. Strains such as *R. corynebacteriorides* (QBTo), *Bacillus subtilis* LAMI008 and *B. pumilus* have been successfully immobilized and employed in hydrocarbon degradation processes. In this book chapter, the use of chitosan and chitosan-based materials in the removal of organic pollutants from water is reviewed.

Keywords: chitin, chitosan, adsorption, organic pollutants, sorbents, oil spill pollution, bioaugmentation, water treatment

1. Introduction

Chitin is a natural polysaccharide and the second most abundant biopolymer on Earth after cellulose [1]. This biopolymer consists of units of β-(1–4)-N-acetyl-D-glucosamine and is the main component of the exoskeleton of arthropods and crustaceans but also can be found in relevant amount in the cell walls of fungi. Chitin is a structural biopolymer whose role is analogous to that of collagen in the higher animals and cellulose in terrestrial plants [2]. In a similar way, plants produce cellulose in their cell walls, and insects and crustaceans synthetize chitin and accumulate it in their shells. Chitin may be regarded as cellulose with hydroxyl at position C2 replaced by an acetamide group (-$CONH_2$). This similarity partly explains some analogies occurring in chitin and cellulose, such as low solubility and low chemical reactivity [3]. Chitin is found in three polymeric forms, α-, β- and γ-chitin, usually found in shrimp and crab shells, squid pen and stomach cuticles of cephalopod, respectively. From the three aforementioned forms of the biopolymer, α-chitin is the most abundant and stable form. α-, β- and γ-chitin correspond to antiparallel, parallel and alternated arrangements of polymer chains, respectively. A hydrogen bond between the acetamide group on the C2 carbons and the secondary alcoholic hydroxyl groups on the C3 carbon is linked through a water molecule with the primary alcoholic hydroxyl groups on a C6 carbon. As a result of this configuration, chitin possesses a strong crystalline structure, which explains the high chemical and solvent stability of the biopolymer. Due to its crystalline structure, chitin exhibits remarkable differences from cellulose in the solubility and reactivity despite of the relatively similar chemical structure [3].

The production of chitin uses basic raw materials of the cuticles of various crustaceans, principally crabs and shrimps. In regular fishery wastes, the biopolymer chitin is associated with proteins, minerals, lipids and pigments [4]. All these substances are considered impurities, and they all have to be quantitatively removed to achieve the required purity of the chitin. The chitin is normally extracted from the carapaces from crustaceans treating the crushed material with acid to achieve complete dissolution of the calcium carbonate structure. After this process, the material is submitted to an alkaline extraction to achieve the solubilization of the proteins. In a later purification step, the material obtained from the deproteinization process follows a decolorization step to remove residues of pigments to yield an almost colorless product [5]. Partial deacetylation of chitin leads to the formation of the polymer chitosan, consisting of units poly(D-glucosamine). A scheme of the production of chitosan from waste crustacean shells following chemical and biological approaches is presented in **Figure 1**.

When the degree of deacetylation of chitin reaches about 50%, the material obtained starts becoming soluble in aqueous acidic media and is called chitosan [5]. The degree of deacetylation is indicative of the amount of amino groups (-NH_2) along the chitosan chain and refers to the degree of removal of acetyl groups (-$COCH_3$) from the amido moieties. The degree of deacetylation and the degree of polymerization (DP), which in turn decides molecular weight of polymer, are two important parameters dictating the use of chitosans for various applications [1]. A scheme of the deacetylation reaction that converts raw chitin into chitosan is presented in **Figure 2**.

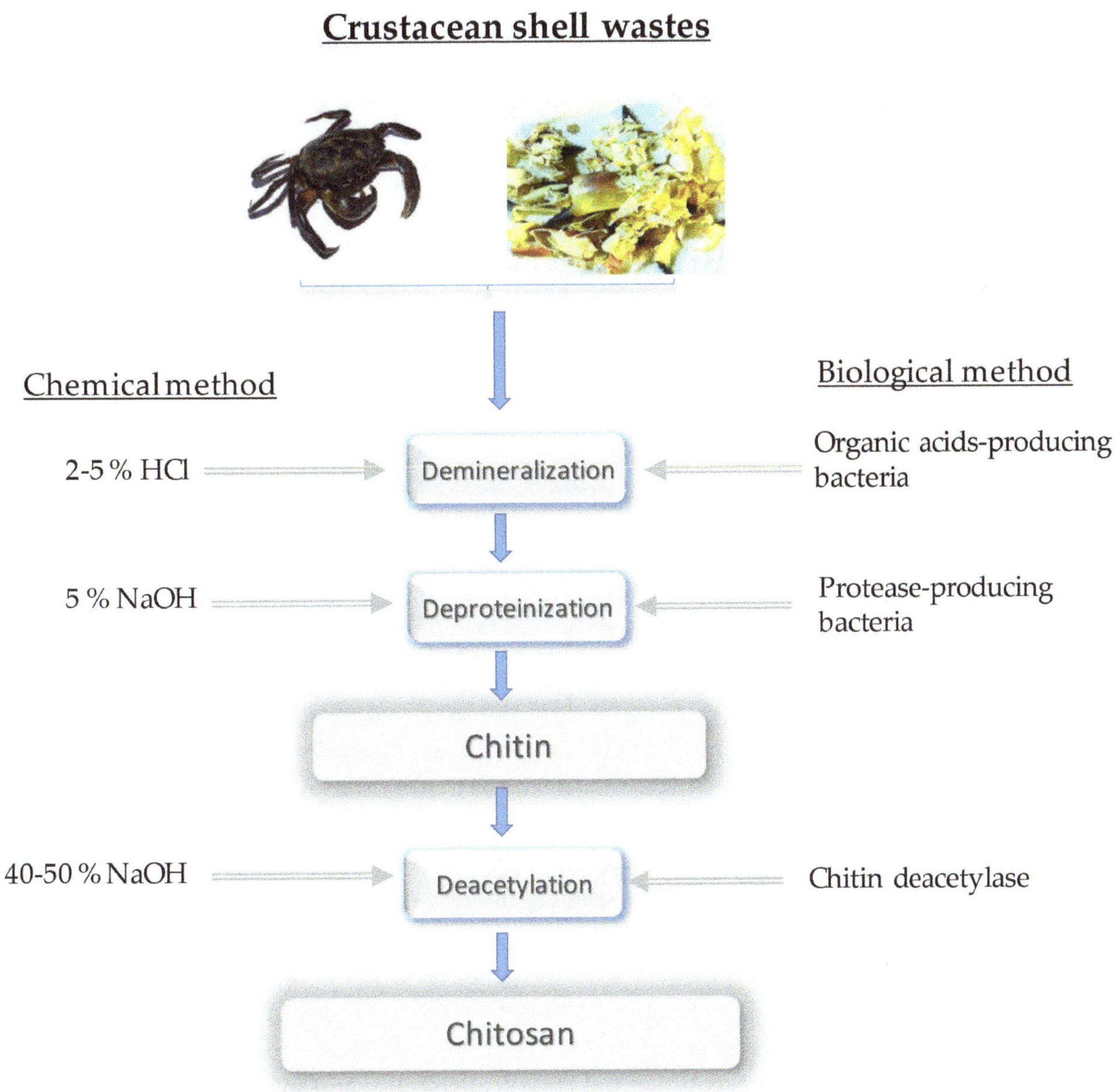

Figure 1. Production of chitin and chitosan using chemical and biological methods (adapted from Jo and co-authors [6]).

Chitosan is insoluble in water, alkali and organic solvents but soluble in most solutions of organic acids when the pH of the solution is lower than 6. Acetic and formic are two of the most widely used acids employed to solubilize chitosan. Some dilute inorganic acids such as nitric acid (HNO_3), hydrochloric acid (HCl), perchloric acid ($HClO_4$) and phosphoric acid (H_3PO_4) can also be used to prepare chitosan solutions but only after prolonged stirring and warming [7]. At low pH chitosan remains as a polycationic species, due to protonation of the amino group according to **Figure 3** [8].

Chitosan exhibits a unique set of properties that makes this polymer a great candidate for the development of water treatment processes. Among them, the most relevant are its high biodegradability, low toxicity, low price and natural availability. The weaknesses exhibited by

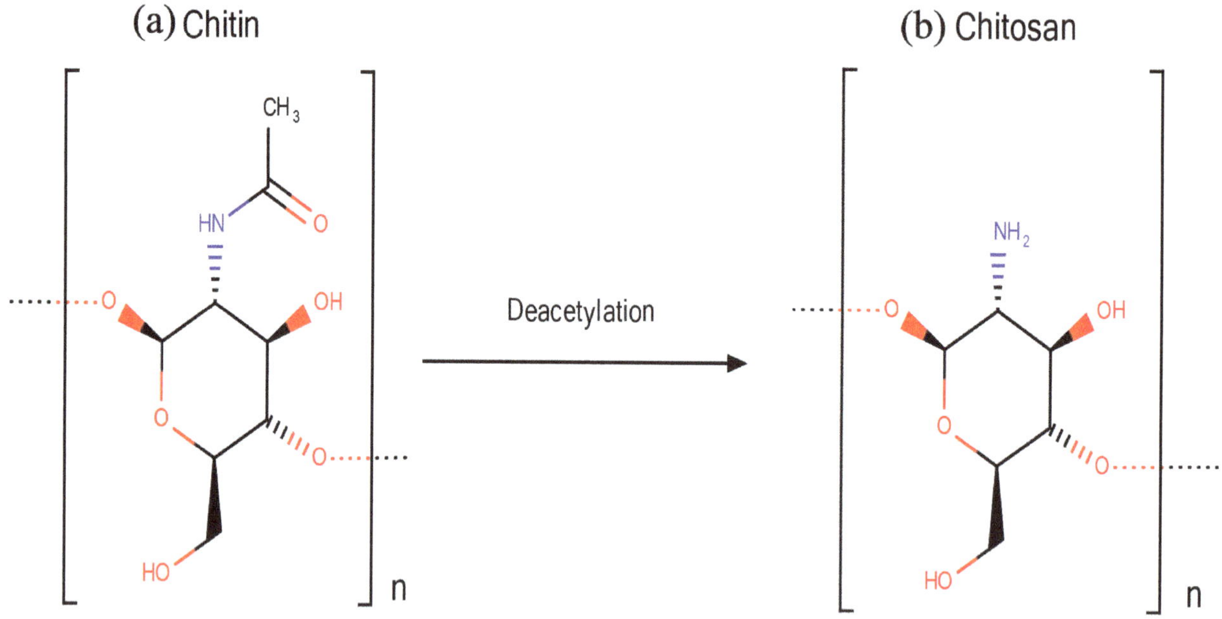

Figure 2. Chemical structure of (a) chitin and (b) chitosan.

this biopolymer however derive from its low acid stability, poor mechanical properties, low thermal stability, resistance to mass transfer, low porosity and surface areas [9, 10]. In order to overcome the drawbacks exhibited by chitosan, a large amount of effort has been devoted to the development of physicochemical modification methods to include different types of functionalization in the polymer. Chemical modifications such as oligomerization, alkylation, acylation, quaternization, hydroxyalkylation, carboxyalkylation, thiolation, sulfation, phosphorylation, enzymatic modifications and graft copolymerization have been carried out, allowing obtaining modified properties for specific end used applications in a large variety of fields [1]. Following physical and chemical methods, researchers have managed to produce a variety of forms of chitosan such as gel beads, membranes, film, fibers, porous frameworks, powders, sponges, hollow fibres and nanoparticles [11–20]. Some chemical modifications have been found capable of enhancing its flexibility and chemical stability and lower its susceptibility to acidic media. Among the most widely employed modifications that contribute modulation of

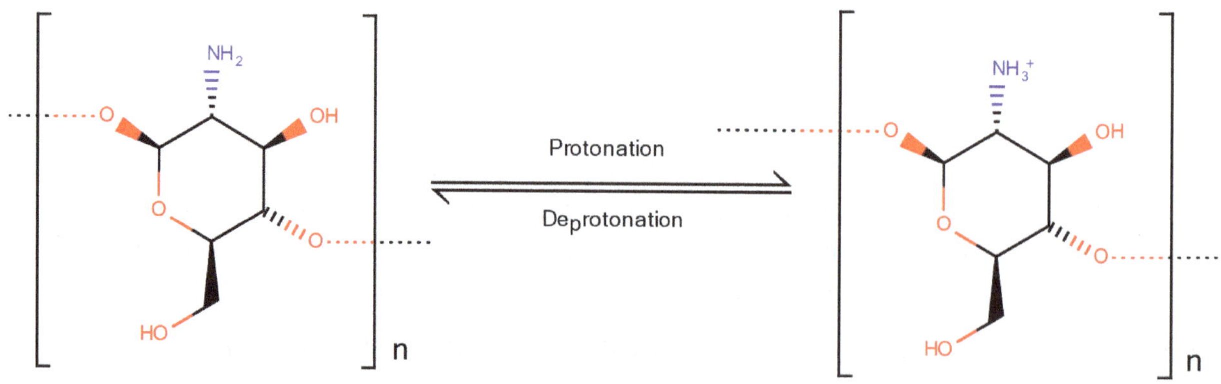

Figure 3. Protonation and deprotonation equilibrium of chitosan.

the functional properties of chitosan while improving the mechanical strength, cross-linking might be found. Several reagents such as glyoxal, glutaraldehyde and epichlorohydrin have been employed to reinforce the structure of chitosan by cross-linking [19, 21–23].

2. Chitosan-based systems for the removal of organic pollutants

Chitosan has revealed a large potential in the detoxification of polluted effluents. This biopolymer on itself and chitosan-based materials have not only shown a high capacity to remove a variety of toxic metals such as Cu(II) [24–27], Pb(II) [28, 29], Cr(VI) [30, 31], As(V) [32], Mo(VI) [33] and Hg(II) [34] but also have demonstrated a large potential to remove other concerning inorganic species from water such as nutrients (NO_3^-, NO_2^- and PO_4^{-3}) [35, 36]. In addition to the abatement of inorganic pollution from water, this biopolymer has been also explored as sorbent against organic pollution. In the next sections, we review the use of raw chitosan and chitosan-based materials in the removal of micropollutants and in the abatement of hydrocarbon pollution associated to oil spills.

2.1. Chitosan-based materials for the removal of organic pollutants

In recent years, chitosan-based composites using metals [37], metal oxides [38] and bimetals [39] have been receiving a large attention as alternative sorbents in water treatment processes. These kinds of materials have been chosen primarily due to their high adsorption capability [40]. Arayne and co-workers in 2011 studied the potential of raw chitosan beads and chitosan beads modified with ZnO nanoparticles (Cs/ZnO NPs) to remove permethrin, an insecticide largely employed in agriculture [41]. Through this study, the authors demonstrated that chitosan beads have an excellent adsorption performance and that the removal efficiency from a 0.1 ppm solution of permethrin increased from 49% when using chitosan beads to 99% when using Cs/ZnO NPs. Despite chitosan demonstrated a high capacity on itself to remove permethrin, including Cs/ZnO NPs in the matrix of the biopolymer, provided a large enhancement that leads to the almost total removal of this pollutant [42]. Saifuddin and co-authors, also in 2011, reported the use of a composite system based on cross-linked chitosan-silver nanoparticles (Cs-Ag NPs) to remove the pesticide atrazine (a very persistant herbicide of the triazine class) [43]. In their study, the kinetics of removal of the pesticide by the developed sorbent was evaluated. The authors reported an equilibrium time at about 65 min for 1, 5, 10, 20 and 25 ppm of atrazine. During their experiments, the authors detected a remarkable increase in the reduction of the pesticide content in water when increasing the sorbent dose. At the dosage of 2.0 g/L of Cs-AgNPs composite, 98% of the initial concentration of atrazine (from 1 ppm solution) was removed [43].

Danalıoğlu and co-authors, in a recent study, developed and tested a novel adsorbent based on a composite magnetic-activated carbon/chitosan system (MACC) for the removal of three widely employed antibiotics: ciprofloxacin, erythromycin and amoxicillin [44]. In their study, MACC nanocomposite demonstrated a good adsorption performance towards the targeted antibiotics [44]. Herein, the initial concentrations of antibiotics tested were 15, 60 and 60 ppm for ciprofloxacin, erythromycin and amoxicillin, respectively. To the initial antibiotic solution,

1 mg of MACC adsorbent was added. As a result, adsorption took place rapidly during the first 30 min, and then the adsorption rate slowed down to reach the equilibrium at about 120 min. The authors performed a set of equilibrium experiments and, by means of the Langmuir isotherm model, managed to calculate the maximum sorption capacity of the material for the different antibiotics (90.01 mg/g for ciprofloxacin, 178.57 mg/g for erythromycin and 526.31 mg/g for amoxicillin [44]). Danalıoğlu et al. also compared the adsorption capacity of MACC for ciprofloxacin to that reported for magnetic alginate-Fe_3O_4 hydrogel fiber and graphene oxide/calcium alginate [44]. Magnetic alginate-Fe_3O_4 hydrogel presented an adsorption capacity for ciprofloxacin ranging from 153 to 555 μg/g [45], and graphene oxide/calcium alginate varied from 18.45 to 39.06 mg/g [46].

In addition to the studies described above, other researchers have reported the removal of phenol and o-chlorophenol using chitosan beads modified with sodium alginate and calcium chloride [47]. These authors reported that such a modification improved the stability of the obtained material as well as the sorption capacity of the beads. The maximum sorption capacity for phenol reported by the authors in this study was 108.69 mg/g, while for o-chlorophenol was 97.08 mg/g. Lie and colleagues investigated the use of raw chitosan, chitosan chemically modified with salicylaldehyde (CS-SA), β-cyclodextrin (CS-CD) and a cross-linked β-cyclodextrin polymer (EPI-CD) in the removal of phenol, p-nitrophenol and p-chlorophenol from aqueous solution [48]. In their study, it was observed that the adsorption capacity of unmodified chitosan for phenol was remarkably lower than that observed for the modified biopolymer. While the chitosan modified by CS-CA was able to achieve a capacity of 8.50, 20.49 and 44.92 mg/g for phenol, p-chlorophenol and p-nitrophenol, respectively, the raw, unmodified chitosan was only able to barely remove about 2 mg/g of the substances. On the other hand, the sorption capacity of chitosan chemically modified by CS-CD was 34.93, 179.73 and 20.562 mg/g for phenol, p-chlorophenol and p-nitrophenol. The last modification by EPI-CD led to the sorption capacities of 131.50 mg/g (phenol), 74.25 mg/g (p-chlorophenol) and 41.11 mg/g (p-nitrophenol) [49].

The removal of phthalate esters (PAEs) [50] by molybdate-impregnated chitosan beads (MICB) in an aqueous solution has also been reported by Chen and co-workers in 2007. The experiments performed by the authors indicated that all PAEs studied were adsorbed by MICB; however, diheptyl phthalate (DHpP) was most efficiently removed, achieving capacity values of 3.01 mg/g and a removal value of 92.5% [51].

2.2. Chitosan-based materials for the removal of oil pollution from water

Among the different types of organic pollution affecting water bodies, a specially concerning type nowadays is the oil pollution. While catastrophic spills such as the Exxon Valdez oil spill in the coast of Alaska (1989) or the BP Deepwater Horizon oil spill in the Gulf of Mexico (2010) caused a very important harm and gathered a large amount of public attention, most of the oil spills are less extraordinary [52]. It is estimated that about 9 million barrels of oil are released globally into the oceans every year. Of this amount, however, more than half come from natural seepage from the ocean floor, and human consumption activities represent the second largest source of oil released into the oceans (about 35%) [52]. Eco-friendly and sustainable approaches to remove oil pollution from water are therefore required to avoid the environmental threat and hazards associated to it.

Different kinds of adsorbents have been explored for the removal of oil droplets from oil-in-water emulsions. For instance, activated carbon, biopolymers, organoclays, sawdust, vermiculite, walnut shell and resins have been tested for this purpose [53–58]. Biopolymer-supported materials have demonstrated being an efficient adsorbent for the removal of several contaminants from aquatic ecosystems. However, some of them have exhibited limitations when facing the scenario of oil removal. Chitosan has demonstrated being one of the most efficient biopolymers for the removal of oil droplets from water. The polymer, in addition to its oil sorption capacity, exhibits a unique structure that is very prone to chemical functionalization, allowing a large versatility in the production of novel sorptive materials with oil-enhanced selectivity and capacity. In addition to this, chitosan has a good biodegradability, biocompatibility, eco-friendliness and low cost [59, 60].

Barros and co-workers [61] investigated the adsorption capacity of chitin flakes, chitin and chitosan powder, chitosan flakes and chitosan solution towards crude oil spilled in seawater. In their study, 5 L of seawater were placed in a plastic container, and 7 g of petroleum were added to it. After 30 min, 50 mL of a 0.5% chitosan solution was sprayed over the oil spill. The results showed that, although chitosan flakes had a better adsorption capacity for oil (0.379 ± 0.030 g oil/g adsorbent) compared to the others, the biopolymer sank after adsorbing the oil. Such a characteristic offered a clear hinder in practical applications. On the other hand, chitosan solution, despite presenting lower adsorption capacity (0.013 ± 0.001 g oil/g adsorbent), did not present the low buoyancy drawback [61]. Elanchezhiyan and colleagues [62] investigated the recovery of oil from oil-in-water emulsion using chitosan/magnesium-aluminum layered double hydroxide hybrid composite (CS-LDHCs) obtained by a single co-precipitation method. CS-LDHC adsorbent was dispersed in 20 mL of deionized water solution containing 4% of oil and the effect of contact time in the oil removal by the CS-LDHCs and the layered double hydroxide hybrid (LDH) was investigated [62]. This was done by varying the contact time from 10 to 120 min at room temperature. Both adsorbents, CS-LDHCs and LDH, reached maximum oil removal saturation at 90 min, and, thus, the authors set 90 min as contact time for both adsorbents in further experiments. The researchers reported an oil removal capacity of 78% for CS-LDHCs, while for the LDH was found to be 30%. Since CS-LDHCs showed a much higher oil adsorption capacity, further studies were performed just targeting this material [62]. The effect of pH was also investigated in their study in the range from 3 to 11. This was done because normally the change in pH of oil-in-water emulsions cause emulsion breaking, which means that demulsification takes place [59, 63]. After their pH study, the authors demonstrated that the adsorption of oil was enhanced in acidic medium (pH 3.0) [62].

Elanchezhiyan and co-workers also studied the effect of contact time on the removal of oil from oil-in-water emulsion using zirconium-chitosan composites (Zr-CS-HC) in time-course experiments at room temperature [59]. The study of the effect of sorbent dosage of Zr-CS-HC indicated that a maximum oil removal percentage of 79% was achieved when exposing 400 mg of sorbent to 25 mL of polluted solution [59]. The authors determined that the maximum oil uptake on chitosan was reached after 4 h contact time. In their study, the authors demonstrated that Zr-CS-HC had a higher removal efficiency compared to chitosan and provided an explanation based on the higher number of vacant sites on the surface of Zr-CS-HC [59].

Grem and co-authors reported that chitosan microspheres produced by ionic gelation of chitosan with sodium tripolyphosphate (STP) were able to separate 90% of the oil from produced water containing 200 ppm oil suspension using packed columns [64].

In another very recent study, Doshi and colleagues [65] studied amphiphilic sodium salt of oleoyl carboxymethyl chitosan (NaO-CMCS) for the removal of oil from a simulated oil spill. Marine diesel was chosen as oil phase for the emulsion studies [65], and in their study, both deionized water and seawater were used to simulate oil-in-water (o/w) emulsions (1:1 v/v). The o/w creamy emulsions were prepared with different dosages of NaO-CMCS (0.5–5 g/L), and a calcium chloride dihydrate solution (0.1%) was also added to that mixture. From the results obtained, the authors concluded that about 75–85 and 19–49% of oil was recovered from the emulsified oil using deionized water and seawater, respectively [65]. The recovery of oil from the polluted aqueous phase was 76% in the case of deionized water containing a 0.5 g/L concentration of NaO-CMCS and 20% from seawater with a 2 g/L dose of NaO-CMCS. It was therefore concluded that this chitosan derivative was an effective material for the removal of oil from spills from polluted seawater [65].

Ummadisingu and co-workers [66] prepared chitosan from seafood industry waste and tested the purified material for the removal of oil from aqueous solutions. The authors explored the effect of contact time, pH, initial concentration and mass of adsorbent. They reported that the sorption equilibrium of oil on chitosan was reached after only 6 min of contact time, and the maximum capacity of chitosan to adsorb oil from oil–water solution was found to be 17.96 g/g of adsorbent [66].

Su and colleagues [67] have recently reported the preparation of a superhydrophobic and superoleophilic chitosan sponge using a freeze-drying method with the assistance of a cross-linking process employing tripolyphosphate/citral followed by octadecanethiol modification. In their paper, the authors describe a procedure that allowed getting a three-dimensional porous structure with large pore volume and good compressive properties. The obtained sponge was able to effectively absorb oil, reaching an absorptive capacity up to 60 times its own weight. The material was able to selectively absorb the emulsified oils in water and achieve continuous oil–water separation. The authors reported an oil–water separation efficiency up to 99% and claimed that the sponge still maintained a highly absorptive capacity after being reused for many cycles while having a good biodegradability.

Bibi and co-workers [68] investigated the adsorption capacity of carbon nanotubes (CNTs) mixed with chitosan (Cs)/poly(vinyl alcohol) (PVA) and cross-linked with silane. In their experimental setup, naphthalene was selected as polycyclic aromatic hydrocarbon (PAH) model, and its removal was studied with two membranes, CM10 and CW with and without CNTs, respectively [68]. A standard solution of 3 mg/L of naphthalene was prepared, and 30 mg of membranes were placed in 40 ml naphthalene-containing solution. The removal percentage of naphthalene with the CW membrane during the first 5 min reported by the authors was 10% and after 150 min 93%. For the CM10 membrane, more than 50% of naphthalene was removed during the first 5 min and 97% after 150 min. These results showed that CM10 membrane had good sorption capacity for naphthalene and that the sorption process took place fast [68]. The correlation between CNT content of the membrane and the removed amount of the naphthalene was 0.16 mg of naphthalene/1 mg CNTs. The authors reported that CW membrane had equilibrium swelling of 217 g/g and CM10 of 162 g/g and concluded that the best membrane was CM10 due to its excellent sorption capacity and fast removal kinetics [68].

2.3. Chitosan-based materials containing immobilized bacteria for the removal of oil pollution

Among the different clean-up actions and materials employed as a first response against oil spills, the use of booms, skimmers, absorbent materials, controlled burning and vacuum and centrifuges might be found. These techniques cannot however achieve a complete clean-up of the polluted area, and their implementation should be done short after the oil spill occurs [69]. In the last years, the use of bioremediation-based techniques has largely attracted the attention of researchers and industrial stakeholders. The use of microorganisms for these decontamination purposes is considered as an effective and environmentally friendly treatment for, i.e., shorelines contaminated as a result of marine oil spills. Most of the compounds present in crude oil and refined products are prone to biodegradation and therefore might be removed from the environment through consumption by microbes [69]. There are mostly two complementary approaches: bioaugmentation and biostimulation. While the first approach involves addition of oil-degrading bacteria to the polluted system, the second approach intends to support the growth of the indigenous hydrocarbon degraders present in the system by the addition of nutrients and/or other growth-limiting substances [69, 70]. A scheme of a biodegradation is presented in **Figure 4**. The most rapid and complete degradation of the majority of organic pollutants is brought about under aerobic conditions. Such a process is normally initiated through an intracellular oxidative attack and the activation of the organic molecule through incorporation of oxygen in a reaction catalyzed by oxygenases and peroxidases [71]. A complete oxidation of the target hydrocarbon would lead to the production of CO_2 and water. The generation of different series of structures corresponding to different transformation products should however not be disregarded when this kind of bioremediation techniques is explored.

In order to improve the performance of the degradation of oil-related pollutants, some researchers have proposed strategies that involve the use of biomass in immobilized systems. One of the preferred entrapment systems for these purposes has been chitosan. In addition to the natural trend of chitosan to absorb oil, chitosan hydrogels have excellent water permeability and mass transfer properties (allowing the required access of the biomass to the nutrients they require). In addition to the aforementioned benefits, chitosan contributes to providing shelter to the valuable biomass while helping preserving the integrity of the culture. The use of microorganisms entrapped in chitosan gel matrices is therefore expected to have a very positive impact in both bioaugmentation- and biostimulation-based decontamination processes.

Dellagnezze and co-workers studied a bacterial consortium composed of four metagenomic clones and *Bacillus subtilis* strain CBMAI 707 (all derived from petroleum reservoirs) entrapped in chitosan beads towards hydrocarbon degradation capacity [72]. Experiments were carried out in mesocosm scale (3000 L) with seawater artificially polluted with crude oil. The compounds present in the oil that were the target of the biodegradation studies were benzo (a) pyrene ($C_{20}H_{12}$), benzo (a) anthracene ($C_{18}H_{12}$) and benzo (K) fluoranthene ($C_{20}H_{12}$) [73–75]. The degradation of hydrocarbons was evaluated in two different treatments: bioaugmentation and control. The authors performed time-course experiments, following the system at days 0, 5, 10, 20 and 30. The researchers demonstrated that degradation ratios increased abruptly from

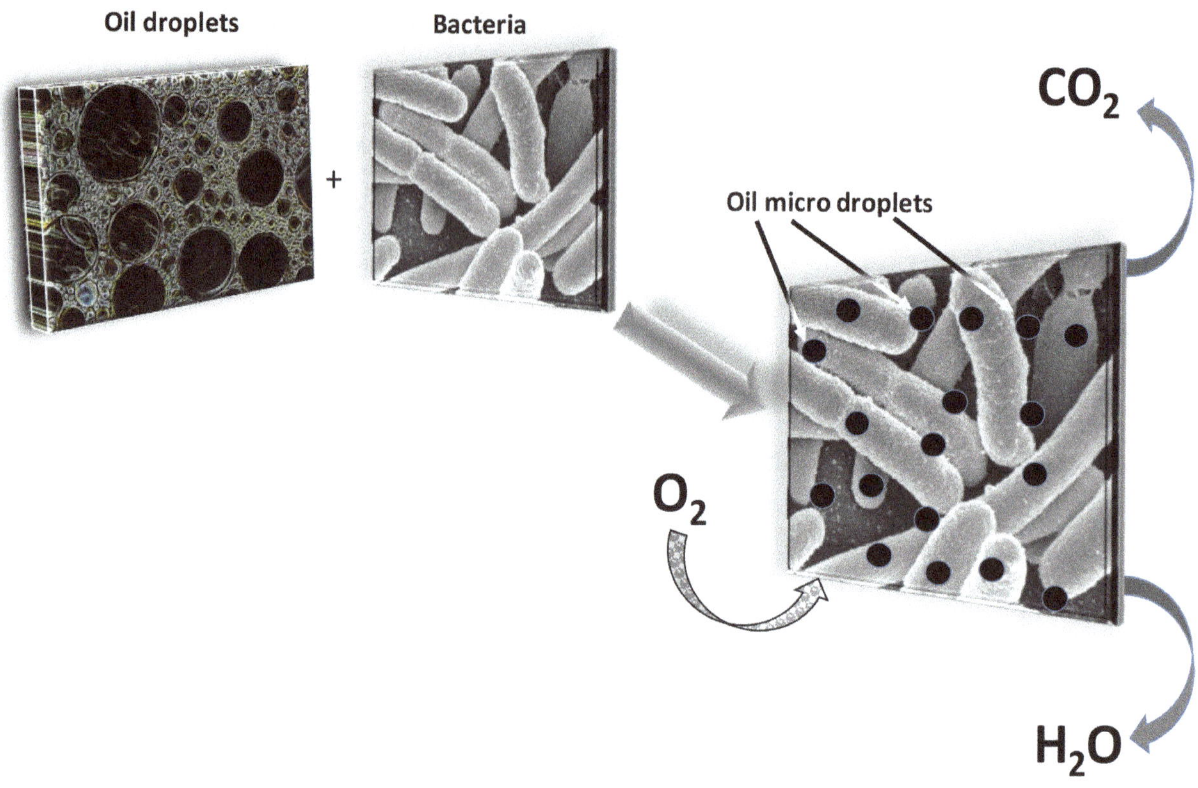

Figure 4. Oil biodegradation scheme under aerobic conditions.

the 5th to the 10th day and then just slightly increased until the end of the experiment. The system remained closed during the first 5 days to allow the acclimation of the bacteria to both treatments, and during this period, the authors did not observe significant biodegradation rates [72]. More than 90% of hydrocarbons' degradation was produced by the 10th day in both treatments. Similar to the 10th day, in day 20 most of the hydrocarbons were totally degraded. After 30 days from the beginning of the assays, the degradation percentages in the bioaugmentation treatment were higher than those observed in the control treatment. For instance, 35% of benzo (k) fluoranthene was degraded in the control treatment, 70% was degraded in the bioaugmentation treatment, and benzo (a) pyrene was almost completely degraded (99%). Likewise, benzo (a) anthracene showed higher degradation percentage, reaching 85% in the bioaugmented series compared to a 68% observed in the control [72].

Sar and Rosenberg [76] studied the biodegradation of n-hexadecane and its biosurfactant recovery. For this purpose, spores of *Bacillus subtilis* LAMI008 were entrapped in 3 mm chitosan beads and cross-linked with 0.3% glutaraldehyde [76]. The authors performed biodegradation assays in 50 mL of mineral medium containing 1% n-hexane (v/v) supplemented with 1% glucose (w/v) and inoculated with 10% of spore suspension (v/v) or with 10% of spore-entrapped chitosan beads (w/v). Both cultures were adjusted to 10^7 CFU/mL [76]. The biodegradation of n-hexadecane by *Bacillus subtilis* LAMI008 entrapped in chitosan beads was compared with that by free cells under similar conditions, and almost 100% of 1% n-hexadecane was degraded within 48 h in both assays.

In another study Gentili and co-authors [77] examined the potential of chitosan flakes as carrier material for the immobilization of *R. corynebacteriorides* (QBTo), a hydrocarbon-degrading bacterial strain. The authors performed biodegradation tests in a crude oil-polluted seawater microcosms [77]. For this purpose, three different microcosm situations were tested. In the microcosms inoculated with QBTo immobilized onto chitosan, 60% of hydrocarbons in the hexane extracts were removed compared with the control microcosms sample and the seawater microcosms. In the control microcosms, QBTo could not produce a significant reduction in the hydrocarbon concentrations [77]. In the seawater microcosms, QBTo was inoculated without a carrier, and a decrease of 30% of hexane extract was obtained. The degradation of hydrocarbons in the microcosms with the strain QBTo immobilized onto chitosan was higher than that obtained in the microcosms without the carrier. The explanation to this improvement is due to the enhancement of the strain survival, since the carrier material and the biofilm structure that the cells developed on it exert a protective effect [77].

Costa and co-authors [78] investigated the potential of the bacterial strain *B. pumilus* entrapped in chitosan in the degradation of hexadecane. The biodegradation assays were performed with free-living and immobilized bacterial cells with 1% hexadecane (v/v). In addition to this, 5 mL of an adjusted culture with 10^9 CFU/mL were used in free-living cell assays [78]. For the immobilized assays with bacterial cells, 5 g of chitosan beads containing the selected strain were used, and the biodegradation was studied at 0, 48, 96 and 144 h. The biodegradation results indicated that after 48 h, the free-living cells removed 81.83% of the hydrocarbon, while the culture of immobilized cell was able to remove only 38.12%. The authors provided an explanation to this based on the fact that in the assays performed with free-living cells, there was approximately 10^4 times more biomass than in the experiments performed with immobilized cells. The authors pointed out that during the immobilization process there was an important loss of cells. Cell counting performed by the authors during the biodegradation experiments showed that the immobilized biomass grew progressively and removed 84.53% of hexadecane after 96 h. These results, obtained in a larger time span, where similar to the 86.28% removed in assays performed with non-immobilized, free-living cells. Moreover, the authors found that the biomass concentration inside the beads was similar to that observed in the free-living cultures at 144 h and also removed hexadecane efficiently [78].

3. Conclusions

Chitosan offers a large potential in the development of sorbent materials for abatement of organic pollution from water. The biopolymer can be used in many different applications on its raw form or included in the preparation of a large variety of materials such as hydrogel beads, nanoparticles, films, membranes and meshes. Among the different scenarios of severe organic pollution, the pollution caused by hydrocarbon spills into water bodies deserves a special mention. Chitosan and materials based on this polymer have demonstrated an enormous potential to efficiently remove hydrocarbons from contaminated water. In addition to the aforementioned, the potential offered by chitosan to develop biological remediation systems deserves a special mention. The excellent biocompatibility of this polymer makes possible development of oil remediation biosystems based on bacteria entrapped into chitosan.

Conflict of interest

The authors certify that they have no conflict of interest.

Author details

Carlos Escudero-Oñate* and Elena Martínez-Francés

*Address all correspondence to: carlos.escudero@niva.no

Norwegian Institute for Water Research (NIVA), Oslo, Norway

References

[1] Mourya VK, Inamdar NN. Chitosan-modifications and applications: Opportunities galore. Reactive and Functional Polymers. 2008;**68**:1013-1051

[2] Pillai CKS, Paul W, Sharma CP. Chitin and chitosan polymers: Chemistry, solubility and fiber formation. Progress in Polymer Science. 2009;**34**:641-678

[3] Wang J, Chen C. Chitosan-based biosorbents: Modification and application for biosorption of heavy metals and radionuclides. Bioresource Technology. 2014;**160**:129-141

[4] Kumari S, Kumar Annamareddy SH, Abanti S, Kumar Rath P. Physicochemical properties and characterization of chitosan synthesized from fish scales, crab and shrimp shells. International Journal of Biological Macromolecules. 2017;**104**:1697-1705

[5] Rinaudo M. Chitin and chitosan: Properties and applications. Progress in Polymer Science. 2006;**31**:603-632

[6] Jo G-H, Park R-D, Jung W-J. Enzymatic production of chitin from crustacean shell waste. In: Kim S-K, editor. Chitin, Chitosan. Oligosaccharides and Their Derivatives. Boca Raton: CRC Press; 2011

[7] Li Q, Dunn ET, Grandmaison EW, Goosen MFA. Applications and properties of chitosan. Journal of Bioactive and Compatible Polymers. 1992;**7**:370-397

[8] Ali A, Ahmed S. A review on chitosan and its nanocomposites in drug delivery. International Journal of Biological Macromolecules. 2018;**109**:273-286

[9] Zhang L, Zeng Y, Cheng Z. Removal of heavy metal ions using chitosan and modified chitosan: A review. Journal of Molecular Liquids. 2016;**214**:175-191

[10] Vakili M, Rafatullah M, Salamatinia B, Abdullah AZ, Ibrahim MH, Tan KB, Gholami Z, Amouzgar P. Application of chitosan and its derivatives as adsorbents for dye removal from water and wastewater: A review. Carbohydrate Polymers. 2014;**113**:115-130

[11] Li C, Lou T, Yan X, Long Y-z, Cui G, Wang X. Fabrication of pure chitosan nanofibrous membranes as effective absorbent for dye removal. International Journal of Biological Macromolecules. 2018;**106**:768-774

[12] Yang J, Lu H, Li M, Liu J, Zhang S, Xiong L, Sun Q. Development of chitosan-sodium phytate nanoparticles as a potent antibacterial agent. Carbohydrate Polymers. 2017;**178**: 311-321

[13] Jara AH, Daza LD, Aguirre DM, Muñoz JA, Solanilla JF, Váquiro HA. Characterization of chitosan edible films obtained with various polymer concentrations and drying temperatures. International Journal of Biological Macromolecules. 2018

[14] Wahba MI. Porous chitosan beads of superior mechanical properties for the covalent immobilization of enzymes. International Journal of Biological Macromolecules. 2017; **105**:894-904

[15] Seyed Dorraji MS, Amani-Ghadim AR, Hanifehpour Y, Woo Joo S, Figoli A, Carraro M, Tasselli F. Performance of chitosan based nanocomposite hollow fibers in the removal of selenium(IV) from water. Chemical Engineering Research and Design. 2017;**117**:309-317

[16] Mirmohseni A, Seyed Dorraji MS, Figoli A, Tasselli F. Chitosan hollow fibers as effective biosorbent toward dye: Preparation and modeling. Bioresource Technology. 2012;**121**: 212-220

[17] Tasselli F, Mirmohseni A, Seyed Dorraji MS, Figoli A. Mechanical, swelling and adsorptive properties of dry–wet spun chitosan hollow fibers crosslinked with glutaraldehyde. Reactive and Functional Polymers. 2013;**73**:218-223

[18] Nguyen S, Escudero C, Sediqi N, Smistad G, Hiorth M. Fluoride loaded polymeric nanoparticles for dental delivery. European Journal of Pharmaceutical Sciences. 2017;**104**:326-334

[19] Riegger BR, Bäurer B, Mirzayeva A, Tovar GEM, Bach M. A systematic approach of chitosan nanoparticle preparation via emulsion crosslinking as potential adsorbent in wastewater treatment. Carbohydrate Polymers. 2018;**180**:46-54

[20] Habiba U, Siddique TA, Lee JJL, Joo TC, Ang BC, Afifi AM. Adsorption study of methyl orange by chitosan/polyvinyl alcohol/zeolite electrospun composite nanofibrous membrane. Carbohydrate Polymers. 2018;**191**:79-85

[21] Li X-Q, Tang R-C. Crosslinking of chitosan fiber by a water-soluble diepoxy crosslinker for enhanced acid resistance and its impact on fiber structures and properties. Reactive and Functional Polymers. 2016;**100**:116-122

[22] Poon L, Wilson LD, Headley JV. Chitosan-glutaraldehyde copolymers and their sorption properties. Carbohydrate Polymers. 2014;**109**:92-101

[23] Kim T-Y, Park S-S, Cho S-Y. Adsorption characteristics of reactive black 5 onto chitosan beads cross-linked with epichlorohydrin. Journal of Industrial and Engineering Chemistry. 2012;**18**:1458-1464

[24] Modrzejewska Z, Rogacki G, Sujka W, Zarzycki R. Sorption of copper by chitosan hydrogel: Kinetics and equilibrium. Chemical Engineering and Processing: Process Intensification. 2016;**109**:104-113

[25] Boamah PO, Huang Y, Hua M, Onumah J, Sam-Amoah LK, Boamah PO, Qian Y, Zhang Q. Sorption of copper onto low molecular weight chitosan derivative from aqueous solution. Ecotoxicology and Environmental Safety. 2016;**129**:154-163

[26] Huang Y, Wu H, Shao T, Zhao X, Peng H, Gong Y, Wan H. Enhanced copper adsorption by DTPA-chitosan/alginate composite beads: Mechanism and application in simulated electroplating wastewater. Chemical Engineering Journal. 2018;**339**:322-333

[27] Dinu MV, Dinu IA, Lazar MM, Dragan ES. Chitosan-based ion-imprinted cryo-composites with excellent selectivity for copper ions. Carbohydrate Polymers. 2018;**186**:140-149

[28] Dinh V-P, Le N-C, Tuyen LA, Hung NQ, Nguyen V-D, Nguyen N-T. Insight into adsorption mechanism of lead(II) from aqueous solution by chitosan loaded MnO_2 nanoparticles. Materials Chemistry and Physics. 2018;**207**:294-302

[29] Qi L, Xu Z. Lead sorption from aqueous solutions on chitosan nanoparticles. Colloids and Surfaces A: Physicochemical and Engineering Aspects. 2004;**251**:183-190

[30] Gokila S, Gomathi T, Sudha PN, Anil S. Removal of the heavy metal ion chromiuim(VI) using chitosan and alginate nanocomposites. International Journal of Biological Macromolecules. 2017;**104**:1459-1468

[31] Kousalya GN, Rajiv Gandhi M, Meenakshi S. Sorption of chromium(VI) using modified forms of chitosan beads. International Journal of Biological Macromolecules. 2010;**47**: 308-315

[32] Gogoi P, Thakur AJ, Devi RR, Das B, Maji TK. A comparative study on sorption of arsenate ions from water by crosslinked chitosan and crosslinked chitosan/MMT nanocomposite. Journal of Environmental Chemical Engineering. 2016;(4):4248-4257

[33] Bertoni FA, González JC, García SI, Sala LF, Bellú SE. Application of chitosan in removal of molybdate ions from contaminated water and groundwater. Carbohydrate Polymers. 2018;**180**:55-62

[34] Gavilan KC, Pestov AV, Garcia HM, Yatluk Y, Roussy J, Guibal E. Mercury sorption on a thiocarbamoyl derivative of chitosan. Journal of Hazardous Materials. 2009;**165**:415-426

[35] Jóźwiak T, Filipkowska U, Szymczyk P, Kuczajowska-Zadrożna M, Mielcarek A. The use of cross-linked chitosan beads for nutrients (nitrate and orthophosphate) removal from a mixture of P-PO_4, N-NO_2 and N-NO_3. International Journal of Biological Macromolecules. 2017;**104**:1280-1293

[36] Jóźwiak T, Filipkowska U, Szymczyk P, Mielcarek A. Sorption of nutrients (orthophos phate, nitrate III and V) in an equimolar mixture of P–PO_4, N–NO_2 and N–NO_3 using chitosan. Arabian Journal of Chemistry. 2016;**17**a:10-957

[37] Gupta A, Chauhan VS, Sankararamakrishnan N. Preparation and evaluation of iron–chitosan composites for removal of As(III) and As(V) from arsenic contaminated real life groundwater. Water Research. 2009;**43**:3862-3870

[38] Zainal Z, Hui LK, Hussein MZ, Abdullah AH. Characterization of TiO_2–chitosan/glass photocatalyst for the removal of a monoazo dye via photodegradation–adsorption process. Journal of Hazardous Materials. 2009;**164**:138-145

[39] Thakre D, Jagtap S, Sakhare N, Labhsetwar N, Meshram S, Rayalu S. Chitosan based mesoporous Ti–Al binary metal oxide supported beads for defluoridation of water. Chemical Engineering Journal. 2010;**158**:315-324

[40] Jia S, Yang Z, Yang W, Zhang T, Zhang S, Yang X, Dong Y, Wu J, Wang Y. Removal of Cu(II) and tetracycline using an aromatic rings-functionalized chitosan-based flocculant: Enhanced interaction between the flocculant and the antibiotic. Chemical Engineering Journal. 2016;**283**:495-503

[41] Arayne MS, Sultana N, Hussain F. Validated RP-HPLC method for determination of permethrin in bulk and topical preparations using UV-vis detector. Journal of Chromatographic Science. 2011;**49**:287-291

[42] Rahmanifar B, Dehaghi SM. Removal of organochlorine pesticides by chitosan loaded with silver oxide nanoparticles from water. Clean Technologies and Environmental Policy. 2014;**16**:1781-1786

[43] Saifuddin N, Nian C, Zhan L, Ning K. Chitosan-silver nanoparticles composite as point-of-use drinking water filtration system for household to remove pesticides in water. Asian Journal of Biochemistry. 2011;**6**:142-159

[44] Danalıoğlu ST, Bayazit ŞS, Kuyumcu ÖK, Salam MA. Efficient removal of antibiotics by a novel magnetic adsorbent: Magnetic activated carbon/chitosan (MACC) nanocomposite. Journal of Molecular Liquids. 2017;**240**:589-596

[45] Konwar A, Gogoi A, Chowdhury D. Magnetic alginate–Fe_3O_4 hydrogel fiber capable of ciprofloxacin hydrochloride adsorption/separation in aqueous solution. RSC Advances. 2015;**5**:81573-81582

[46] Wu S, Zhao X, Li Y, Zhao C, Du Q, Sun J, Wang Y, Peng X, Xia Y, Wang Z. Adsorption of ciprofloxacin onto biocomposite fibers of graphene oxide/calcium alginate. Chemical Engineering Journal. 2013;**230**:389-395

[47] Nadavala SK, Swayampakula K, Boddu VM, Abburi K. Biosorption of phenol and o-chlorophenol from aqueous solutions on to chitosan–calcium alginate blended beads. Journal of Hazardous Materials. 2009;**162**:482-489

[48] Li J-M, Meng X-G, Hu C-W, Du J. Adsorption of phenol, p-chlorophenol and p-nitrophenol onto functional chitosan. Bioresource Technology. 2009;**100**:1168-1173

[49] Barbusiński K, Salwiczek S, Paszewska A. The use of chitosan for removing selected pollutants from water and wastewater—Short review. Architecture Civil Engineering Environment. 2016;**9**:107-115

[50] Barreto R, Hissa D, Paes F, Grangeiro T, Nascimento R, Rebelo L, Craveiro A, Melo V. New approach for petroleum hydrocarbon degradation using bacterial spores entrapped in chitosan beads. Bioresource Technology. 2010;**101**:2121-2125

[51] Chen C-Y, Chung Y-C. Removal of phthalate esters from aqueous solution by molybdate impregnated chitosan beads. Environmental Engineering Science. 2007;**24**:834-841

[52] Cohen MA. Water Pollution from Oil Spills A2. In: Shogren, Jason F, editors. Encyclopedia of Energy, Natural Resource, and Environmental Economics. Waltham: Elsevier; 2013. pp. 121-126

[53] Zhang J, Xue Q, Pan X, Jin Y, Lu W, Ding D, Guo Q. Graphene oxide/polyacrylonitrile

fiber hierarchical-structured membrane for ultra-fast microfiltration of oil-water emulsion. Chemical Engineering Journal. 2017;**307**:643-649

[54] Ahmad A, Sumathi S, Hameed B. Residual oil and suspended solid removal using natural adsorbents chitosan, bentonite and activated carbon: A comparative study. Chemical Engineering Journal. 2005;**108**:179-185

[55] Cambiella A, Ortea E, Rios G, Benito JM, Pazos C, Coca J. Treatment of oil-in-water emulsions: Performance of a sawdust bed filter. Journal of Hazardous Materials. 2006;**131**:195-199

[56] da Silva UG, Melo MAF, da Silva AF, de Farias RF. Adsorption of crude oil on anhydrous and hydrophobized vermiculite. Journal of Colloid and Interface Science. 2003;**260**:302-304

[57] Srinivasan A, Viraraghavan T. Removal of oil by walnut shell media. Bioresource Technology. 2008;**99**:8217-8220

[58] Franco CA, Cortés FB, Nassar NN. Adsorptive removal of oil spill from oil-in-fresh water emulsions by hydrophobic alumina nanoparticles functionalized with petroleum vacuum residue. Journal of Colloid and Interface Science. 2014;**425**:168-177

[59] Elanchezhiyan SS, Sivasurian N, Meenakshi S. Enhancement of oil recovery using zirconium-chitosan hybrid composite by adsorptive method. Carbohydrate Polymers. 2016;**145**:103-113

[60] Farzana MH, Meenakshi S. Exploitation of zinc oxide impregnated chitosan beads for the photocatalytic decolorization of an azo dye. International Journal of Biological Macromolecules. 2015;**72**:900-910

[61] Barros FCF, Vasconcellos LCG, Carvalho TV, Nascimento RFD. Removal of petroleum spill in water by chitin and chitosan. Orbital-The Electronic Journal of Chemistry. 2014;**6**:70-74

[62] Elanchezhiyan SS, Meenakshi S. Synthesis and characterization of chitosan/Mg-Al layered double hydroxide composite for the removal of oil particles from oil-in-water emulsion. International Journal of Biological Macromolecules. 2017

[63] Schulz P, Rodriguez M, Del Blanco L, Pistonesi M, Agullo E. Emulsification properties of chitosan. Colloid & Polymer Science. 1998;**276**:1159-1165

[64] Grem ICS, Lima BNB, Carneiro WF, Queirós YGC, Mansur CRE. Chitosan microspheres applied for removal of oil from produced water in the oil industry. Polímeros. 2013;**23**:705-711

[65] Doshi B, Repo E, Heiskanen JP, Sirviö JA, Sillanpää M. Sodium salt of oleoyl carboxymethyl chitosan: A sustainable adsorbent in the oil spill treatment. Journal of Cleaner Production. 2018;**170**:339-350

[66] Ummadisingu A, Gupta S. Characteristics and kinetic study of chitosan prepared from seafood industry waste for oil spills cleanup. Desalination and Water Treatment. 2012;**44**:44-51

[67] Su C, Yang H, Zhao H, Liu Y, Chen R. Recyclable and biodegradable superhydrophobic and superoleophilic chitosan sponge for the effective removal of oily pollutants from water. Chemical Engineering Journal. 2017;**330**:423-432

[68] Bibi S, Yasin T, Hassan S, Riaz M, Nawaz M. Chitosan/CNTs green nanocomposite membrane: Synthesis, swelling and polyaromatic hydrocarbons removal. Materials Science and Engineering: C. 2015;**46**:359-365

[69] Nikolopoulou M, Pasadakis N, Kalogerakis N. Evaluation of autochthonous bioaugmentation and biostimulation during microcosm-simulated oil spills. Marine Pollution Bulletin. 2013;**72**:165-173

[70] Nikolopoulou M, Kalogerakis N. Biostimulation strategies for enhanced bioremediation of marine oil spills including chronic pollution. In: Timmis KN, editor. Handbook of Hydrocarbon and Lipid Microbiology. Berlin Heidelberg: Springer-Verlag; 2010. pp. 2521-2529

[71] Das N, Chandran P. Microbial degradation of petroleum hydrocarbon contaminants: An overview. Biotechnology Research International. 2011;**2011**:1-13

[72] Dellagnezze B, Vasconcellos S, Angelim A, Melo V, Santisi S, Cappello S, Oliveira V. Bioaugmentation strategy employing a microbial consortium immobilized in chitosan beads for oil degradation in mesocosm scale. Marine Pollution Bulletin. 2016;**107**:107-117

[73] Phale PS, Basu A, Majhi PD, Deveryshetty J, Vamsee-Krishna C, Shrivastava R. Metabolic diversity in bacterial degradation of aromatic compounds. Omics: A Journal of Integrative Biology. 2007;**11**:252-279

[74] Kanaly RA, Harayama S. Biodegradation of high-molecular-weight polycyclic aromatic hydrocarbons by bacteria. Journal of Bacteriology. 2000;**182**:2059-2067

[75] Jones MD, Singleton DR, Carstensen DP, Powell SN, Swanson JS, Pfaender FK, Aitken MD. Effect of incubation conditions on the enrichment of pyrene-degrading bacteria identified by stable-isotope probing in an aged, PAH-contaminated soil. Microbial Ecology. 2008;**56**:341-349

[76] Sar N, Rosenberg E. Emulsifier production by Acinetobacter calcoaceticus strains. Current Microbiology. 1983;**9**:309-313

[77] Gentili AR, Cubitto MA, Ferrero M, Rodriguéz MS. Bioremediation of crude oil polluted seawater by a hydrocarbon-degrading bacterial strain immobilized on chitin and chitosan flakes. International Biodeterioration & Biodegradation. 2006;**57**:222-228

[78] Costa SP, Angelim AL, de Queiroz MFV, Melo VMM. Vegetative cells of Bacillus pumilus entrapped in chitosan beads as a product for hydrocarbon biodegradation. International Biodeterioration & Biodegradation. 2014;**87**:122-127

2

Chitosan's Wide Profile from Fibre to Fabrics

Xue Luo and Li Li

Abstract

Textile has a high structure capacity, is adaptive to multiple situations and is applied in food, energy, environmental, construction and medical industries. Its stable and flexible characteristics are sure to attract even more attention. Biofunctional textile is one of the most important categories of functional textile, taking up 7% of the total amount, and is expected to be the most promising section of growth. Due to the restrict requirement of fibre production, chitosan is one of the few materials that can be spun into pure fibre. The pure chitosan fibre can be blend with other fibres and produce durable functional fabric suitable for medical as well as daily use. This article also reviewed existed modification on chitosan material prepared for fibre spinning and technology related to chitosan-based textile production and discussed the difficulties and possible solutions in chitosan yarn spinning and possible ways of fabric forming.

Keywords: chitosan, textile, fibres, fabrics, biofunctional materials

1. Introduction

The textile industry accounts for 2% of the world's gross domestic product and is the seventh largest economic sector, according to a recent report by McKinsey & Company [1]. The capabilities and stability of textiles have advanced considerably. Entanglement of fibres in yarns can be designed and fabricated by changing numerous technical factors during production. Furthermore, diverse methods are used for forming fabrics. Knitting and weaving structures differ in elasticity, density and air and liquid permeability. There are numerous methods of enriching textile structures by combining and layering existing structures without adherence between components. Moreover, textiles are considerably more stable than common film structures because of the elasticity embedded in their structure. Textiles are durable under conditions subjecting them to impact forces and are resistant to abrasion. Therefore, the structure of textiles is favoured for both industrial use and apparel.

Advances in material development offer opportunities for cross discipline studies to further improve the mechanical and functional performance of textiles. Textiles are applied in the food, energy, environmental, construction and medical industries, meaning they are not only a domestic product. Their stable and flexible characteristics have attracted considerable attention. Functional textiles account for 27% of worldwide fabric production, with the functional textile market expected to be worth US$175 billion by 2020.

Biofunctional textiles are among the most valuable categories of functional textiles, accounting for up to 7% of total functional textiles. Developing countries involving growing textile industries, such as China, have lost global textile and apparel market share. The textile market has shifted to more value-added products, namely technical textiles. Biofunctional textiles are categorised according to their usage: in vitro and in vivo use. Biofunctional textiles for in vivo use emphasise biocompatibility and biostability because they come into direct contact with cells and biological fluids. One method of endowing these capabilities in a textile is by coating a textile polymer with functional molecules, which is sometimes enhanced through chemical bonds. However, the coating is thin and vulnerable to abrasion and other physical movements, even when the environment of application is in vivo. Another means of developing biofunctional textiles is by spinning fibres with pure functional materials. Candidate materials for spinning fibre are extremely limited because of both process requirements and mechanical properties of the fibre required for yarn spinning.

The two main methods of forming fibres are melt and wet spinning. Melt spinning is typically applied to thermoplastic polymers that transform to a liquid form under heat and recover to a solid and flexible form after cooling. The mechanical and functional properties of the fibre should not be affected by the high temperatures required in melt spinning.

Such a requirement of melt spinning is hardly achieved by biofunctional fibre materials, such as chitosan and alginate, because they are not thermoplastic and also degrade when heated at elevated temperatures. In such cases, low-temperature wet spinning or gel spinning is used. In wet spinning, the polymer is dissolved in a solvent and then extruded through a spinneret into a nonsolvent in a coagulation bath in which the fibres are precipitated and solidified. The filaments are then washed to remove the remaining solvents and nonsolvents, drawn, dried and lubricated before being wound on a bobbin. Other solvent removal methods exist. Instead of precipitating the fibres in a coagulation bath, dry spinning involves solidifying the fibres by evaporating the solvent in a stream of hot air or hot inert gas. The solvent is recovered and reused in such a method.

The strict requirements of fibre spinning and the fragility of biofunctional materials limit the material choices for pure biofunctional fibres. Typical pure biofunctional fibres are made from corn, bamboo, milk, microorganism products and animal shells. The first four material sources are developed mainly as a substitute for oil-based manmade fibres, whereas the fifth material source has special biofunctions other than biocompatibility and biostability.

Chitosan and chitin fibres are the products of such technology, which inherent the antibacterial and wound-healing characteristics of chitosan and chitin materials while holding adequate mechanical strength.

2. Chitosan textile production

Traditionally, chitosan is mainly treated on a yarn's or fabric's surface [2–6] and is exposed to harm from abrasion and other movements. Chitosan peals when it wears, and the biofunctions are vulnerable to physical and chemical damage. Producing a pure fibre from chitosan is a reasonable means of improving stability and durability.

Chitosan fibres are derived mainly from wet spinning. Dissolution, deaeration and filtration of chitosan are performed before spinning [7]. The semi-finished fibres are then refined, dried and post-finished. Factors in dissolution, deaeration and drying are vital to control properties of chitosan fibre, including solvent, pH and concentration during dissolution, method and agent of drying, etc. A study [8] shows that methanol drying yielded chitosan fibre has superior mechanical properties to fibres dried using other methods and agent.

The three main methods are available for producing a fabric from fibres. A nonwoven method entails entangling fibres by using water or air jets. This process is similar to wool felting and requires no yarn spinning. The nonwoven method is fast and cheap, and it is suitable for producing cheap, disposable products. Woven and knit fabrics are more durable and are more suitable for daily use because they can be reused hundreds of times, withstanding frequent washing and abrasion [9, 10] (**Figures 1–5**).

Chitosan fibres are weaker than cotton. Typical chitosan fibres have a dry tensile strength of 2.09 cN/dtex, which is close to the minimum dry tensile strength value for cotton fibres (1.9–3.1 cN/dtex). Their wet tensile strength (1.8 cN/dtex) is lower than that of cotton fibres (2.23.1 cN/dtex) (**Table 1**). In a spinning factory, the humidity is usually set to a high level (70–80%), under which conditions chitosan becomes adhesive and weak, causing problems during the spinning process, especially when yarns have a high chitosan fibre ratio.

Chitosan's weakness also affects end-product usage. A weak textile does not withstand tearing, bursting or abrasion. Therefore, the strength and weakness must be balanced by mixing and strengthening the fibre with other materials. The price of chitosan and fibres is expensive at up to $100 per kilogramme, which is comparable to cashmere. The blending of two or more types of fibre solves both the mechanical strength and cost problems.

Blending two or more biofunctional materials or blending biofunctional materials with non-biofunctional materials can be achieved during fibre spinning, yarn spinning or fabric formation. Blending in or before fibre spinning requires property consistency between the two or more biofunctional materials in the spinning process; the ratio of biofunctional materials is settled and less flexible regarding the production process. Blending after yarn production is likely to cause an uneven distribution of the biofunctional materials and influence the performance. A balanced method of forming blended fabric requires the blending of pure fibres during yarn spinning.

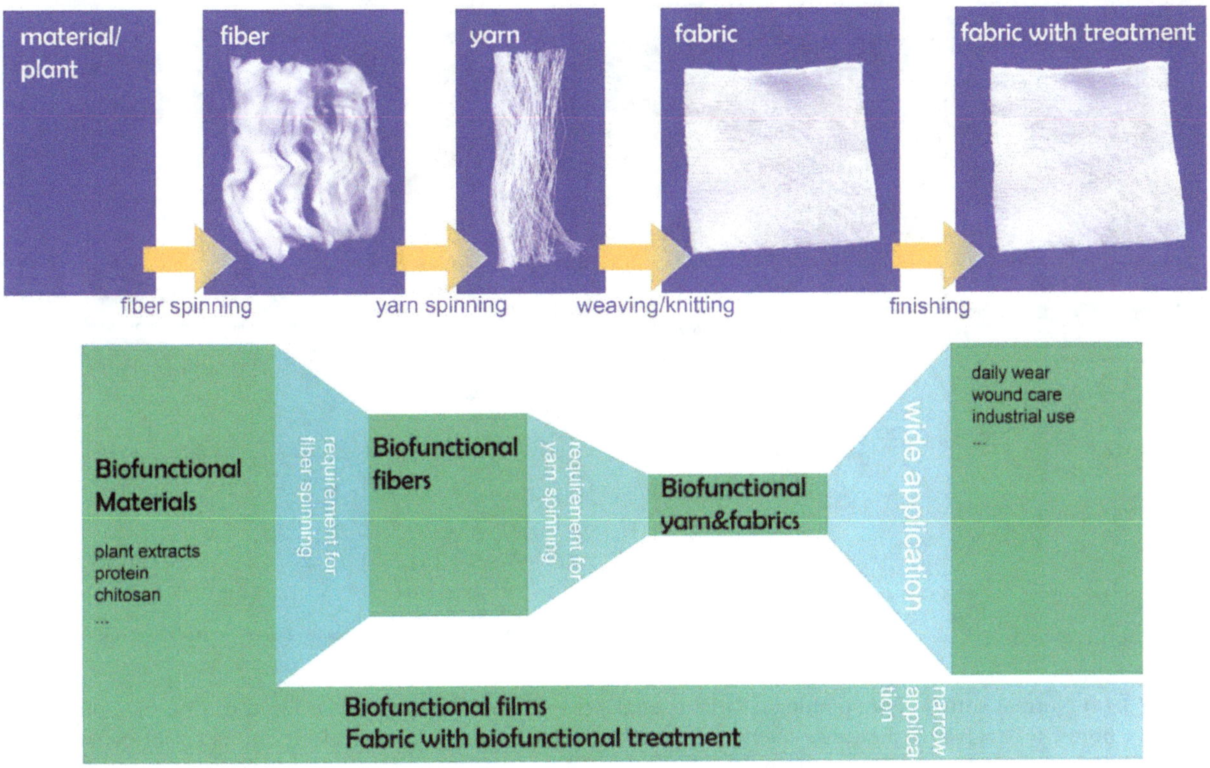

Figure 1. Process of textile production (upper); the strict requirements of textile production heavily restrict the applicable material list (chitosan is an example). The rich structural capability and stability of biofunctional textiles render them versatile.

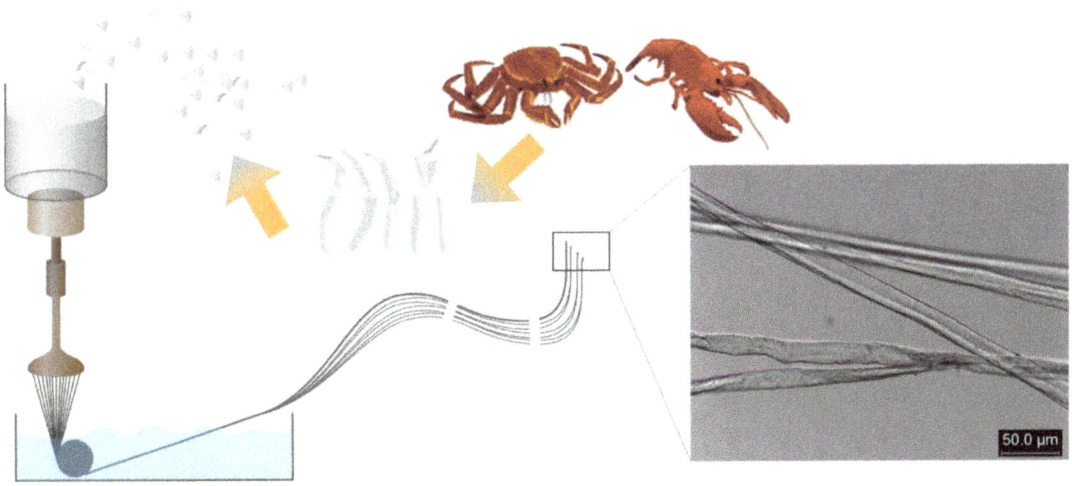

Figure 2. Spinning process of chitosan fibre (left) and chitosan fibre under microscope (right).

Blending biofunctional fibres during yarn spinning may cause problems. Biofunctional fibres normally have weaker mechanical properties than normal fibres, but they may have special properties. Chitosan fibres easily adhere to rubber rollers, which not only causes difficulty in spinning yarns with a high chitosan content but also causes material loss, raising production costs. Therefore, the mechanical properties and biofunction of yarns must be carefully balanced to ensure sufficient strength and comfort handle and maintain stable biofunctions while controlling the production difficulty and costs.

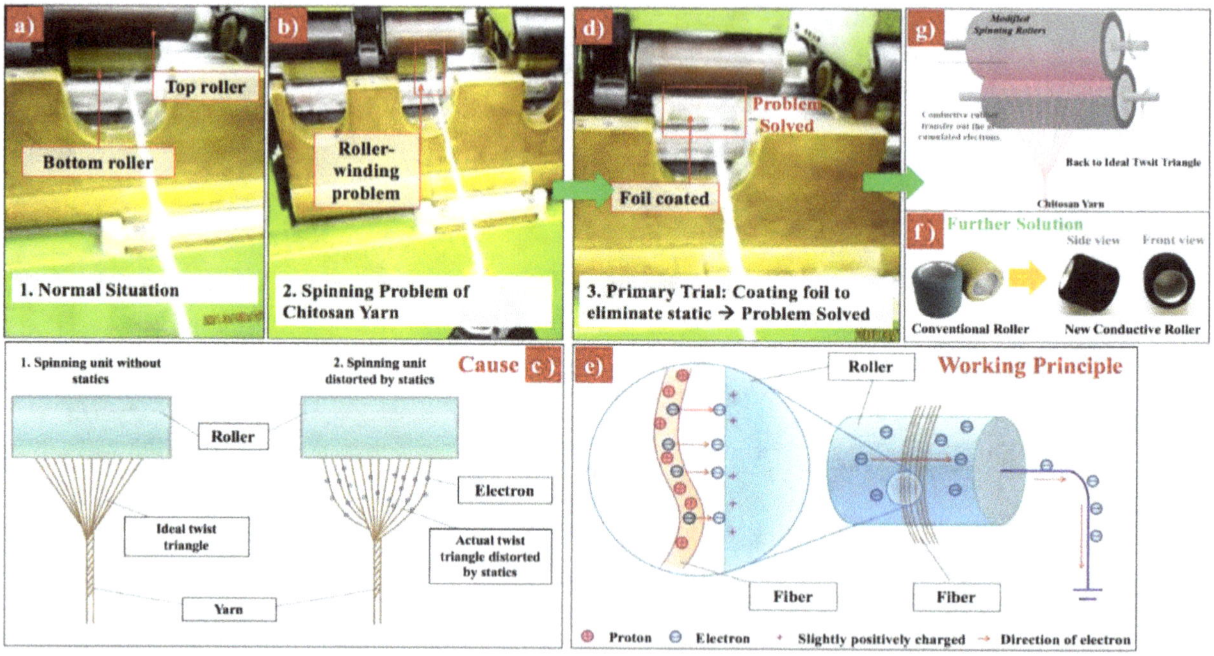

Figure 3. (a) Normal situation of spinning a twist triangle; (b) roller-winding problem faced during the spinning of chitosan fibres; (c) analysing the cause of the problem: adhesion caused by static electricity is easily generated by abrasion (due to the extremely low electronic work function of chitosan fibres); (d) primary trial to solve the problem by using a layer of foil to remove (transfer) accumulated electrons; (e) working principle of the primary trial; (f) further trial: modified conductive roller to remove the static electricity; and (g) a twist triangle turns back to an ideal situation when the modified rollers are applied.

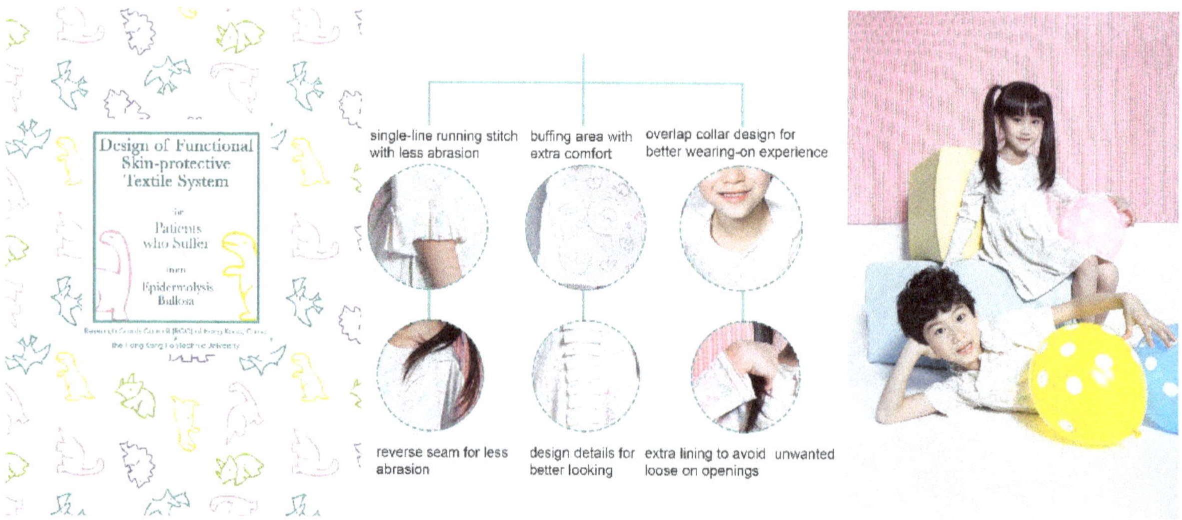

Figure 4. Example of product design targets epidermolysis bullosa patients.

Studies [11–13] have shown that factors associated with the blending process, such as blending time (fibre or sliver blending) and spinning factors, may influence the distribution of yarn fibres. Regarding the ratios of biofunctional fibres, variables must be examined not only for academic purposes but also for industrial and domestic application. Efficient and precise testing methods must be developed to facilitate the regulation of biofunctional textile production, testing and usage.

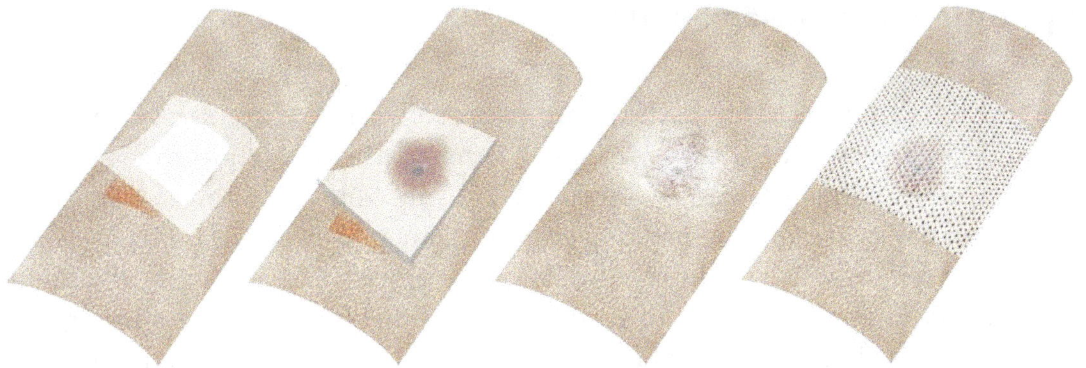

Figure 5. Four common biofunctional materials related to wound care: Band-Aid, gel dressing (artificial skin), powder or cream and textile.

Parameters	Values	Cotton*
Dry tensile strength (cN/dtex)	2.09	1.9–3.1
Dry tensile strength ratio (%)	15.1	7–10
Wet tensile strength (cN/dtex)	1.80	2.2–3.1
*Source from www.intechopen.com		

Table 1. Parameters of chitosan fibre.

Besides blending after fibre spinning, blending before fibre spinning and chitosan material modification are other common ways for researchers to reinforce the weak fibres especially in wet and acidic environment. Blending chitosan with other polymers can strengthen the fibre but compatibility needs to be improved, and thus reactive compatibilisers such as epoxy-functionalised LDPE has been introduced to increase the strength of the blend [14].

Biofunctions of chitosan/chitosan-treated fabric can be improved by adding functional groups on chitosan material or adding other ingredients. Some approaches make reaction faster by improving chitosan water solubility by acylation, alkylation, pegylation, hydroxyalkylation, carboxyalkylation and depolymerisation, but the durability is sure to be shortened. Some others improve cationic properties by deacetylation, quaternisation and addition of cationic moieties. Concerning non-toxic requirement of chitosan textile for medical use, some safe additives include neem [15].

Cross-linked chitosan fibre shows equal [16] or better [17] antibacterial effect and possibly has better mechanical property due to the reduction of crystallinity [18]. Stronger antibacterial function is found in N,N,N-trimethyl chitosan fibres [19] and succinic anhydride-modified fibre [20]. N-carboxyethyl chitosan fibres [21], quaternisation-functionalised chitosan fibre [22] and succinic anhydride-modified fibre [20] have better liquid absorption capacity than normal chitosan fibre, but tensile strength and elongation at break are lower. There are few studies that clearly report improvement in both mechanical and biofunction properties.

3. Application of chitosan textile

Chitosan textiles can be used for medical care to provide safe and continuous protection for patients and caregivers. For patients with skin trauma, particularly those who are bedbound and at risk of developing bedsores, or those with low immunity, chitosan textiles can be used as materials for daily dressing and bedding or as fundamental materials for long-term wound care [23]; this is because such textiles control bacterial growth and accelerate wound healing. Moreover, chitosan textiles can avoid or reduce the side effects of long-term antibiotic or silver fungicide use. The antibacterial function can be controlled by the blending ratio of the chitosan fibres to other fibres to meet different needs.

Chitosan textiles are also required in space and military activities. The extreme environment in space renders the balance of microorganisms even more vital. Excessive or insufficient microorganisms can pose a risk. An excessively high number of microorganisms pose an infection risk. Furthermore, the growth and mutation of bacteria in space are uncontrollable, which may pose a threat to humans and other on-board animals. An excessively low microbial population may cause immune disorders or impaired immune function. Using chitosan textiles in ship interiors and clothing can continuously maintain the number of microorganisms, which is beneficial to the health of shipborne personnel.

In addition to special uses, chitosan textiles can be used for ordinary clothing, especially underwear. The market demand for antibacterial underwear is considerably high. In relatively hostile market environments, functional underwear can be marketed as having advantageous properties over products of competing enterprises, in order to prevent the prices from being trapped at low levels. In contrast to chitosan antibacterial agents coated on the surface of normal fabrics, blended chitosan-knitted fabrics do not lose their biofunctions after friction or washing. The antibacterial performance of the fabrics is still stable after being washed hundreds of times. Therefore, such fabrics are particularly suitable for end products such as underwear, which must be washed frequently. The safety of chitosan-knitted fabrics is also superior to that of traditional silver-based antibacterial fabrics, because they can come into direct contact with the skin.

Chitosan also have antifungal effects [24–29], which render chitosan textile particularly useful for bedding, home decor and personal hygiene products. However, chitosan fibres are not bright or white, which limits the consumer acceptance of chitosan textiles.

4. Conclusion

Further development of relevant technologies will give chitosan textiles a more crucial role in functional textiles in the future. Producing daily wear from a blended fabric with a low proportion of chitosan will ensure that the price is acceptable to consumers. However, chitosan fabrics with special properties generally require a high proportion of chitosan fibres. This requires breakthroughs in spinning technologies, because chitosan fibres are sticky and easily generate static electricity. The upper and lower limits of the functions of chitosan fabrics deserve attention. Because of the high price of chitosan fibres, increasing their proportion should engender greater functional improvement.

Tuning the biological function of chitosan and increasing the ratio of chitosan in fabrics both require a systematic study of blending technologies. Even in the traditional spinning process, numerous parameters may affect the distribution, feather, yarn thickness and compactness of the final fibre. For chitosan yarns, these parameters may affect wet gas absorption, comfort, biological function and the mechanical properties of yarns and fabrics. Results from studies on the textile industry could be used to strengthen the function of chitosan textiles. Meanwhile, factors influencing the structure and effects of chitosan textiles can be eliminated in yarn spinning and during knitting, weaving or felting processes of nonwoven fabric to improve the effects of the final product.

Relevant test methods and standards for chitosan textiles must be further standardised. Chitosan fibre-blended textiles are different from textiles treated with traditional antibiotics or silver fungicides. In contrast to treated fabrics, bactericidal components are insoluble in chitosan fibre-blended textiles. Furthermore, the antibacterial properties of chitosan fibres may not be detected by certain test methods such as those that entail testing the zone of inhibition.

Meanwhile, the method proposed in most studies for testing the antibacterial effects of chitosan in a liquid form after it has been dissolved in an acetic acid solution is not suitable for chitosan textiles, because this situation would not occur in practical use. After existing test methods and determination methods are compared, a comprehensive test system should be developed for chitosan textiles. Furthermore, the antibacterial principle of chitosan fibres may be different from the antiseptic principle of the acetic acid solution for chitosan raw materials. Therefore, studying their differences and similarities is necessary. Moreover, the production of chitosan raw materials causes pollution problems that must be solved. Chitosan textiles have been continuously refined. It is expected that many new findings and technologies will be discovered and developed in this cross domain.

Author details

Xue Luo and Li Li*

*Address all correspondence to: li.lilly@polyu.edu.hk

Institute of Textiles and Clothing, The Hong Kong Polytechnic University, China

References

[1] Amed I, Berg A, Brantberg L, Hedrich S. The State of Fashion. McKinsey & Company. 2016. Available from: www.mckinsey.com [Retrieved: January 10, 2018]

[2] Enescu D. Use of chitosan in surface modification of textile materials. Roumanian Biotechnological Letters. 2008;**13**(6):4037

[3] Lim SH, Hudson SM. Review of chitosan and its derivatives as antimicrobial agents and their uses as textile chemicals. Journal of Macromolecular Science, Part C: Polymer Reviews. 2003;**43**(2):223-269

[4] Lim SH, Hudson SM. Application of a fiber-reactive chitosan derivative to cotton fabric as an antimicrobial textile finish. Carbohydrate Polymers. 2004;**56**(2):227-234

[5] Shanmugasundaram OL. Chitosan coated cotton yarn and its effect on antimicrobial activity. Journal of Textile and Apparel, Technology and Management. 2006;**5**(3):1

[6] Zhang Z, Chen L, Ji J, Huang Y, Chen D. Antibacterial properties of cotton fabrics treated with chitosan. Textile Research Journal. 2003;**73**(12):1103-1106

[7] East GC, Qin Y. Wet spinning of chitosan and the acetylation of chitosan fibers. Journal of Applied Polymer Science. 1993;**50**(10):1773-1779

[8] Knaul J, Hooper M, Chanyi C, Creber KA. Improvements in the drying process for wet-spun chitosan fibers. Journal of Applied Polymer Science. 1998;**69**(7):1435-1444

[9] Lam NY, Zhang M, Yang C, Ho CP, Li L. A pilot intervention with chitosan/cotton knitted jersey fabric to provide comfort for epidermolysis bullosa patients. Textile Research Journal. 2018;**88**(6):704-716

[10] Wawro D, Skrzetuska E, Włodarczyk B, Kowalski K, Krucińska I. Processing of Chitosan Yarn into Knitted Fabrics. Fibres & Textiles in Eastern Europe. 2016;**24**(6):120-152

[11] Lam NYK, Zhang M, Guo HF, Ho CP, Li L. Effect of fiber length and blending method on the tensile properties of ring spun chitosan–cotton blend yarns. Textile Research Journal. 2017;**87**(2):244-257

[12] Guo HF, Yang C, Li L. Study on the dynamics of chitosan/cotton fiber in an airflow around two rotating cylinders. Textile Research Journal. 2017. DOI: 0040517517715082

[13] Liu S, Hua T, Luo X, Yi Lam N, Tao XM, Li L. A novel approach to improving the quality of chitosan blended yarns using static theory. Textile Research Journal. 2015;**85**(10): 1022-1034

[14] Prasanna K, Sailaja RRN. Blends of LDPE/chitosan using epoxy-functionalized LDPE as compatibilizer. Journal of Applied Polymer Science. 2012;**124**:3264-3275

[15] Chandrasekara S, Vijayakumara S, Rajendran R. Application of chitosan and herbal nanocomposites to develop antibacterial medical textile. Biomedicine & Aging Pathology. 2014;**4**:59-64

[16] Li XQ, Tang RC. Crosslinked and dyed chitosan fiber presenting enhanced acid resistance and bioactivities. Polymer. 2016;**8**(4):119

[17] Yang Q, Dou F, Liang B, Shen Q. Studies of cross-linking reaction on chitosan fiber with glyoxal. Carbohydrate Polymers. 2005;**59**(2):205-210

[18] Yang Q, Dou F, Liang B, Shen Q. Investigations of the effects of glyoxal cross-linking on the structure and properties of chitosan fiber. Carbohydrate Polymers. 2005;**61**(4):393-398

[19] Zhou Z, Yan D, Cheng X, Kong M, Liu Y, Feng C, Chen X. Biomaterials based on N, N, N-trimethyl chitosan fibers in wound dressing applications. International Journal of Biological Macromolecules. 2016;**89**:471-476

[20] Xia G, Lang X, Kong M, Cheng X, Liu Y, Feng C, Chen X. Surface fluid-swellable chitosan fiber as the wound dressing material. Carbohydrate Polymers. 2016;**136**:860-866

[21] Yang S, Dong Q, Yang H, Liu X, Gu S, Zhou Y, Xu W. N-carboxyethyl chitosan fibers prepared as potential use in tissue engineering. International Journal of Biological Macromolecules. 2016;**82**:1018-1022

[22] Zhou Y, Yang H, Liu X, Mao J, Gu S, Xu W. Potential of quaternization-functionalized chitosan fiber for wound dressing. International Journal of Biological Macromolecules. 2013;**52**:327-332

[23] Lam NYK. Application in fashion design with chitosan based yarn development [Doctoral dissertation]; The Hong Kong Polytechnic University; 2017

[24] Martínez-Camacho AP, Cortez-Rocha MO, Ezquerra-Brauer JM, Graciano-Verdugo AZ, Rodriguez-Félix F, Castillo-Ortega MM, et al. Chitosan composite films: Thermal, structural, mechanical and antifungal properties. Carbohydrate Polymers. 2010;**82**:305-315

[25] Tajdini F, Amini MA, Nafissi-Varcheh N, Faramarzi MA. Production, physiochemical and antimicrobial properties of fungal chitosan from *Rhizomucor miehei* and *Mucor racemosus*. International Journal of Biological Macromolecules. 2010;**47**:180-183

[26] Li K, Xing R, Liu S, Qin Y, Meng X, Li P. Microwave-assisted degradation of chitosan for a possible use in inhibiting crop pathogenic fungi. International Journal of Biological Macromolecules. 2012;**51**:767-773

[27] Ai H, Wang F, Xia Y, Chen X, Lei C. Antioxidant, antifungal and antiviral activities of chitosan from the larvae of housefly, *Musca domestica* L. Food Chemistry. 2012;**132**:493-498

[28] Hongpattarakere T, Riyaphan O. Effect of deacetylation conditions on antimicrobial activity of chitosan prepared from carapace of black tiger shrimp (*Penaeus monodon*). Songklanakarin Journal of Science and Technology. 2008;**30**:1-9

[29] Tsai G-J, Su W-H, Chen H-C, Pan C-L. Antimicrobial activity of shrimp chitin and chitosan from different treatments and applications of fish preservation. Fisheries Science. 2002;**68**:170-177

3

Sewage Polluted Water Treatment via Chitosan

Thomas Hahn and Susanne Zibek

Abstract

Due to the increasing scarcity of water, wastewater treatment and water conditioning are one of the major future issues. Together with the need to apply highly accessible abundant materials and the demand to replace fossil-based chemicals with sustainable compounds from renewable resources, chitosan (CS) provides some of the solutions to obtain these goals and combines both, abundance and sustainability. Hence, the focus of this review is on the application of CS in wastewater treatment providing advantages and drawbacks in using CS in contrast to chitin. We herewith present the application of CS for coagulation/flocculation purposes, whether as native compound, as functionalized molecule or as blend, respectively, composite. The heavy metal, respectively, dye removal is an additional theme to be addressed in the body of the text. The third topic of this review contains the application of CS blends or composites in order to prepare membrane materials for water purification or conditioning. Together with a summary of the recent study, we discuss these findings and possible consequences for future works. In addition, we provide some theoretical background of the processes that CS is involved in and state some mechanistic insights.

Keywords: adsorption, anionic dye, blends, composites, coagulation/flocculation, heavy metal removal, membrane, wastewater treatment, water conditioning

1. Introduction

There are different kinds of sewages derived from industrial production, agriculture, or directly emerging from the households. As a consequence, the three sectors generate high volumes of wastewaters containing inorganic and organic compounds of every description: dyes, heavy metal ions, antibiotics, hormones, feces, colloids, and further contaminants with a broad structural variety. The removal of these pollutants from wastewater and the conditioning of the water are thus of major interest for the water conservancy. The major parameters that need to be considered and measured in municipal wastewater are as follows: total solids, suspended solids, dissolved solids, total organic carbon, chemical oxygen demand, conductivity, alkalinity, pH, nitrogen, phosphorus, and sulfate content [1]. Special wastewaters, for example, from pharmaceutical or mining industry, require further quantifications such as hormone or heavy metal cation content. According to this, the methods applied for water conditioning mainly depend on the kind of pollution that needs to be separated. In general, processes to purify wastewater include physical, mechanical, chemical, and biological methods with various technologies ranging from adsorption, filtration, biodegradation, oxidation and reduction, UV irradiation, and coagulation/flocculation [2]. However, at the end, qualitative and quantitative criteria required must be met after purification, whereas sustainability as modern key performance is pushed into the foreground. A further aspect concerns the availability and the efficiency of the materials used during conditioning of the wastewater. Chitin, or especially the acid-soluble derivative CS, as the second abundant polysaccharide after cellulose, is one of the polymers to be applied in the wastewater meeting the two requirements. CS has the benefit that it originates from fishery waste and is biocompatible, biodegradable as well as nontoxic and has thus a versatile application portfolio [3]. In contrast to the most naturally occurring polysaccharides, which are either anionic or neutral, such as cellulose, starch, or alginate, CS is a polycation. The free amine groups, which are responsible for the polycationic character, bear nonbinding electrons providing donor properties suitable for coupling to electrophiles, for example, for the formation of imines or amides [4, 5].

One question always arises concerning the application in wastewater treatment: Rather use CS than chitin? Chitin is a glycosidic polymer that has a high degree of polymerization, forms numerous intra- and intermolecular hydrogen bonds, is semi-crystalline, and almost completely acetylated. Chitin manufacturing benefits from known chemical or biotechnological production routes starting at fishery waste that provides a cheap and renewable source. Furthermore, the purification of chitin results in calcium carbonate- or protein-rich side streams that can be further processed supporting economics of chitin production [6]. CS is a derivative of chitin with commonly lower molecular weight and lower acetylation degree, predominantly produced by the chemical conversion of chitin. The conversion reaction includes the application of high concentrated alkaline solution at increased temperature providing thus a nonsustainable process. The harsh conditions are prerequisite due to the fact that the acetamido groups are arranged in a trans-configuration with regard to the hydroxyl group at C3 [7]. The enzymatic deacetylation of chitin at mild conditions is nowadays the content of numerous extensive studies so that the development of prospective economic conversion processes is expected. The prospect of an economic production process contributes to the prediction that CS has a more promising future than chitin in the wastewater treatment [8]. Younes and Rinaudo accordingly stated that CS has a wider range of application areas in comparison to chitin [9]. This is due to the high availability and accessibility of the amine groups, the lower intermolecular forces, and the solubility in acidic aqueous media. Chitin takes advantage of the poor solubility resulting in decreased leaching and thus repeated application in all media. On the other hand, there are processes and functionalizations described increasing

CS stability in acidic media: CS is applicable as powder, flakes, and gel, such as membranes or beads. Gel preparation processes comprise freeze-drying, ionotropic gelation, neutralization, crosslinking, and solvent evaporation method. At least the latter four methods include concluding steps resulting in a more robust CS derivative, composite, or blend [10]. However, CS could be applied as a concentrated solution in acidic media, and this offers the opportunity for homogeneous modification reactions or heavy metal adsorption at the free amine functionalities. The bulky acetyl groups of the chitin provide steric hindrance disabling the proper approximation of reagents to the nitrogen. The inefficient adsorption of heavy metal ions by chitin was already demonstrated [11]. Further, the higher the deacetylation degree, as in CS, the higher the density of the available primary amine groups mainly responsible for the electrostatic interaction. The higher proportion of amorphous regions increases accessibility, sorption capacity and makes it furthermore suitable to act as a flexible linker of different colloids as required for the application in coagulation/flocculation techniques [12].

Flocculation/coagulation is an abundant, efficient, cheap and thus one of the most important processes for the treatment of effluents [13]. The removal of the suspended solids or colloids is mandatory in order to perform the succeeding purification steps. This is one of the processes carried out with inorganic metal salts and/or polyelectrolytes, as CS is. The majority of the CS publications and patents in wastewater treatment focuses on this field of application. Hence, section 2.1 of this chapter elucidates the CS usage in coagulation and flocculation. The specific removal of heavy metals or anionic dyes is a topic that needs to be considered in special effluents as derived from mining or textile industries. CS showed promising results within this field of application and is thus content of section 2.2 of this chapter. The last section has the application of chitin and CS in membrane materials for a theme. Membrane-assisted approaches become more and more important due to their efficiency, improved process controls, and the opportunity for a directed compounding of different materials. Within these applications, CS takes the role of not only a structural substance but also a functional compound as it is the case for adsorption as well as coagulation and flocculation.

2. Main body

2.1. Chitosan as coagulating and flocculating agent

Water quality is commonly diminished by the presence of colloids or smaller organic substances resulting in a high chemical oxygen demand (COD) and high turbidity. The removal of the majority of these compounds is predominantly performed via coagulation and flocculation. Though coagulation and flocculation are used interchangeably, they represent two distinct processes: coagulation is the act of destabilizing a suspension, whereas flocculation means either the spontaneous or polymer-induced formation of large agglomerates succeeding destabilization [14]. The technique transforms colloidal particles or solutes with settling times of years to flocculated or precipitated particles with settling times between seconds and hours. Furthermore, the moisture content of the resulting slurry from up to 99% can be reduced to 65–85% enabling succeeding processing [15]. In general, different organic and inorganic coagulants and organic flocculants were used. Often, ferric chloride, lime, or alum as agents of choice are applied as the primary coagulant for the destabilization of suspensions due to their availability, efficacy, and their little cost [16]. However, residual concentrations of alum bear environmental risks to the whole aquatic biota [17, 18]. Synthetic polymeric flocculants, as polyacrylamide, are currently used based on the ability to improve flocculation despite their application is associated with lack of biodegradability and the release of degradation products [19]. A decrease or substitution of these reagents by the addition of biopolymers would be beneficial regarding sustainability, ecology, and health. One of the polymers focused on within the investigations was CS. An overview of the recent studies concerning the usage of CS for coagulation and flocculation purposes with original wastewaters rather than model solutions can be obtained from **Table 1**.

The experiments summarized in the table were carried out with different kinds of wastewaters and with native crab shell CS rather than the modified one. Economic viability would be greatly decreased if CS modification is essential to ensure a high efficiency of such a low-cost unit operation even if several investigations were performed with grafted or modified CS [27]. The results revealed that the CS could significantly reduce COD (>60%) and turbidity of the treated solutions (>80%) applied to all kinds of wastewaters and at different conditions. Regarding this, the authors stated that CS has approved to be efficient, concerning the coagulation of suspended matter in wastewater even at low temperatures and low doses [28].

Wastewater from	Effect	Notes	Refs.
Textile treatment	72.5% COD and 94.9% of nephelometric turbidity units (NTO) reduction	Optimum performance at pH 4–6 and 30 mg/L chitosan	[20]
Pulp and paper mill industry	90% turbidity reduction and 60% COD reduction	$FeCl_3$ as coagulant, CS as auxiliary, improves sedimentation and compaction, heterogeneous photocatalysis succeeded	[21]
Olive oil and wine production	80–94% turbidity reduction, 81–94% decrease of total suspended solids, and 73% COD decrease	Best performance at actual pH of wastewater (olive oil wastewater) or no significant influence of pH (wine production wastewater)	[22]
Catfish farming	>99% of microalgae removal efficiency, significant NTU reduction and 80% of the microalgae recovered	CS could be used for both removal recovery of biomass and the overall reduction of the microbial community	[23]
Cardboard industry	Turbidity lowered more than 85% and COD more than 80%	Higher efficiencies than polyaluminum chloride; additionally decreases heavy metal, removes colored compounds	[24]
Tequila industry	CS was the most efficient biopolymer removing 84% of COD	Catalytic oxidation succeeds coagulation/ flocculation process	[25]
Rural domestic water treatment	Turbidity reduction of at maximum 99%	Unmodified chitin was used in comparison to alum and sago	[26]

Table 1. An overview about latest results obtained applying CS for coagulation/flocculation purposes with "real" effluents.

The independence of the CS flocculation efficacy from different environment conditions, as temperature, is mandatory since the proposed application is likely to be carried out in non-temperated outdoor sewage plants or in open aquatic systems. The application in aquatic systems for fish or shrimp farming reasons on the requirements to reduce the high nutrient loading and to prevent the accumulation of toxic substances. Synthetic polymers or inorganic compounds mentioned before are not adequate to perform this since they are commonly not biocompatible and biodegradable. For the same reason, CS was successfully applied in accelerating the sedimentation of microalgae. The authors reported differing flocculation efficiencies depending on the pH and dosage but approved that CS can be successfully applied to harvest microalgae [29]. Hence, the application of CS to induce biofloc formation with microorganisms for clarification purposes or reducing harmful nutrient contents would be suitable. This could on one hand increase the water quality in aquatic systems, on the other hand, the resulting flocs can be easily removed or uptaken by the marine animals [23, 30]. However, it can be expected that the growth of different bacterial strains is decreased due to the antibacterial characteristics of the CS constraining the application to eukaryotic microorganisms that are used for this process [31]. Since the antibacterial property is dependent on the amount of amine functionalities, chitin usage in aquatic systems is worth trying out. Native chitin was also successfully applied to decrease COD and turbidity of surface water, but it has to be considered that high concentrations of chitin (>0.1 g/L) were applied. At the same time, surface water contains a lower amount of contaminants in comparison to industrial or municipal effluents [26]. This reduces the amount of flocculant or "active" sites necessary, possibly enabling the application of chitin.

In contrary to the findings of the authors summarized in **Table 1**, it was proposed several years ago that the effectiveness of flocculants based on biopolymers is low compared to synthetic polymers [27]. It is also obvious that a combination of CS with other techniques seems to be favored rather than the direct application as flocculant [21]. The addition of CS could significantly lower the amount of coagulant required or enhance floc formation, approved for the flocculation of a model system with clay or bentonite, the conditioning of groundwater or for clarification of pulp and paper mill wastewater [21, 32, 33]. Such a process is exemplary provided for the synergistic action with Fe salts shown in **Figure 1**.

In **Figure 1**, ferric chloride is applied as a primary coagulant to destabilize the system, and the so-called perikinetic flocculation due to Brownian motion under vigorous stirring succeedingly occurs. This results in a formation of smaller flocs. The polyelectrolyte activity initiating a large floc formation (orthokinetic flocculation) is not only based on one mechanism. For cationic polymers, such as CS, two main mechanisms can be postulated which occur coincidently: (a) Bridging by the adsorption of one polymer to adjacent colloids (bridging model). This applies also to polymers bearing the same charge as the colloid. (b) Reduction of the electronic repulsion between adjacent colloids by electrostatic interaction, adsorption, and charge neutralization of a polymer with opposite charge (patch mechanism) [14]. The interaction of the CS with the colloids leads to the formation of flocs having low settling times. The performance of floc formation and turbidity removal is withal dependent from the CS dosage, thereby approving the contribution of the patch mechanism. First, increasing CS dosage fosters turbidity decrease; further CS addition concludes in a contrary effect based on the

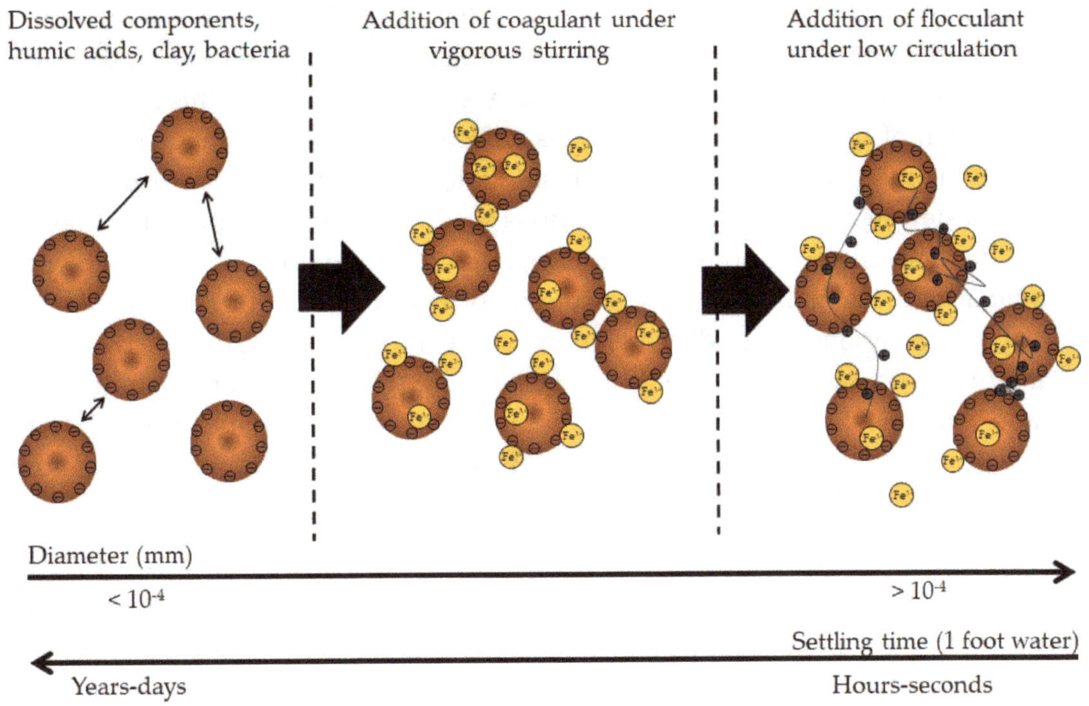

Figure 1. Mechanism for coagulation/flocculation of colloids or solutes with a ferric chloride-CS system. The mechanism is exemplarily shown for Fe^{3+} salts (yellow circles) as primary coagulant and CS (gray line) as flocculant auxiliary. The size relations are not representative.

repulsion of the excess CS adsorbed to the colloids [34]. The size or molecular weight of the polysaccharide provides a further influencing factor. Investigations revealed a higher efficacy of the turbidity removal with increasing CS molecular weight. This approves that bridging also contributes to the effect, but only in tap water, determining the importance of the ionic strength for the mechanism [14, 35]. Guibal et al. stated that the effect of the deacetylation degree was not very significant except at nearly neutral pH values and low ionic strength suspensions. This confirms that a variance analysis with only one independent factor to optimize the flocculation process with CS is not expressive without coincident consideration of other relevant parameters [36].

Based on the results, CS was approved to be an efficient flocculant auxiliary but needs to be applied at an optimum dosage and as a polymer bearing physical-chemical characteristics suitable for the application. Furthermore, the performance is greatly influenced by the pH of the reaction medium [37]. The pH dependency of the CS is linked to the charge density, as confirmed by several authors, providing a double-edged sword [38, 39]. The pH of an effluent can scarcely be adapted due to the commonly high volumes of wastewater generated. This applies not only to the application as flocculant but the more for the usage as adsorbent which is highly sensitive for pH changes.

2.2. Chitosan as sorbent

The coagulation/flocculation mechanism is also effective with regard to the removal of anionic compounds and positively charged heavy metal ions via solid–liquid separation. Besides the destabilization of dispersions and the removal of the formed floccules, CS can also act as a specific adsorbent although both processes, coagulation/flocculation and adsorption, cannot be investigated separately and occur in general simultaneously. In this section, we focus on the adsorption of the mentioned contaminants present in the effluents. This is especially the case for effluents generated by metal finishing, textile dyeing, or board manufacturing, resulting in wastewaters with high concentrations of toxic heavy metal ions and anionic dyes. Native and derivatized CS demonstrated to separate both compounds with a high effectivity [40]. The capacity of the native CS to adsorb dyes or heavy metals ions is in general dependent from various parameters as deacetylation degree, the particle size, the physical state of the CS, the pH value, and the temperature [41–44]. According to different studies, deacetylation grade is the most relevant parameter and thus the primary amine groups [7]. It has to be stated that the total amount is not relevant but the accessible amount of amine groups is, depending on crystallinity and diffusional properties [45].

Native CS is able to interact with other compounds via free primary alcohol or free primary amine groups depending on the system conditions. In comparison to the application as flocculant, it is widely accepted to modify or combine the CS in order to modulate the stability, rigidity, and viscosity [46]. Crosslinking as one of the most prominent modification procedures prevents leaching of CS at acidic pH and gives additionally the opportunity to recycle, respectively, reuse, the resin [40]. Furthermore, modification is carried out to increase sorption capacity as well as selectivity to adsorb specific compounds as can be inferred from **Table 2**. There are two general modification processes described here for CS: the linkage to reactive molecules and thus the insertion of functional groups, named as grafting, or the crosslinking reactions to form a dense network of CS chains conferring stability to the resin [47]. For derivatized CSs, crosslinking method, crosslinking grade, and the kind of derivatization are crucial for the performance. There are several techniques, covalent and ionic, to crosslink the CS [48]. A third opportunity besides the crosslinking and grafting is the formation of composites or blends to combine the benefits of CS and other materials or better, to develop synergistic effects [40, 49]. Here, the type of compound used and the content greatly alter the functionality and efficiency.

Common to all experiments is that the majority of compounds applied in combination or used to derivatize CS are non-sustainable materials (polyacrylamide and epichlorhydrin). To become the benefits important and to pursue a holistic sustainable approach, materials from renewable resources have to be applied in combination with CS, focusing investigations concerning the removal efficiency with different effluents. All investigations, summarized in **Table 2**, were performed with aqueous solutions spiked with model substances. Although this is only a short overview of the current study, it is obvious that there is a lack of experiments carried out with "real" wastewaters as it is shown in **Table 1** for the coagulation/flocculation studies. Studies with model solutions are suitable for an estimation of the prospective potential and application field but cannot substitute the experiments with effluents. This bases on an activity and stability reduction that has to be expected in a complex matrix. However, results revealed the effective removal of heavy metal ions or dyes from model solutions. Especially the dye removal based on the polycationic character of CS seems to be promising indicated by the high removal efficiencies.

Substrate	Agents	Adsorption characteristics	Notes	Refs.
Synthetic heavy metal solutions	CS dianhydride (ChD)and ChD amine (ChDA)	>66% removal of Cu^{2+}, Pb^{2+}, Ca^{2+}, Ni^{2+}, Zn^{2+}, Cd^{2+}, and Cr^{3+}	Binding constants from 10^4 to 10^5 M^{-1}; ChDA sedimentation velocity 20* higher than native CS	[50]
	Native CS, polyethylene-imine-CS	146 mg/g Cu^{2+} capacity	Desorption with alkaline EDTA; capacity of native CS lower	[51]
	Carboxymethyl CS-hemicellulose	Up to 909 mg/g Cd^{2+} and 333 mg/g Cu^{2+}	Desorption with EDTA; covalent bond via thermal crosslinking	[52]
	Glycine or chloroacetic acid reacted CS	280 mg/g Cu^{2+}, 99% removal of Cu^{2+} and Co^{2+}	Modification does not significantly enhance properties; pH 9 adsorption works best	[53]
Synthetic dye and heavy metal solution	Carrageenan/ CS-microspheres	212 mg/g for dyes and 20 mg/g for Cu^{2+}	Ampholytic microspheres with magnetic Fe_3O_4 core were applied	[54]
	CS-lignin composites	>95% removal and 111 mg/g for RBBR, >95% removal and 20 mg/g of Cr(VI)	The composites exhibit better performance than CS and lignin, no significant pH effect	[55]
Synthetic dye solutions	Carboxymethyl CS grafted polyacrylamide	>93% removal of dyes	Effective removal of anionic and cationic dyes also results in flocculation of dyes	[56]
	CS crosslinked with sulfonates and epichlorohydrin	>90% removal of dyes	Ampholytic character enables the removal of different dyes with one crosslinked CS	[57]
	CS/PVA-blends	130 mg/g capacity for RR	Desorption by pH increase, Langmuir model fits best	[49]
	Nano-ZnO/CS composite beads	76% removal, adsorption capacity of 190 mg/g for RB5	Langmuir model fits best, pH 4 suitable for adsorption	[58]

RBBR: Remazol Brilliant Blue R; RB5: Reactive Black 5; RR: Reactive Red.

Table 2. Abstract of the current research and results concerning the application of native and derivatized CS in dye and heavy metal removal.

2.2.1. *Adsorption of dyes*

Dyes as adsorbate are usually classified with regard to their charge, succeeding dissolution in water. There are cationic (basic) dyes, reactive (acidic dyes), and non-ionic (dispersed) dyes. The adsorption of anionic dyes is a property originally derived by native CS due to the cationic character at low pH values. Electrostatic interactions play thus the major role with regard to the adsorption of the dyes. The modification of the CS is commonly carried out to improve stability or to extend the adsorbate spectrum. Herrera-González et al., for example, crosslinked the CS with epichlorohydrin and subsequently grafted the modified polymer with sulfonates resulting in the ability to capture positive-charged dyes [57]. In general, to capture a broader spectrum of dyes and to overcome the hurdle that they show inert properties, CS composites, as CS/bentonite, CS/montmorillonite, or CS/activated clay, were stated as promising materials. As an additional feature, the materials provide an increased stability at low pH values [40]. The compounds to form the composite resins implement new properties resulting in a variety of further interactions between dyes and adsorbent and thus in stronger bonds [55]. Bond strength between adsorbent and adsorbate can be assessed by thermodynamic measurements. The adsorption of reactive dyes with crosslinked CS or CS composites revealed enthalpic values from −53 to 46 kJ/mol. This determines that the enthalpy values are highly dependent on the crosslinking agent and the other compounds the CS is applied with. This is not the case for the Gibbs energy showing low negative values for all investigations and thus exhibiting a spontaneous reaction [58–60].

Rashid et al. bridged the gap between heavy metal removal and dye adsorption. They applied a mixed Fe^{3+}- and Cu^{2+}-CS complex for the efficient removal of Reactive Black 5 [61]. The separation of both, dyes and heavy metal ions, is the content of several investigations using composites. It has to be stated that the adsorption capacity of the materials is decreased (~20 mg/g) in respect to materials developed for the recovery of one compound class alone even if the simultaneous separation was not in the focus of these experiments. Hence, the adsorption of heavy metals should be focused using other CS grafts, blends, or composites.

2.2.2. *Remediation of heavy metal pollutants/contaminants*

Heavy metal contaminations constitute a severe risk for the environment and also for humans. Not least because they are at the top of the food chain, humans will inevitable uptake and accumulate heavy metals released. An adsorptive removal directly at the source of formation would thus be advantageous in minimizing exposition potential. Based on many studies, CS provides an adsorbent to accomplish this task.

Since the term "heavy metal" is inconsistently defined in the scientific literature [62], we refer to toxic metals with a high density and their oxyanions. The majority of investigations thus focuses on Al^{3+}, Cu^{2+}, Pb^{2+}, Hg^{2+}, Ni^{2+}, Zn^{2+}, Cd^{2+}, Cr^{3+}, and Fe^{3+}, demonstrating the successful application of CS for the removal of these metal ions. On the other hand, CS does not tightly adsorb alkali and alkaline earth metals according to the HSAB (hard soft acid base) principle [63].

As it is the case for the dye removal, the combination of CS with other compounds was in the scope of recent investigations: the hydroxyapatite/CS nanostructures were efficient concerning the removal of Pb^{2+} from wastewater, revealing a higher adsorption capacity than comparable sorbents [64]. The same applies to CS-tannic acid modified biopolymers being able to adsorb Pb^{2+} and Al^{3+} [65]. Blending lignin and CS provides a sustainable material for the removal of metal ions, whereas not only interpolymeric interactions exist but also synergistic effects to capture the adsorptive. The authors state several adsorption sites for one adsorptive based on the interactions derived from hydrogen bridge bonds [55]. All together is that the CS additionally provides a backbone for modifications with functional molecules, improving the chelation of metal ions. However, the results also indicate that the CS itself significantly contributes to the adsorption of these compounds [53].

Investigations approved the high affinity for different metal cations, as, for example, Fe^{3+} to CS by an equilibrium constant of 9.49×10^5 M^{-1} determining that the equilibrium is strongly shifted toward the CS-heavy metal complex [66]. These findings were confirmed by thermodynamic measurements revealing highly negative Gibbs free energy values (−37 kJ/mol) and negative enthalpic values (−41 kJ/mol) [67]. In contrast to that, Negm et al. claimed that the adsorption of copper and cobalt ions is endothermic and entropy-driven, likely reasoning on the different conditions and the functionalization of shrimp shell CS with glycine/chloroacetic acid greatly altering the sorption mechanism [53]. Summarizing the CS-metal ion equilibrium systems revealed that particularly the Langmuir isotherm was the isotherm of choice to analyze the equilibrium data in over 30 systems since they provided a very good fit to the data. However, the authors mentioned that there is a lack of comparable results with other isotherms [68].

Although the detailed mechanism of metal sequestration remains unclear, in general, the removal of metal ions by the action of CS can occur via coprecipitation, chelation, coordination of amine groups as well as ligand exchange or electrostatic interactions with protonated amine groups [69, 70]. The majority of the complexation and chelation studies were performed using Cu^{2+} which is particularly reasoned by the fact that copper ions provide the highest affinity towards CS (up to 1.2 mmol/g) [43]. According to the HSAB model of Pearson, nitrogen as the hard base is appropriate, donating electrons to a borderline acid such as Cu^{2+} [71].

In contrast to the removal of anionic dyes, it is stated that the adsorption of heavy metal ions decreases due to protonation. A chelation process would be efficient at increased pH values since the adsorption of different metal cations is mainly attributed to the unprotonated amine groups of the CS acting as ligands of the metal ion [72]. In the year 1986, the bridge model was one of the first trials to propose a coordination geometry emphasizing the relevance of the amine groups for adsorption [73]. Schlick suggested a square planar structure of the CS-Cu^{2+} complex with four nitrogen ligands derived from the primary amine groups of CS. An octahedral coordination geometry (coordination number: 6) is formed by the arrangement of axial water molecules (see **Figure 2a**) [74]. However, there are investigations suggesting the C3-hydroxyl group as further ligand for complexation substantiated by thermodynamic data that led to the development of refined models (**Figure 2b**) [67]. Further investigations supposed a neutral complex (see **Figure 2c**) consisting of two nitrogen ligands and two hydroxide ions coordinating the copper ion occurring at high Cu^{2+} loadings and pH values of >5.5 [75, 76]. The structures illustrated in **Figures 2a–c** can be described as "bridge model "since they involve the chelation of the Cu^{2+} by nitrogen atoms from different glucosamine units in an inter- or intramolecular fashion [77]. The structure depicted in **Figure 2d** is called the "pendant model "and is up to date the most prominent model describing the interaction between Cu^{2+} and CS supported by potentiometry and circular dichroism data [78]. The model developed by Ogawa et al. in the year 1993 based on X-ray studies assumes the chelation of different heavy metal cations by one amino group only [79]. On the other hand, several authors stated that at least a degree of polymerization of four is required to affect an efficient chelation, the consequence being that not only one glucosamine unit is responsible for chelation [80]. However, the truth probably lies somewhere in between these models based on the dependency of the heavy-metal ion-CS ratio and pH [75]. Some amine functions may be inaccessible for chelation, and others suffer from steric hindrance to form a regular coordination geometry. This is based on the fact that CS provides a natural polysaccharide with all its heterogeneities.

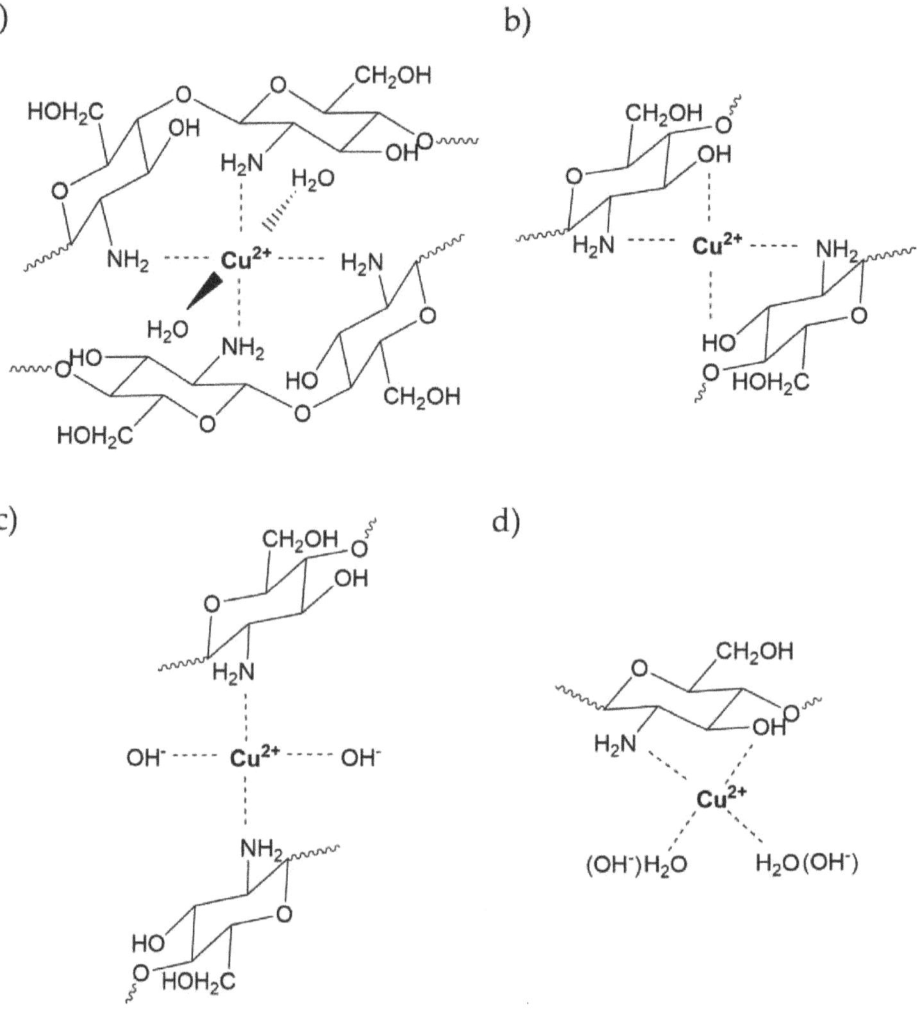

Figure 2. Different chelation mechanisms between CS and metal ions using the example of copper ions: (a) octahedral coordination geometry of the CS-Cu^{2+} complex with four amine and two axial water ligands; (b) Cu^{2+} chelation mechanism utilizing the C3-hydroxyl groups of two glucosamine units; (c) neutral complex with two nitrogen ligands and two hydroxide ions coordinating the Cu^{2+}; (d) "pendant model", chelation of Cu^{2+} by one amino group and the C3-hydroxyl group of one glucosamine unit.

The desorption of metal cations succeeding the adsorption onto CS is scarcely described in the studies. The desorption of the metal ions was either performed by the application of ammonium chloride, potassium iodide, or by EDTA [81]. Shifting pH represents the most suitable method for elution. For example, a pH decrease resulted in a 94% cadmium ion desorption, whereas 8.3 mM H^+ per gram of beads was adsorbed to substitute the bound metal ion [82].

Oxyanions like chromate or vanadate can be adsorbed by protonated amine groups at lower pH values due to electrostatic interactions resulting in an exothermic reaction [67, 83]. The simultaneous recovery of oxyanions and metal cations is described in a further study. In common, protonation reduces the adsorption capacity for metal cations but increases the effectivity of metal anion adsorption. The authors performing the experiments take advantage of the distribution of deprotonated and protonated amine groups in a pH range of 5–6. Deprotonated amine groups enable the formation of chelate complexes with Ni^{2+}, Cu^{2+}, and Fe^{2+}, the metal anions were attracted by electrostatic interaction [84]. The optimum pH removing majority of metal anions by electrostatic attraction is in the range of 2 and 4. At lower pH values, competitive pressure by other anions derived from the acid for pH adjsutement for binding sites on CS is drastically increasing [77]. The decreased adsorption capacity of metal anions in the presence of, for example, high chloride, sulfate, or nitrate concentrations is based on the same effect. A further option to remove toxic oxyanions as selenite or arsenite was provided by Yamani et al. The group demonstrated that the chelation of copper ions leads to the formation of electron-accepting "anion" adsorption sites which did not exist previously, enabling the directed binding of these oxyanions. The bimetallic complex connected via oxygen linker offers the separation of both even in the presence of phosphate at high concentrations [76]. According to this, it was approved that Fe-crosslinked CS enables the adsorption of chromate. The adsorption mechanism is suggested to be a ligand exchange substituting an anion by the chromate in the coordination sphere of the iron ion, resulting in an uptake of Cr(VI) (295 mg/g at pH 4.8) [85].

Resuming the study and investigation results, CS is a valuable sorbent for dyes and heavy metals. However, an efficient simultaneous removal of both compound classes with native CS is unlikely due to the pH dependency of both processes. A successful removal can be expected if the solute containing media has a suitable pH for adsorption. CS solubilization can be prevented by crosslinking, functionalization, or blending, additionally resulting in an increased performance of the resin, widening pH range for optimum sorption and creating synergies between the compounds as for CS-functionalized membranes.

2.3. Chitosan-functionalized membranes

The previously mentioned wastewater treatments, adsorption and coagulation/flocculation, require the direct physical-chemical interaction of the effective agent with the contaminant in the effluent. The intention is to remove the compounds from the bulk solution to achieve effluents for further processing steps. By way of contrast, membrane-assisted applications are commonly applied to enable the purification or conditioning of water with regard to physical rather than chemical properties. The composition of the membrane and the quantity of the materials contained are of great importance to provide selectivity for membrane permeation. In common use, membrane materials consist of synthetic polymers and their composites or blends. Green and sustainable compounds as membrane components are highly demanded for well-known reasons. According to Dobosz et al., biopolymers could additionally reduce biofouling which is crucial for the lifetime, increasing the time span between sanitization cycles of the membrane [86]. Especially CS with its antibacterial activity is thus predestined for the production of membranes sensitive to fouling. Carboxymethyl CS membranes, for example, were applied during a 6-week protein separation process within which no fouling or deterioration in the membrane flux was recorded [87]. Weng et al. approved the antimicrobial activity of a cellulose/CS membrane against *Escherichia coli* in disc diffusion experiments [88]. Studies concerning CS-coated polyacrylonitrile hollow fiber membranes approved the antimicrobial and antibiofouling effect in respect to Gram-positive and Gram-negative bacteria [89, 90]. Another aspect to consider is the hydrophilicity of the membranes, which is a major requirement in water-conditioning applications to obtain membrane permeability. CS provides a high hydrophilicity allowing especially water from aqueous solutions to permeate. Together with a high salt rejection efficiency, this is also the relevant property for the main application fields of CS membranes which can be obtained from **Table 3**.

As can be inferred from the table, CS is content of reverse osmosis (RO), forward osmosis (FO) and nanofiltration membranes. Nanofiltration membranes differ from the other mentioned in the ability to separate particles in the size of 2–5 nm and thus enable permeation of minerals, which would be separated by osmosis membranes. In contrast to FO, RO processes work against the osmotic potential demanding membranes produced to resist high pressure. Both together have the need for semi-permeable membranes revealing high salt rejection grades and high water permeability, approving high efficiencies at moderate costs.

Research and innovation activities concerning the utilization of CS in the three membrane processes are mainly rooted in the countries in North Africa and the Arabian Peninsula. This originates from the access to seawater, resulting in a high availability of crustacean-derived chitin/CS and water deficiency leading to an increased demand for conditioning of water. Seawater is scarce, whereas saltwater is ubiquitous. Hence, in the majority of studies, synthetic monovalent salt solutions were applied, and the results revealed that removal by the CS-containing membranes is

Application	Membrane material	Effect/notes	Refs.
Reverse osmosis (RO) with NaCl solutions	Polyamide-6/CS	Salt rejection increases with the addition of CS as additive up to 52%	[91]
	CS crosslinked graphene oxide (GO)/ titania hybrid lamellar membrane	Rejection rate is ~30% at 7.2–14.3 LMH/bar	[92]
	Thin film composite RO membranes covalently linked to CS	>90% salt rejection at LMH/bar >3	[93]
Forward (FO) and pressure-retarded RO with NaCl and Na_2SO_4 solutions	Polymerization of CS with trimesoyl chloride on the surface of SPES/PES support layer	95% NaCl rejection with up to 4.6 LMH/bar permeability, higher CS concentration revealed lower water permeabilities, but higher salt rejections	[94]
FO with synthetic salt and sucrose solution	Membrane consisting of layer-by-layer assembly of CS and GO nanosheets on a sulfonated polyethersulfone (SPES)/ PES support layer	>90% Na_2SO_4 rejection at 3 LMH/ bar, 2–4 orders of magnitude higher water flux compared to polyamide membrane on sulfonated PES/PES support	[95]
Nanofiltration with synthetic mineral saltwater	CS crosslinked buckypaper membranes	Rejection rates: 80–95% $MgCl_2$, 21–63% for NaCl, 18–37% $MgSO_4$ and 6–14% for Na_2SO_4; water permeability of 0.2–0.9 LMH/bar	[96]
Nanofiltration, retention of basic, neutral and acidic dyes	Polyurethane foam membrane filled with humic acid-CS crosslinked gels	Rejection efficiency >60% for all dye types, >99% for the anionic dyes, adsorption of dyes at the membrane	[97]
Nanofiltration, synthetic emulsion and dye solution	Coating layer with CS and silver nanoparticles, alginate nanofibers as midlayer and nonwoven as mechanical support	Nanoparticle retention >98%, oil removal >93%	[98]
Nanofiltration, humic acid retention	Carboxymethyl CS was blended with polyvinylidene fluoride	>97% humic acid retention; low irreversible fouling, good permeability, durability and stability	[99]

Table 3. An overview of the application fields of chitinous membrane materials and the thereof obtained results.

successfully carried out but with swaying rejection rates (30–90%) [91–93]. Separation efficiencies of nanofiltration membranes are significantly higher considering low-molecular weight organic compounds (>60%) grounding on larger molecular weight differences between the molecules to be separated and thus higher selectivity. Hence, it is not necessary to build high-density networks, which can be produced by utilizing the amine functions of CS as anchor points for modifications. In common, CS is not applied as native but as crosslinked polymer embedded in a matrix or coated on a support layer in order to introduce and combine the advantages of several compounds, or to compensate their weaknesses. Researchers report on the incorporation of modern and innovative components in CS, such as metal-organic frameworks, developing synergies of the two materials, resulting in $MgCl_2$ rejection efficiencies of 93% [100]. As already mentioned, the hydrophilicity of the membrane is the relevant factor for the water flux commonly determined by measuring the contact angle. CS coating of membranes indicated a significant higher water flux than the native membranes and thus resulted in a decreased pressure and energy demand in the process [101].

Swelling of CS is one of the properties to be compensated for the adequate application in membrane technology. As swelling is tantamount to a high water content, this greatly affects the water permeability as well as the mechanical strength of the membrane. Investigations revealed that the ability to swell must be controlled to create membranes that enable a selective separation of water and salt whatever, simultaneously guaranteeing a high water flux [102]. Decreasing the swelling of CS-based membranes means to constrain the movement of the CS chains especially in solvents in which the CS can be solubilized [103]. Further properties affecting the swelling behavior are the pH of the medium and the resulting electrostatic repulsion of CS chains at low pH values [104]. Crosslinking is thus a suitable tool to control this phenomenon as several researchers reported a significant decrease in swelling by crosslinking or modification of the CS up to 300% [96, 105].

Finally, CS seems to be a promising material with regard to the application as a membrane component. Its antibacterial activity in combination with the functionality of the amine groups provides a suitable tool to prevent fouling and coincidently adapts the network to the substrates to be filtrated. The challenges to be mastered are the reduction of swelling of these membranes while approving a high water permeability predominantly in drinking water purification. In addition, further investigations concerning osmosis membranes have to test real saltwater samples not lacking all other natural occurring compounds than sodium chloride as is the case for the synthetic model solutions.

3. Conclusion/summary

Within this publication, we reviewed the purification of effluents with native and modified CS as well as the application of CS-containing membranes for filtration purposes. Crosslinking, derivatization, and the production of composites or blends with other natural and synthetic polymers as well as low-molecular weight compounds are the main type of application described in the study rather than the usage of native CS. It seems to be appropriate to introduce new functionalities, to prevent leaching, or to foster the beneficial properties. Due to these manifold-positive properties in combination with other compounds, the preconditions are favorable for the implementation of CS in wastewater treatment. CS represents a compound to be effective as coagulant/flocculant in the native state, whereas the addition as flocculant auxiliary decreasing the amount of inorganic coagulants required for the process seems to be promising. The good overall performance adsorbing heavy metal ions and dyes is stated in several investigations, enabling CS to be applied in the treatment of special wastewaters, such as textile and mining effluents. The simultaneous adsorption of both is limited due to the necessity to adjust the pH in order to protonate/deprotonate the CS, effectively removing the single compounds, respectively. Investigations concerning CS-containing membranes showed that a biopolysaccharide could also contribute to more sophisticated water conditioning processes. The water permeability and the selectivity have to be evaluated especially considering the swelling behavior. It can be assumed that CS can also be implemented to produce switchable membranes, which means membranes altering the properties due to pH shifts. However, the investigations concerning the application of CS and its derivatives suffer from several drawbacks not adequately addressed in the past, aggravating the market accessibility (1) cost factors were not considered; (2) experiments limited to lab scale; (3) only batch experiments were carried out; (4) mechanical strength should be increased; (5) studies have to be performed with actual wastewaters; (6) regeneration of the materials was not investigated; (7) swelling behavior of CS needs to be limited, and (8) the heterogeneity of different CS batches is not considered, yet [47, 106]. The heterogeneity particularly derives from the origin of the CS, the crab shells being exposed to varying environmental conditions. Not only the CS derived from fishery waste can be applied for wastewater treatment, but also the CS isolated from fungi or insects for the provision of more homogeneous batches. Since the production of fungal biomass or insect-based protein is already industrially established, a higher quantity and quality of fungal- or insect-based CS can be assumed and could be applied in a prospective effluent purification. For example, Adnan et al. successfully applied commercial fungal CS to purify a synthetic kaoline solution and palm oil mill effluent. They stated that in contrast to crab shell and shrimp shell CS, the fungal CS is available all over the year. Further, it has a narrow molecular weight based on the controlled production process, opposing the argument that the heterogeneity of CS limits its application. The authors assume therefore that prospective works will focus on fungal- and insect-based CS, increasing the opportunity to develop profitable products with improved properties [107].

Conflict of interest

The authors declare that there is no conflict of interest.

Appendices and nomenclature

CS	chitosan
FO	forward osmosis
LMH	liters per m^2 per hour
RO	reverse osmosis

Author details

Thomas Hahn and Susanne Zibek*

*Address all correspondence to: susanne.zibek@igb.fraunhofer.de

Fraunhofer-Institute of Interfacial Engineering and Biotechnology, Stuttgart, Germany

References

[1] Von Sperling M. Wastewater Characteristics, Treatment and Disposal. 1st ed. London: IWA Publishing; 2007. 296 p

[2] Rajasulochana P, Preethy V. Comparison on efficiency of various techniques in treatment of waste and sewage water – A comprehensive review. Resource-Efficient Technologies. 2016;**2**:175-184. DOI: 10.1016/j.reffit.2016.09.004

[3] Shukla SK, Mishra AK, Arotiba OA, Mamba BB. Chitosan-based nanomaterials: A state-of-the-art review. International Journal of Biological Macromolecules. 2013;**59**:46-58. DOI: 10.1016/j.ijbiomac.2013.04.043

[4] Tondwal R, Singh M. Chitosan functionalization with a series of sulfur-containing -amino acids for the development of drug-binding abilities. Journal of Applied Polymer Science. 2018;**135**. DOI: 10.1002/app.46000

[5] Cumpstey I. Chemical modification of polysaccharides. ISRN Organic Chemistry. 2013;**2013**:27. DOI: 10.1155/2013/417672

[6] Yan N, Chen X. Don't waste seafood waste. Nature. 2015;**524**:155-157. DOI: 10.1038/524155a

[7] Rhazi M, Tolaimate A, Habibi Y. Interactions of chitosan with metals for water purification. In: Polysaccharide Building Blocks. John Wiley & Sons, Inc.; 2012. pp. 127-141. DOI: 10.1002/9781118229484.ch4

[8] Rangel-Mendez JR, Barrios VAE, Davila-Rodriguez JL. Chitin based biocomposites for removal of contaminants from water: A case study of fluoride adsorption. In: Elnashar M, editor. Biopolymers. InTech; 2010

[9] Younes I, Rinaudo M. Chitin and chitosan preparation from marine sources. Structure, properties and applications. Marine Drugs. 2015;**13**:1133-1174. DOI: 10.3390/md13031133

[10] Krajewska B. Membrane-based processes performed with use of chitin/chitosan materials. Separation and Purification Technology. 2005;**41**:305-312. DOI: 10.1016/j.seppur.2004.03.019

[11] Inoue K, Baba Y. Chitosan: A versatile biopolymer for separation, purification, and concentration of metal ions. In: Senguta AK, editor. Ion Exchange and Solvent Extraction. Vol. 18. Boca Raton: CRC Press; 2007

[12] Ioelovich M. Crystallinity and hydrophility of chitin and chitosan. Research and Reviews: Journal of Chemistry. 2014;**3**:7-14

[13] Teh CY, Budiman PM, Shak KPY, Wu TY. Recent advancement of coagulation–flocculation and its application in wastewater treatment. Industrial & Engineering Chemistry Research. 2016;**55**:4363-4389. DOI: 10.1021/acs.iecr.5b04703

[14] Bratby J. Coagulation and Flocculation in Water and Wastewater Treatment. 3rd ed. London: IWA Publishing; 2016

[15] Zemmouri H, Mameri N, Lounici H. Chitosan use in chemical conditioning for dewatering municipal-activated sludge. Water science and technology : a journal of the International Association on Water Pollution Research. 2015;**71**:810-816. DOI: 10.2166/wst.2014.532

[16] Lin J, Couperthwaite SJ, Millar GJ. Effectiveness of aluminium based coagulants for pretreatment of coal seam water. Separation and Purification Technology. 2017;**177**:207-222. DOI: 10.1016/j.seppur.2017.01.010

[17] Agarwal S, Tyagi I, Gupta VK, Dehghani MH, Ghanbari R. Investigating the residual aluminum elimination from conventional and enhanced coagulation by phosphate compounds in wastewater treatment process. Journal of Molecular Liquids. 2016;**221**:673-684. DOI: 10.1016/j.molliq.2016.06.051

[18] Smeltzer E, Kirn RA, Fiske S. Long-term water quality and biological effects of alum treatment of Lake Morey. Vermont. Lake and Reservoir Management. 1999;**15**:173-184. DOI: 10.1080/07438149909354115

[19] Lee CS, Robinson J, Chong MF. A review on application of flocculants in wastewater treatment. Process Safety and Environmental Protection. 2014;**92**:489-508. DOI: 10.1016/j.psep.2014.04.010

[20] Ariffin M, Hassan M, Pei Li T, Zainon Noor Z. Coagulation and flocculation treatment of wastewater in textile industry using chitosan. International Journal of Engineering and Applied Sciences (IJEAS). 2009;**4**:43-53

[21] Rodrigues AC, Boroski M, Shimada NS, Garcia JC, Nozaki J, Hioka N. Treatment of paper pulp and paper mill wastewater by coagulation-flocculation followed by heterogeneous photocatalysis. Journal of Photochemistry and Photobiology A: Chemistry. 2008;**194**:1-10. DOI: 10.1016/j.jphotochem.2007.07.007

[22] Rizzo L, Lofrano G, Belgiorno V. Olive mill and winery wastewaters pre-treatment by coagulation with chitosan. Separation Science and Technology. 2010;**45**:2447-2452. DOI: 10.1080/01496395.2010.487845

[23] Mohd Yunos FH, Nasir NM, Wan Jusoh HH, Khatoon H, Lam SS, Jusoh A. Harvesting of microalgae (Chlorella sp.) from aquaculture bioflocs using an environmental-friendly chitosan-based bio-coagulant. International Biodeterioration & Biodegradation. 2017;**124**:243-249. DOI: 10.1016/j.ibiod.2017.07.016

[24] Renault F, Sancey B, Charles J, Morin-Crini N, Badot P-M, Winterton P, et al. Chitosan flocculation of cardboard-mill secondary biological wastewater. Chemical Engineering Journal. 2009;**155**:775-783. DOI: 10.1016/j.cej.2009.09.023

[25] Ferral-Pérez H, Torres Bustillos LG, Méndez H, Rodríguez-Santillan JL, Chairez I. Sequential treatment of tequila industry Vinasses by biopolymer-based coagulation/flocculation and catalytic Ozonation. Ozone: Science & Engineering. 2016;**38**:279-290. DOI: 10.1080/01919512.2016.1158635

[26] Saritha V, Srinivas N, Srikanth Vuppala NV. Analysis and optimization of coagulation and flocculation process. Applied Water Science. 2017;**7**:451-460. DOI: 10.1007/s13201-014-0262-y

[27] Yang R, Li HJ, Huang M, Yang H, Li AM. A review on chitosan-based flocculants and their applications in water treatment. Water Research. 2016;**95**:59-89. DOI: 10.1016/j.watres.2016.02.068

[28] Kos L. Use of chitosan for textile wastewater decolourization. Fibres & Textiles in Eastern Europe. 2016;**24**:130-135. DOI: 10.5604/12303666.1196623

[29] Gerchman Y, Vasker B, Tavasi M, Mishael Y, Kinel-Tahan Y, Yehoshua Y. Effective harvesting of microalgae: Comparison of different polymeric flocculants. Bioresource Technology. 2017;**228**:141-146. DOI: 10.1016/j.biortech.2016.12.040

[30] Xu Y, Purton S, Baganz F. Chitosan flocculation to aid the harvesting of the microalga Chlorella sorokiniana. Bioresource Technology. 2013;**129**:296-301. DOI: 10.1016/j.biortech.2012.11.068

[31] Divya K, Vijayan S, George TK, Jisha MS. Antimicrobial properties of chitosan nanoparticles: Mode of action and factors affecting activity. Fibers and Polymers. 2017;**18**:221-230. DOI: 10.1007/s12221-017-6690-1

[32] Hu C-Y, Lo S-L, Chang C-L, Chen F-L, Wu Y-D, Ma J-I. Treatment of highly turbid water using chitosan and aluminum salts. Separation and Purification Technology. 2013;**104**:322-326. DOI: 10.1016/j.seppur.2012.11.016

[33] Zemmouri H, Drouiche M, Sayeh A, Lounici H, Mameri N. Chitosan application for treatment of Beni-Amrane's water dam. Energy Procedia. 2013;**36**:558-564. DOI: 10.1016/j.egypro.2013.07.064

[34] Bhalkaran S, Wilson LD. Investigation of self-assembly processes for chitosan-based coagulant-flocculant systems: A mini-review. International Journal of Molecular Sciences. 2016;**17**:1-21. DOI: 10.3390/ijms17101662

[35] Li J, Jiao S, Zhong L, Pan J, Ma Q. Optimizing coagulation and flocculation process for kaolinite suspension with chitosan. Colloids and Surfaces A: Physicochemical and Engineering Aspects. 2013;**428**:100-110. DOI: https://doi.org/10.1016/j.colsurfa.2013.03.034

[36] Guibal E, Van Vooren M, Dempsey BA, Roussy J. A review of the use of chitosan for the removal of particulate and dissolved contaminants. Separation Science and Technology. 2006;**41**:2487-2514. DOI: 10.1080/01496390600742807

[37] Feng B, Peng J, Zhu X, Huang W. The settling behavior of quartz using chitosan as flocculant. Journal of Materials Research and Technology. 2017;**6**:71-76. DOI: 10.1016/j.jmrt.2016.09.004

[38] Chung YC. Improvement of aquaculture wastewater using chitosan of different degrees of deacetylation. Environmental Technology. 2006;**27**:1199-1208. DOI: 10.1080/09593332708618734

[39] Huang C, Chen S, Ruhsing Pan J. Optimal condition for modification of chitosan: A biopolymer for coagulation of colloidal particles. Water Research. 2000;**34**:1057-1062. DOI: 10.1016/S0043-1354(99)00211-0

[40] Wan Ngah WS, Teong LC, Hanafiah MAKM. Adsorption of dyes and heavy metal ions by chitosan composites: A review. Carbohydrate Polymers. 2011;**83**:1446-1456. DOI: 10.1016/j.carbpol.2010.11.004

[41] Piccin JS, Vieira MLG, Goncalves JO, Dotto GL, Pinto LAA. Adsorption of FD&C Red No. 40 by chitosan: Isotherms analysis. Journal of Food Engineering. 2009;**95**:16-20. DOI: 10.1016/j.jfoodeng.2009.03.017

[42] Vakili M, Rafatullah M, Salamatinia B, Abdullah AZ, Ibrahim MH, Tan KB, et al. Application of chitosan and its derivatives as adsorbents for dye removal from water and wastewater: A review. Carbohydrate Polymers. 2014;**113**:115-130. DOI: 10.1016/j.carbpol.2014.07.007

[43] Rhazi M, Desbrières J, Tolaimate A, Rinaudo M, Vottero P, Alagui A, et al. Influence of the nature of the metal ions on the complexation with chitosan.: Application to the treatment of liquid waste. European Polymer Journal. 2002;**38**:1523-1530. DOI: 10.1016/S0014-3057(02)00026-5

[44] Pestov A, Bratskaya S. Chitosan and its derivatives as highly efficient polymer ligands. Molecules. 2016;**21**. DOI: 10.3390/molecules21030330

[45] Vieira RS, Beppu MM. Chitosan as adsorbent for heavy metal ions: Performance and adsorption mechanism. Water Resources Research Progress. 2008:149-175

[46] Pillai SK, Ray SS. Chitosan-based nanocomposites. In: Maya JJ, Sabu T, editors. Natural Polymers. Vol. 2. Nanocomposites. London: RSC Publishing; 2012. 17 p

[47] Kyzas GZ, Bikiaris DN. Recent modifications of chitosan for adsorption applications: A critical and systematic review. Marine Drugs. 2015;**13**:312-337. DOI: 10.3390/md13010312

[48] Yong SK, Shrivastava M, Srivastava P, Kunhikrishnan A, Bolan N. Environmental applications of chitosan and its derivatives. In: Whitacre DM, editor. Reviews of Environmental Contamination and Toxicology. Vol. 233. Cham: Springer International Publishing; 2015. pp. 1-43. DOI: 10.1007/978-3-319-10479-9_1

[49] Mahmoodi NM, Mokhtari-Shourijeh Z. Preparation of PVA-chitosan blend nanofiber and its dye removal ability from colored wastewater. Fibers and Polymers. 2015;**16**:1861-1869. DOI: 10.1007/s12221-015-5371-1

[50] Martínez-Quiroz M, López-Maldonado EA, Ochoa-Terán A, Pina-Luis GE, Oropeza-Guzman MT. Modification of chitosan with carbamoyl benzoic acids for testing its coagulant-flocculant and binding capacities in removal of metallic ions typically contained in plating wastewater. Chemical Engineering Journal. 2018;**332**:749-756. DOI: 10.1016/j.cej.2017.09.042

[51] Wang B, Zhu Y, Bai Z, Luque R, Xuan J. Functionalized chitosan biosorbents with ultra-high performance, mechanical strength and tunable selectivity for heavy metals in wastewater treatment. Chemical Engineering Journal. 2017;**325**:350-359. DOI: 10.1016/j.cej.2017.05.065

[52] Wu S-P, Dai X-Z, Kan J-R, Shilong F-D, Zhu M-Y. Fabrication of carboxymethyl chitosan–hemicellulose resin for adsorptive removal of heavy metals from wastewater. Chinese Chemical Letters. 2017;**28**:625-632. DOI: https://doi.org/10.1016/j.cclet.2016.11.015

[53] Negm NA, El Sheikh R, El-Farargy AF, Hefni HHH, Bekhit M. Treatment of industrial wastewater containing copper and cobalt ions using modified chitosan. Journal of Industrial and Engineering Chemistry. 2015;**21**:526-534. DOI: 10.1016/j.jiec.2014.03.015

[54] Liang X, Duan J, Xu Q, Wei X, Lu A, Zhang L. Ampholytic microspheres constructed from chitosan and carrageenan in alkali/urea aqueous solution for purification of various wastewater. Chemical Engineering Journal. 2017;**317**:766-776. DOI: https://doi.org/10.1016/j.cej.2017.02.089

[55] Nair V, Panigrahy A, Vinu R. Development of novel chitosan-lignin composites for adsorption of dyes and metal ions from wastewater. Chemical Engineering Journal. 2014;**254**:491-502. DOI: 10.1016/j.cej.2014.05.045

[56] Yang Z, Yang H, Jiang Z, Cai T, Li H, Li H, et al. Flocculation of both anionic and cationic dyes in aqueous solutions by the amphoteric grafting flocculant carboxymethyl chitosan-graft-polyacrylamide. Journal of Hazardous Materials. 2013;**254-255**:36-45. DOI: https://doi.org/10.1016/j.jhazmat.2013.03.053

[57] Herrera-González AM, Peláez-Cid AA, Caldera-Villalobos M. Adsorption of textile dyes present in aqueous solution and wastewater using polyelectrolytes derived from chitosan. Journal of Chemical Technology & Biotechnology. 2017;**92**:1488-1495. DOI: 10.1002/jctb.5214

[58] Çınar S, Kaynar ÜH, Aydemir T, Çam Kaynar S, Ayvacıklı M. An efficient removal of RB5 from aqueous solution by adsorption onto nano-ZnO/Chitosan composite beads. International Journal of Biological Macromolecules. 2017;**96**:459-465. DOI: 10.1016/j.ijbiomac.2016.12.021

[59] Chiou MS, Li HY. Adsorption behavior of reactive dye in aqueous solution on chemical cross-linked chitosan beads. Chemosphere. 2003;**50**:1095-1105. DOI: 10.1016/S0045-6535(02)00636-7

[60] Hasan M, Ahmad AL, Hameed BH. Adsorption of reactive dye onto cross-linked chitosan/oil palm ash composite beads. Chemical Engineering Journal. 2008;**136**:164-172. DOI: 10.1016/j.cej.2007.03.038

[61] Rashid S, Shen C, Chen X, Li S, Chen Y, Wen Y, et al. Enhanced catalytic ability of chitosan-Cu-Fe bimetal complex for the removal of dyes in aqueous solution. RSC Advances. 2015;**5**:90731-90741. DOI: 10.1039/C5RA14711E

[62] Duffus John H. "Heavy metals" a meaningless term? (IUPAC Technical Report). Pure and Applied Chemistry. 2002. p. 793

[63] Guibal E, Vincent T, Navarro R. Metal ion biosorption on chitosan for the synthesis of advanced materials. Journal of Materials Science. 2014;**49**:5505-5518. DOI: 10.1007/s10853-014-8301-5

[64] Mohammad AM, Salah Eldin TA, Hassan MA, El-Anadouli BE. Efficient treatment of lead-containing wastewater by hydroxyapatite/chitosan nanostructures. Arabian Journal of Chemistry. 2017;**10**:683-690. DOI: 10.1016/j.arabjc.2014.12.016

[65] Badawi MA, Negm NA, Abou Kana MTH, Hefni HH, Abdel Moneem MM. Adsorption of aluminum and lead from wastewater by chitosan-tannic acid modified biopolymers: Isotherms, kinetics, thermodynamics and process mechanism. International Journal of Biological Macromolecules. 2017;**99**:465-476. DOI: 10.1016/j.ijbiomac.2017.03.003

[66] McIlwee HA, Schauer CL, Praig VG, Boukherroub R, Szunerits S. Thin chitosan films as a platform for SPR sensing of ferric ions. The Analyst. 2008;**133**:673-677. DOI: 10.1039/B717736D

[67] Monteiro OAC, Airoldi C. Some thermodynamic data on copper–chitin and copper–chitosan biopolymer interactions. Journal of Colloid and Interface Science. 1999;**212**:212-219. DOI: 10.1006/jcis.1998.6063

[68] Gerente C, Lee VKC, Le Cloirec P, McKay G. Application of chitosan for the removal of metals from wastewaters by adsorption – Mechanisms and models review. Critical Reviews in Environmental Science and Technology. 2007;**37**:41-127. DOI: 10.1080/10643380600729089

[69] Guibal E, Vincent T, Blondet FP. Biopolymers as support for heterogeneous catalysis: Focus on chitosan, a promising aminopolysaccharide. In: Ion Exchange and Solvent Extraction. Vol. 18. Sengupta AK, editor. Boca raton: CRC Press; 2007

[70] Miretzky P, Cirelli AF. Hg(II) removal from water by chitosan and chitosan derivatives: A review. Journal of Hazardous Materials. 2009;**167**:10-23. DOI: 10.1016/j.jhazmat.2009.01.060

[71] Pearson RG. Recent advances in the concept of hard and soft acids and bases. Journal of Chemical Education. 1987;**64**:561-567. DOI: 10.1021/ed064p561

[72] Lü R, Cao Z, Shen G. Comparative study on interaction between copper (II) and chitin/chitosan by density functional calculation. Journal of Molecular Structure: THEOCHEM. 2008;**860**:80-85. DOI: 10.1016/j.theochem.2008.03.013

[73] Schlick S. Binding-sites of Cu^{2+} in chitin and chitosan – An electron-spin-resonance study. Macromolecules. 1986;**19**:192-195. DOI: 10.1021/ma00155a030

[74] Oliveira M, Simoni JA, Airoldi C. Chitosan metal-crosslinked beads applied for n-alkylmonoamines removal from aqueous solutions – A thermodynamic study. The Journal of Chemical Thermodynamics. 2014;**73**:197-205. DOI: 10.1016/j.jct.2013.12.030

[75] Rhazi M, Desbrières J, Tolaimate A, Rinaudo M, Vottero P, Alagui A. Contribution to the study of the complexation of copper by chitosan and oligomers. Polymer. 2002;**43**:1267-1276. DOI: 10.1016/S0032-3861(01)00685-1

[76] Yamani JS, Lounsbury AW, Zimmerman JB. Towards a selective adsorbent for arsenate and selenite in the presence of phosphate: Assessment of adsorption efficiency, mechanism, and binary separation factors of the chitosan-copper complex. Water Research. 2016;**88**:889-896. DOI: 10.1016/j.watres.2015.11.017

[77] Guibal E. Interactions of metal ions with chitosan-based sorbents: A review. Separation and Purification Technology. 2004;**38**:43-74. DOI: 10.1016/j.seppur.2003.10.004

[78] Domard A. pH and c.d. Measurements on a fully deacetylated chitosan: Application to CuII – Polymer interactions. International Journal of Biological Macromolecules. 1987;**9**:98-104. DOI: 10.1016/0141-8130(87)90033-X

[79] Ogawa K, Oka K, Yui T. X-ray study of chitosan-transition metal complexes. Chemistry of Materials. 1993;**5**:726-728. DOI: 10.1021/cm00029a026

[80] Shahgholi M, Callahan JH, Rappoli BJ, Rowley DA. Investigation of copper-saccharide complexation reactions using potentiometry and electrospray mass spectrometry. Journal of Mass Spectrometry. 1997;**32**:1080-1093. DOI: 10.1002/(sici)1096-9888(199711)32:10<1080::aid-jms562>3.3.co;2-2

[81] Lalmi S, Kameche M, Innocent C, Haddou B, Derriche Z, Pourcelly G. Adsorption of biodegradable polyelectrolyte onto cotton for fixation of copper and lead: Comparison to a cation exchange textile. Separation Science and Technology. 2013;**48**:156-165. DOI: 10.1080/01496395.2012.676138

[82] Tzu Yang H, Yu Ling L. Desorption of cadmium from porous chitosan beads. In: Ning RY, editor. Advancing Desalination. Rijeka: InTech; 2012. Ch. 08. DOI: 10.5772/50142

[83] Gérente C, Andrès Y, McKay G, Le Cloirec P. Removal of arsenic(V) onto chitosan: From sorption mechanism explanation to dynamic water treatment process. Chemical Engineering Journal. 2010;**158**:593-598. DOI: 10.1016/j.cej.2010.02.005

[84] Mende M, Schwarz D, Steinbach C, Boldt R, Schwarz S. Simultaneous adsorption of heavy metal ions and anions from aqueous solutions on chitosan—Investigated by spectrophotometry and SEM-EDX analysis. Colloids and Surfaces A: Physicochemical and Engineering Aspects. 2016;**510**:275-282. DOI: https://doi.org/10.1016/j.colsurfa.2016.08.033

[85] Zimmermann AC, Mecabo A, Fagundes T, Rodrigues CA. Adsorption of Cr(VI) using Fe-crosslinked chitosan complex (Ch-Fe). Journal of Hazardous Materials. 2010;**179**:192-196. DOI: 10.1016/j.jhazmat.2010.02.078

[86] Dobosz KM, Kolewe KW, Schiffman JD. Green materials science and engineering reduces biofouling: Approaches for medical and membrane-based technologies. Frontiers in Microbiology. 2015;**6**:1-8. DOI: 10.3389/fmicb.2015.00196

[87] Chakrabarty T, Shahi VK. Modified chitosan-based, pH-responsive membrane for protein separation. RSC Advances. 2014;**4**:53245-53252. DOI: 10.1039/C4RA05314A

[88] Weng RG, Chen LH, Lin S, Zhang H, Wu H, Liu K, et al. Preparation and characterization of antibacterial cellulose/chitosan nanofiltration membranes. Polymer. 2017;**9**. DOI: 10.3390/polym9040116

[89] Mukherjee M, De S. Investigation of antifouling and disinfection potential of chitosan coated iron oxide-PAN hollow fiber membrane using gram-positive and gram-negative bacteria. Materials Science and Engineering: C. 2017;**75**:133-148. DOI: https://doi.org/10.1016/j.msec.2017.02.039

[90] Shanthana Lakshmi D, Jaiswar S, saxena M, Tasselli F, Raval HD. Preparation and performance of biofouling resistant PAN/chitosan hollow fiber membranes. 3Biotechnology. 2017;**7**:224. DOI: 10.1007/s13205-017-0798-2

[91] El-Gendi A, Deratani A, Ahmed SA, Ali SS. Development of polyamide-6/chitosan membranes for desalination. Egyptian Journal of Petroleum. 2014;**23**:169-173. DOI: https://doi.org/10.1016/j.ejpe.2014.05.003

[92] Deng H, Sun PZ, Zhang YJ, Zhu HW. Reverse osmosis desalination of chitosan cross-linked graphene oxide/titania hybrid lamellar membranes. Nanotechnology. 2016;**27**:8. DOI: 10.1088/0957-4484/27/27/274002

[93] Raval HD, Gondaliya MD. A novel high-flux thin film composite reverse osmosis membrane modified by polysaccharide supramolecular assembly. Journal of Applied Polymer Science. 2017;**134**:7. DOI: 10.1002/app.45026

[94] Shakeri A, Salehi H, Rastgar M. Chitosan-based thin active layer membrane for forward osmosis desalination. Carbohydrate Polymers. 2017;**174**:658-668. DOI: https://doi.org/10.1016/j.carbpol.2017.06.104

[95] Salehi H, Rastgar M, Shakeri A. Anti-fouling and high water permeable forward osmosis membrane fabricated via layer by layer assembly of chitosan/graphene oxide. Applied Surface Science. 2017;**413**:99-108. DOI: https://doi.org/10.1016/j.apsusc.2017.03.271

[96] Alshahrani AA, Al-Zoubi H, Nghiem LD, Panhuis MIH. Synthesis and characterisation of MWNT/chitosan and MWNT/chitosan-crosslinked buckypaper membranes for desalination. Desalination. 2017;**418**:60-70. DOI: 10.1016/j.desa1.2017.05.031

[97] Yang H-C, Gong J-L, Zeng G-M, Zhang P, Zhang J, Liu H-Y, et al. Polyurethane foam membranes filled with humic acid-chitosan crosslinked gels for selective and simultaneous removal of dyes. Journal of Colloid and Interface Science. 2017;**505**:67-78. DOI: https://doi.org/10.1016/j.jcis.2017.05.075

[98] Mokhena TC, Luyt AS. Development of multifunctional nano/ultrafiltration membrane based on a chitosan thin film on alginate electrospun nanofibres. Journal of Cleaner Production. 2017;**156**:470-479. DOI: https://doi.org/10.1016/j.jclepro.2017.04.073

[99] Ekambaram K, Doraisamy M. Fouling resistant PVDF/Carboxymethyl chitosan composite nanofiltration membranes for humic acid removal. Carbohydrate Polymers. 2017;**173**:431-440. DOI: https://doi.org/10.1016/j.carbpol.2017.06.017

[100] Ma XH, Yang Z, Yao ZK, Xu ZL, Tang CYY. A facile preparation of novel positively charged MOF/chitosan nanofiltration membranes. Journal of Membrane Science. 2017;**525**:269-276. DOI: 10.1016/j.memsci.2016.11.015

[101] Mehta BB, Joshi RN, Raval HD. A novel ultra-low energy reverse osmosis membrane modified by chitosan with glutaraldehyde crosslinking. Journal of Applied Polymer Science. 2018;**135**:1-7. DOI: 10.1002/app.45971

[102] Geise GM, Lee H-S, Miller DJ, Freeman BD, McGrath JE, Paul DR. Water purification by membranes: The role of polymer science. Journal of Polymer Science Part B: Polymer Physics. 2010;**48**:1685-1718. DOI: 10.1002/polb.22037

[103] Tanabe T, Okitsu N, Tachibana A, Yamauchi K. Preparation and characterization of keratin-chitosan composite film. Biomaterials. 2002;**23**:817-825. DOI: 10.1016/s0142-9612(01)00187-9

[104] Qu X, Wirsen A, Albertsson AC. Novel pH-sensitive chitosan hydrogels: Swelling behavior and states of water. Polymer. 2000;**41**:4589-4598. DOI: 10.1016/s0032-3861(99)00685-0

[105] Tsai HS, Wang YZ. Properties of hydrophilic chitosan network membranes by introducing binary crosslink agents. Polymer Bulletin. 2008;**60**:103-113. DOI: 10.1007/s00289-007-0846-x

[106] Liu C, Bai R. Recent advances in chitosan and its derivatives as adsorbents for removal of pollutants from water and wastewater. Current Opinion in Chemical Engineering. 2014;**4**:62-70. DOI: https://doi.org/10.1016/j.coche.2014.01.004

[107] Adnan O, Abidin ZZ, Idris A, Kamarudin S, Al-Qubaisi MS. A novel biocoagulant agent from mushroom chitosan as water and wastewater therapy. Environmental Science and Pollution Research. 2017;**24**:20104-20112. DOI: 10.1007/s11356-017-9560-x

4

Overview of Electrospinned Chitosan Nanofiber Composites for Wound Dressings

Claudia A. Vega-Cázarez,
Dalia I. Sánchez-Machado and
Jaime López-Cervantes

Abstract

Chitosan has a medical application because of its natural origin and properties of biodegradability, biocompatibility, nontoxicity, and antimicrobial capacity. Electrospinning produces non-woven nanofibers to wound dressing with high specific surface area and small pores. These properties are favorable for absorption of exudates and prevent the penetration of bacteria, thus promoting wound healing. For this reason, chitosan blends are used to produce nanofiber dressings, and the characterization of the structural, mechanical, and biological properties is very promising for further studies. Nowadays, the researchers are seeking for biomaterials that provide modern dressings with many qualities, which are designed to promote wound healing. In this chapter, the electrospinning parameters that affect the nanofiber properties based on chitosan to prepare wound dressings are highlighted.

Keywords: electrospinning, chitosan, dressings, nanofibers, biomaterials

1. Introduction

Chitin is a polymer composed of *N*-acetylglucosamine that has been recognized as one of the most abundant natural polysaccharides and has been found in crustaceans, fungi, and insects [1]. To improve its solubility in water, it is partially deacetylated and converted into chitosan, which is a linear polymer of glucosamine and *N*-acetylglucosamine [2]. Due to its properties, chitosan has many applications in food preservation, medicine, biotechnology, agriculture, and water treatment. Among the properties of interest in biomedicine are its biodegradability, biocompatibility, nontoxicity, and high antimicrobial activity [3]. Thus, chitosan preparations have been used as gels, microparticles, films, and coating agents. Medical products based on chitosan have been studied as dietary supplements, wound healing agents, hemostatic devices, and drug delivery [4].

The appropriate selection of dressings depends on the characteristics of the wound and its mode of action. The purposes of the dressings are to facilitate the healing process, control symptoms, achieve an esthetic healing, keep the wound moist, absorb exudate without leakage, prevent infection, avoid trauma when dressing needs to be changed, not present toxic skin irritation, and maintain gaseous exchange [5]. Currently, it is necessary to find these conditions in a single product, with chitosan and other biomaterials being the most studied [6].

The electrospinning process is a technique that emerged in the 1970s with the purpose of producing fibers with a size smaller than 100 nm called nanofibers [7, 8]. Specifically, the electrospinning system is recognized for its ability to produce nanofibers from a wide variety of polymers of natural or synthetic origin and natural proteins [9].

The electrospinning system faces several problems during the elaboration of nanofibers for biomedical applications, particularly, the optimization of the process conditions for each polymer [10]. This chapter presents a detailed review of the influence of electrospinning parameters on the properties of the biocomposites of chitosan in mixture with other polymers to produce dressings, as well as some in vitro and in vivo assays of their application as wound dressings.

2. Electrospinning technique for production of nanofibers

The electrospinning process is a recognized system for obtaining polymer fibers by creating an electric field. To do this, a jet of the polymer is injected through a charged needle, and the solvent where the polymer is dissolved evaporates, thus allowing the deposition of solid polymer fibers on the collector. As a result, the electrospinning system equipment consists of a high voltage source, syringe pump, a syringe loaded with the polymer solution, and a collector (**Figure 1**).

During the process, high-voltage loads are applied to the polymer solution that is injected. The increase in the electric field formed causes the repulsion interactions between the same charges and attraction between the oppositely charges in the liquid. Whereas the collector that is grounded exerts forces of attraction on the injected drop, once the electric field increases, it will reach a point where the electrostatic forces reach the equilibrium with the surface tension, deforming the cone to a drop, which it is called Taylor cone [11].

Many polymers and operating parameters have been reported for the electrospinning process. The electrospinning system is affected by polymer characteristics such as concentration, viscosity, molecular weight, and surface tension. Other parameters that influence are the components and processing parameters of the system, specifically the applied voltage, injection flow, distance between the needle and the tip, and the type of collector used [12], presented in **Table 1**.

2.1. Characteristics of chitosan polymer for electrospinning nanofiber process

According to Kai et al. [12], electrospinning requires a very concentrated polymer solution. The low concentrated solutions cause the formation of weak chains, as well as insufficient electrostatic

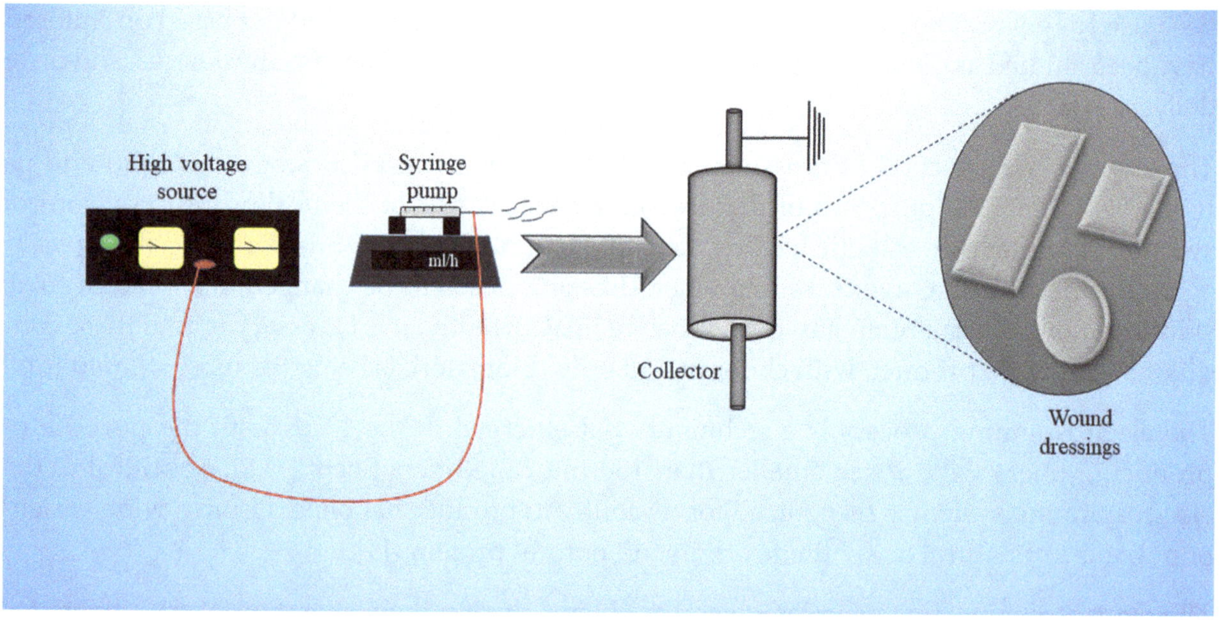

Figure 1. Components of the electrospinning system.

repulsive forces, giving rise to the appearance of droplets in the collector and the incomplete formation of the fibers.

Surface tension and viscosity are parameters of the polymer solution that mainly affect the diameter of the fiber [10]. These authors agree with Sill and von Recum [11], who mentioned that if the solution is very diluted the polymeric fiber will break into droplets before reaching the collector due to surface tension. When the solution is very concentrated does not complete the fiber formation is not obtained due to the high viscosity, which makes difficult the flow during injection.

The molecular weight of the polymer is another parameter that affects electrospinning. Low molecular mass polymers lead to bead formation, while high molecular weight polymers produce fibers with larger diameter [12]. Likewise, Cui et al. [24] found that to produce pectin nanofibers by electrospinning the molecular weight of the polymer should be in the range of 200–950 kDa.

2.2. Components and processing parameters of the electrospinning technique

Haider et al. [25] reported that the critical amount of applied voltage varies according to the polymer. Also, the formation of nanofibers with small diameters is due to the increase in the applied voltage, attributed to the stretching of the Taylor cone, and the formation of the jet (**Figure 2**). However, the increase in the critical voltage value forms beads in the nanofibers, which is caused by the decrease in the size of the Taylor cone and an increase in the flow velocity.

One of the electrospinning parameters with greater influence on the mechanical properties is the type of collector. During the electrospinning process, the fibers are deposited in a lower potential electrode known as a rotating or static collector. Thus, the fibers can be guided through the electric field formed between the tip of the needle loaded with the polymer solution and the high voltage source [26].

Polymer	Electrospinning parameters	Collector	References
PVA	AV (14.5–17.5 kV) TCD (125 mm) FR (6 μL/min)	Stainless steel collector	[13]
CS/PVA	AV (18 kV) TDC (15 cm) FR (0.35 ml/h)		[14]
CS/HA	AV (15 kV) TCD (15 cm) FR (1.2 ml/h)	Rectangular 6 × 2 cm aluminum collecting plate	[15]
PVA/SA	AV (15 kV) TCD (15 cm)		[16]
CS/PVA	AV (15 kV) TCD (15 cm)	Copper plate wrapped with aluminum foil	[17]
SA/PEO	AV (20 kV) TCD (20 cm)		[18]
CS/PEO	AV (15–35 kV) TCD (15 cm) FR (0.1–2 mL/h)	Aluminum foil attached to a drum collector	[19]
SA/PVA	AV (17 kV) TCD (5 cm) FR (0.2 ml/h)	Aluminum target	[20]
SA/PVA	AV (17 kV) TCD (10 cm) FR (2 μl/min)	Rectangular nickel collector with a glass microscope slide taped to its surface	[21]
Ge	AV (12 kV) TCD (10 cm) FR (0.003 ml/min)	Stainless steel mandrel (30 cm length and 5 cm in diameter) and rotational speeds (1700, 2400, 3100, 3800, and 4500 rpm)	[22]
PL/CS	AV (5 kV) TCD (12.5 cm) FR (0.6 ml h^{-1})		[23]

PVA, Poly(vinyl alcohol); CS, chitosan; HA, hydroxyapatite; PEO, poly(ethylene oxide); SA, sodium alginate; PL, polycaprolactone; Ge, gelatin; AV, applied voltage; TCD, tip-to-collector distance; FR, flow rate

Table 1. Polymers and parameters used in electrospinning.

The collectors of the first electrospinning equipment lacked movement, but over time they have been modified in order to improve the alignment of the fibers and modify their surface. The use of rotating cylindrical collectors has helped in the alignment of the fibers. At the same

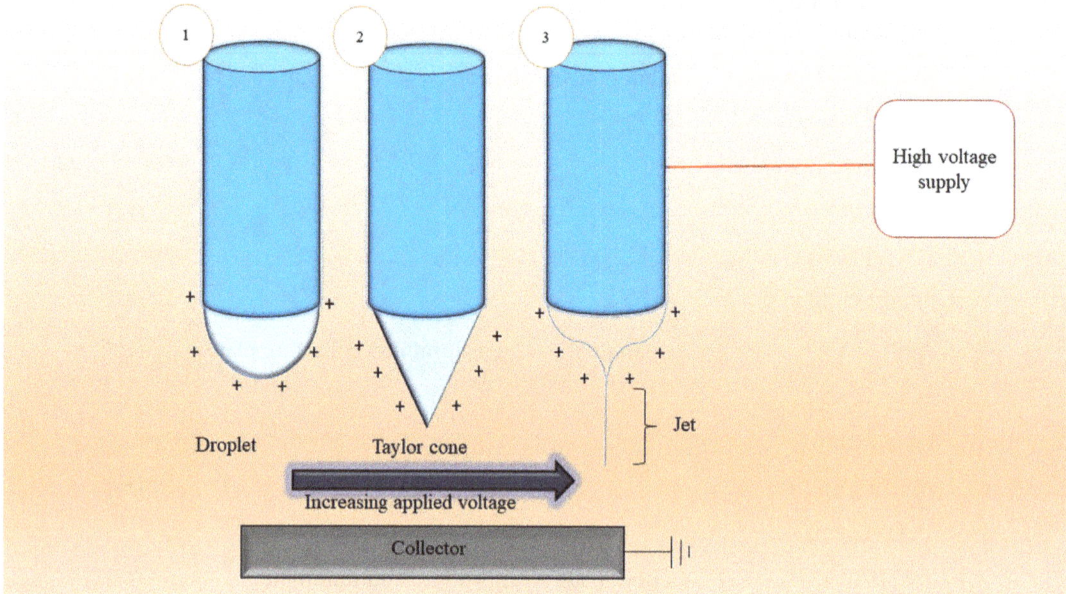

Figure 2. Taylor cone formation stages during voltage application.

time, Huang et al. [27] used a thin circular collector with a sharp edge and found that with this modification electric field is concentrated, favoring the alignment and collection of the fibers.

The collectors can be adjusted to various specifications from a stationary plate to a rotating cylinder. When a static collector is used, the fibers will have a random order, while aligned fibers will be generated with the rotary collector [11].

Huang et al. [27] used collectors with movement to prepare gelatin hydrofibers. They confirmed that the use of rotating collectors of cylindrical shape with high speed orientates the nanofibers. In addition, they reported that when the collector speed is lower than the alignment speed, the hydrofibers will have a random order, while high speed causes instability in the injection jet, affecting the surface morphology of the fibers.

In addition, the position of the collector influences the morphology of the fibrous materials. Haider et al. [25] reported that the diameter increases at short distances between the needle and the collector, while the diameter decreases at long distances between the needle and the collector. Also, Sill and von Recum [11] explained that the distance between the needle and the collector affects the continuous and complete formation of the fibers. Small distances are not enough to achieve the evaporation of the solvent before being deposited in the collector, resulting in the formation of loose and weak fibers that tend to stick together and make it difficult to remove from the collector.

With the intention of providing greater guidance in continuous nanofibers, Dabirian et al. [28] incorporated two injection syringes with different charges and positioned in a triangular form with respect to the initial needle of the process. Both syringes are attracted to each other, and the nanofibers are twisted and collected continuously in the form of a yarn.

3. Characteristics parameters of chitosan biopolymer

Chitosan is a linear polysaccharide consisting of *N*-acetyl-D-glucosamine (GlcNAc) and D-glucosamine (GlcN), which is produced by alkaline deacetylation of crustacean chitin [29]. It is soluble in dilute acid solutions of acetic, lactic, malic, formic, or succinic acid. It is polycationic at pH 6 and easily interacts with negatively charged molecules such as proteins, fatty acids, anionic polysaccharides, bile acids, and phospholipids and, therefore, can be used as blended solutions [30].

This polysaccharide has antibacterial, antifungal, mucoadhesive, analgesic, hemostatic, nontoxicity, biodegradable, and biocompatible activity (**Figure 3**) [31]. These biological properties have favored its topical application and implantation. Another recent medical application is controlled drug delivery systems [32]. Baldrick [33] reported the possible oral use of chitosan nanoparticles with drug carrier, as gel beads of chitosan can be degraded between 14 and 28 days after being implanted. Mansouri et al. [34] propose to chitosan as a candidate to be used in gene therapy due to its ability to bind and form complexes with DNA by the electrostatic attraction created.

Chitosan has been shown to be a nontoxic polymer, the reason why the Food and Drug Administration (FDA) has approved its use in wound dressings [35]. Likewise, in the clinical reports with chitosan-based biomaterials, inflammation or allergic reactions have not been reported [2]. Waibel et al. [36] detailed the first study of chitosan as bandages in soldiers with shellfish allergy, where all soldiers tolerate the chitosan bandage without reaction.

A very important property is biodegradation, which can be chemical or enzymatic degradation. Chemical degradation refers to acid-catalyzed degradation, such as the gastric juices of the stomach. Likewise, chitosan can be degraded by enzymes, which are responsible for

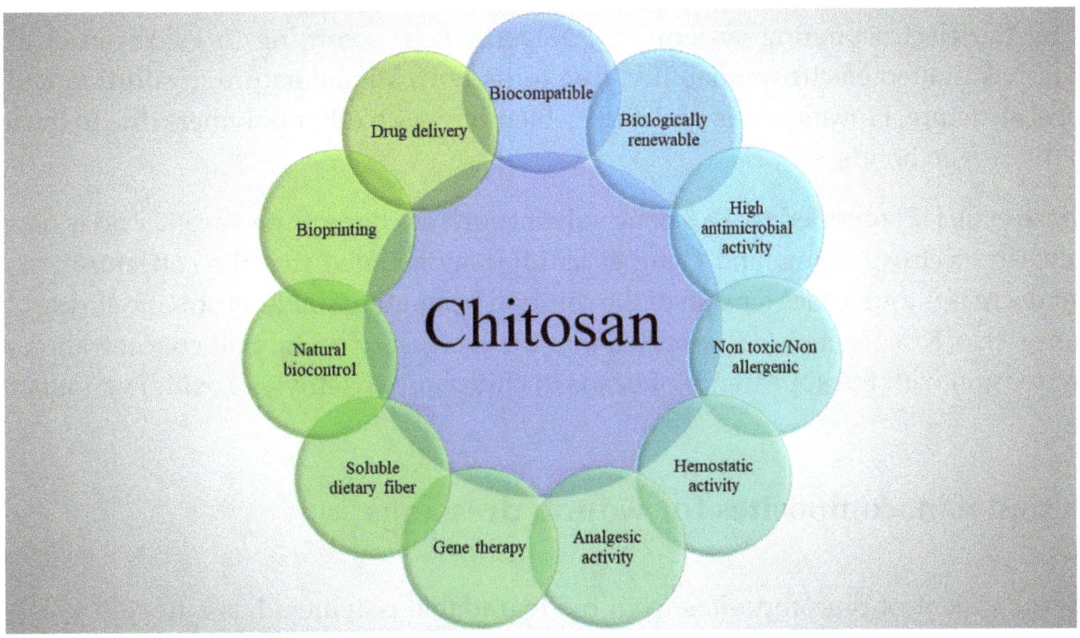

Figure 3. Promising properties of chitosan for medical use.

hydrolyzing the bonds between glucosamine-glucosamine, glucosamine-*N*-acetyl-glucosamine, and *N*-acetyl-glucosamine-*N*-acetyl-glucosamine [35]. In addition, the degradation rate of chitosan depends mainly on the molecular weight and the degree of deacetylation. However, the degradation of chitosan is not fully understood.

Ng et al. [37] reported a new emerging technology that can provide 3D nanostructures named bioprinting and highlighting for chitosan, due to its poor printability. They prepared a polyelectrolyte complex blending chitosan and gelatin. The hydrogels were customized and fabricated considering the wound size.

Due to its cationic properties, the chitin and its derivatives have demonstrated that are a potent elicitor in low concentrations. Thus can activate defense mechanisms against pathogen nematodes. Also, in plants can be used to control herbivorous insect and viral diseases [38].

Biocompatibility according to Baldrick [33] is attributed to glucosamine, which is the most abundant component in chitosan and produced in the human body from glucose. Also, it is responsible for producing glycosaminoglycan which forms a cartilage tissue in the body. Chitosan by its positive charge can bind to free fatty acids and bile salt from dietary lipid during the absorption in the gut.

The hemostatic and antimicrobial activity of chitosan is related to its polycationic nature. Therefore, it is involved in the agglutination of red cells that stimulates the formation of clots [39]. The antimicrobial activity is attributed to the electrostatic interactions between the polycationic structure of chitosan and the anionic components of the surface of microorganisms [40].

Xia et al. [41] mentioned that chitosan has free amino groups which are able to neutralize gastric juices and form a protective barrier in the stomach. On this basis, they have proposed it as a safe material with biomedical applications in the treatment of peptic ulcers, for wound healing.

In the preparation of materials for wound healing, chitosan has been used to develop nanofibers by the electrospinning system [29]. Nevertheless, according to Pakravan et al. [19], chitosan has limited electrospinnability, due to its polycationic nature in solution and rigid chemical structure. However, it is possible by blending with other polymers due to the formation of hydrogen bonds.

De Vrieze et al. [42] reported that a strong solvent such as concentrated acetic acid is necessary for chitosan electrospinning, and Geng et al. [3] demonstrated that the concentration of the solvent decreases the surface tension of the solution generating stable jet stream during electrospinning. Also, Kriegel et al. [43] reported that the increase in acetic acid concentration affects surface tension with the appearance of beads in chitosan/poly(ethylene oxide) nanofibers.

4. Myriad nanocomposites for wound dressings

The development of biocomposites from biodegradable polymers has attracted great interest due to their capacity to completely degrade and not produce toxic effects [44]. Also, the main challenge in science of biomaterials and tissue engineering is to create matrices either of natural origin, synthetic, or blends that are suitable for the development of medical prototypes with improved properties.

In biomedical area, they are interested in developing polymers as biomaterials, because more complex structures need to be achieved on the requirements for their different applications [45].

According to the National Institute of Health (NIH), biomaterials are defined as "any substance (other than a drug) or combination of substances synthetic or natural in origin, which can be used for any period of time, as a whole or part of a system which treats, augments, or replaces tissue, organ, or function of the body " [46].

4.1. Natural and synthetic polymeric nanofibers

In biomedicine, the natural polymeric nanofibers are polysaccharides, collagen, keratin, silk, tubulin, actin, cellulose, chitin [47]. Likewise, there is great variety polymers such as poly(lactic acid-co-glycolic acid) (PLGA), poly(lactic acid) (PLLA), polycaprolactone (PCL), poly(ethylene oxide), (PEO), and poly(vinyl alcohol) (PVA). All these have been used in wound dressings [7, 45], as shown in **Table 1**.

4.2. Wound dressing theory

Traditional dressings used on wound treatment were made from natural or synthetic materials. In the past, the main function of the dressings was to keep the wound dry through the evaporation of the exudates and to avoid the introduction of dangerous microorganisms to the wound [48]. However, over time this idea has been modified, according to Newman et al. [49]; nowadays, an ideal wound dressing must have the objective of providing an optimum moisture, accelerate the healing process, absorb large amounts of exudates, and prevent tissue maceration around the wound, which would cause a second injury. Also, Caló and Khutoryanskiy [50] reported that the dressings are designed with the purpose of maintaining a moist environment between the wound and the dressing, favoring the healing of wounds. The healing process is a dynamic process which allows a complex sequence of events, which include homeostasis, inflammation, proliferation, and remodeling [25].

A wound is defined as a defect or a break in the skin resulting from trauma or medical/physiological conditions. Traditionally, wounds can be classified depending on the number of layers of skin and affected area. When only the epidermis is damaged, the wound is superficial. If, epidermis and deeper layers are affected, is named partial thickness. Moreover, when the subcutaneous fat and deeper tissue have been affected, the wound is named full thickness [50].

On the other hand, Zahedi et al. [51] mentioned that the dressings can be classified from other aspects. Therefore, these authors classified them into three groups called passive, interactive, and bioactive dressings. Passive dressings are ordinarily used for common wound coverage such as cotton gauze. Interactive dressings are characterized for being transparent and permeable. Bioactive dressings have the advantage that requires long periods to be removed and are made from biopolymers such as collagen, chitosan, alginate, and elastin. In the **Figure 4** are presented commercial dressing such as hydrocolloids, alginates, collagens and hydrofibers.

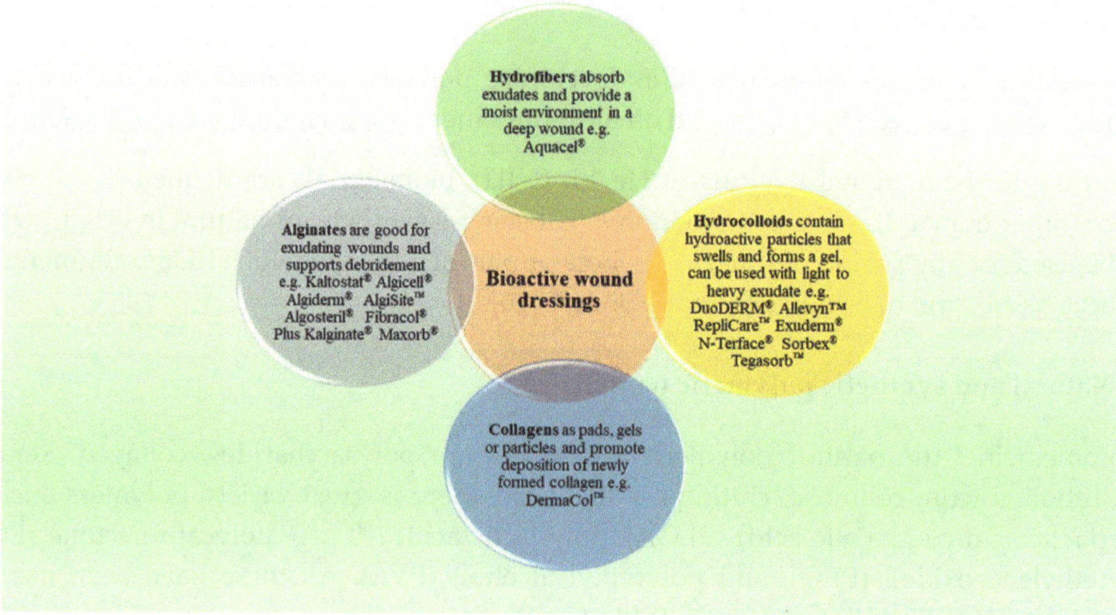

Figure 4. Commercial products of bioactive products.

According to Newsom et al. [52], hydrogels are effective for necrotic wounds because they hydrate wounds and reduce postoperative pain. However, they have disadvantages in wounds with large amounts of exudates and hemorrhages; that is why the use of alginates is recommended. Hydrocolloids promote a moist environment, autolytic tissue debridement, also as protection against microorganisms and are used in postoperative wounds. Hydrofibers are recognized for accelerating the wound healing process and improving the appearance of the final healing.

This wide variety of modern dressings has the advantage of creating an optimal environment that allows the proliferation of epithelial cells and improves the treatment of wounds [48].

5. Properties of electrospinned chitosan nanofibers/composites for wound dressings

The properties of the dressings are optimized through the modification of the electrospinning parameters. Most studies mentioned that the composition of the polymer solution such as solvent, concentration, viscosity, and surface tension affects the morphological structure and mechanical properties of nanofibers [12].

5.1. Superficial morphology

The structural morphology is visually evaluated with scanning electron microscopy (SEM), which also allows the measurement of the diameter of the nanofibers.

According to Bhardwaj and Kundu [8], the viscosity of the polymer solution determines the size and surface morphology of the nanofibers. Also, Bhattarai et al. [53] indicated that electrical conductivity is a physical property that modifies the final diameter of nanofibers.

Haider et al. [54] mentioned that the changes in the surface of nanofibers are attributed to the parameters of electrospinning and concentration of the solution and solvent (trifluoroacetic acid). Haider et al. [25] mentioned that selection of the solvent is very important for the formation of fibers with smooth and bead-free surface formation, considering the solvent capacity to dissolve the polymer and moderate boiling temperature. Also, Geng et al. [3] mentioned that a higher concentration of acetic acid decreases the surface tension and increases the charge density of the jet, improving the morphology of the nanofiber.

Pakravan et al. [19] prepared chitosan nanofibers with poly(ethylene oxide) and found that chitosan solutions, due to the polycationic nature, are more conductive compared to poly(ethylene oxide). Specifically, poly(ethylene oxide) reduces the protonation of the chitosan and avoid the formation of hydrogen bonds between amino groups of chitosan and other groups. The diameters of the chitosan and poly(polyethylene) oxide nanofibers are in the range of 60 to 120 nm.

Similarly, Bhattarai et al. [53] worked with chitosan and poly(ethylene oxide). They reported cylindrical nanofibers with a diameter of 75 nm and mentioned that viscosity is the parameter that modifies the electrospinning and the morphology of the nanofiber. They also indicated that the viscosity is attributed to the interactions between the chains of both polymers.

Lu et al. [55] prepared nanofibers of poly(ethylene oxide) with a low concentration of sodium alginate, showing smooth surfaces with a diameter of 250 nm; this was achieved by increasing the viscosity and decreasing the electrical conductivity of the solutions. Likewise, Shalumon et al. [20] prepared sodium alginate nanofibers blended with poly(vinyl alcohol) adding zinc oxide nanoparticles. The nanoparticles increased the electrical conductivity improving the electrospinning process to produce 190–240 nm fibers with smooth and uniform surface. **Table 2** shows the diameter of the nanofibers prepared by electrospinning with various polymers.

5.2. Porosity and pore distribution

The porosity of biomaterials such as nanofibers is advantageous for their application in wound healing, because they provide more structural space to enhance cell seeding, in addition to facilitating cell proliferation and migration [61]. It also improves oxygen exchange, nutrient delivery, and absorption of exudates. In addition, small pores in dressing nanofibers reduce wound infections and dehydration during the healing process [12].

The porosity is analyzed with water vapor transmission rate (WVTR). Archana et al. [62] mentioned that an ideal dressing should maintain water loss from the skin at an adequate rate, between 2000 and 2500 g^{-2} day^{-1}, indicating that higher values dry wounds quickly and retard healing, finding that for a chitosan dressing with pectin and nanosized titanium dioxide (TiO_2) particles the WVTR was from 1950 to 2050 g^{-2} day^{-1}.

However, Ziabari et al. [63] mentioned that there is no specific literature to measure the size of the pores and their distribution along the fiber. The techniques currently used are indirect and not very precise. Therefore, image analysis techniques are used, and these tools are more precise and direct. Coimbra et al. [64] used scanning electron microscopy (SEM) to observe the morphology of the chitosan/pectin wound dressing, which formed porous shapes from sheetlike to fibrous-like structures. Additionally, Ninan et al. [65] used micro computed

Polymeric solution	Average diameter (nm)	References
CS/PVA	279.843	[14]
CS/HA	227.8 ± 154.3	[15]
PVA/SA	500–50	[16]
CS/PVA	75–400	[17]
SA/PEO	130 ± 51	[18]
CS/PEO	60–120	[19]
SA/PVA	190–240	[20]
SA/PVA	250 ± 30	[21]
CS/PEO	40	[53]
SA/PEO	250	[54]
CMCTS/CS/PEO	50–300	[56]
COL/CS	434–691	[57]
CMC/HA	15–35	[58]
CA/PVA	98.1–191.5	[59]
PEO/CS/PC/O	86	[60]

PVA, poly(vinyl alcohol); CS, chitosan; HA, hydroxiapatitha; SA, sodium alginate; PEO, poly(ethylene oxide); PL, polycaprolactone; PC, poly(ε-caprolactone); O, olive oil; CA, calcium alginate; CMCTS, carboxymethyl chitosan; COL, collagen.

Table 2. Diameter of nanofibers by electrospinning.

tomography (micro-CT) in pectin/carboxymethylcellulose/microfibrillated cellulose dressings (15–280 μm) obtaining 3D images and the quantity of the number of pixels in the pore image. Kumar et al. [66] measured the porosity of the chitosan/pectin/calcium carbonate ($CaCO_3$) dressings based on the empty spaces present (41.8%) as a fraction of the total volume by liquid displacement method. Also, Liuyun et al. [67] used liquid displacement method to measure porosity (77.8%) and pore diameter, between 100 and 500 μm, in nanohydroxyapatite/chitosan/carboxymethylcellulose dressings. Besides, these authors reported the presence of interconnected pores that favor the administration of nutrients.

Recently, Sarhan et al. [68] reported the use of an automatic analyzer called mercury porosimetry to measure the size of the pores; however, it only measures sizes in the range of 0.0018–400 μm, finding that in chitosan nanofibers/honey/PVA the diameter of the pores is 140 μm.

5.3. Water absorption capacity

One of the properties of dressings is the ability to absorb exudates in the wound and provide a moist environment. However, it affects the efficiency of oxygen and nutrient transfer [62].

According to Ninan et al. [65], the swelling analysis and the water uptake provide information about the ability of the dressing to transport the nutrients inside the pores, in addition to avoiding dehydration of the wound.

Choi et al. [69] mentioned that the hydrophilicity of chitosan is due to a modification in its structure, which improves its solubility in water at physiological pH. Archana et al. [62] evaluated the swelling of the dressings of chitosan/pectin and TiO_2 in phosphate-buffered sodium (PBS), finding the highest values (1215%) at pH 2.0 and the lowest (900%) at pH 7.0, due to the osmotic effect of the dressing due to the absence of ionized amino groups.

Liuyun et al. [67] reported the loss in weight, 30% in 30 days, by degradation of the dressings of chitosan/nanohydroxyapatite/carboxymethylcellulose, caused by intermolecular interaction between the components of the system and the microstructure, which favors the growth of the cells for the regeneration of the bone tissue.

Zarghami et al. [60] reported that cross-linking agents modify nanofibers and their swelling behavior. For chitosan nanofibers with poly(ethylene oxide) cross-linked with glutaraldehyde vapors, the swelling decreases from 300 to 75%, due to the aldimine bonds (—CH=N—) formed between the free amino groups of chitosan and the aldehydes of glutaraldehyde. This increases the diffusivity resistance of water and decreases its water absorption capacity.

Duan et al. [70] also used glutaraldehyde vapors as a cross-linking agent, causing a decrease in swelling from 328.7 to 146.2% and shrinkage of poly(lactide-co-glycolide)/chitosan/PVA dressings. Similarly, Chen et al. [71] used glutaraldehyde vapors to reduce the water solubility of chitosan nanofibers with collagen and poly(ethylene oxide).

5.4. Mechanical properties

The mechanical properties are evaluated through Young's modulus, tensile strength, and elongation at break by tensile tests with a universal mechanical testing machine [8]. Parameters such as the composition of the polymer solution, the interaction between its components, and the cross-linking agents affect the mechanical properties of fibrous materials during electrospinning [55].

In some studies, it has been reported that obtaining nanofibers has great advantages for the development of medical prototypes, and the precise measurement of their mechanical properties is very important, especially in dressings which must be able to withstand the forces exerted by the growing tissue or during physiological and biomechanical activities [8].

The composition of the mixture influences the mechanical behavior of the fiber. Chitosan is a fragile and rigid polymer, while poly(lactic acid) (PLA) is resistant; it is possible to blend it to form nanofibers, which show greater tensile strength at rupture with decreases in elongation at break.

Archana et al. [62] studied the chitosan and pectin blend with nanoparticles of titanium dioxide (TiO_2). They found that by increasing the pectin content and the presence of the nanoparticles, the resistance increases significantly compared to those with a lower proportion of pectin and without the presence of TiO_2. Cui et al. [24] reported that pectin nanofibers with poly(ethylene oxide) (PEO) were rigid with values of Young's modulus (192.3 MPa) and tensile strength (14.6 MPa); this is attributed to the orientation of the polymer chains of pectin during electrospinning.

Cross-linking agents have been used with the intention of improving or adding properties; some of the most used cross-linking agents in electrospinning systems are glutaraldehyde vapors [60, 70, 71] and genipin [15].

Shalumon et al. [20] proposed cross-linking agents with glutaraldehyde vapors because it has less or no toxic effect. Chen et al. [71] elaborated by electrospinning nanofibers of chitosan/collagen/poly(ethylene oxide) cross-linked with glutaraldehyde vapors reported Young's modulus increase from 0.29 to 0.65 MPa and the decrease in tensile stress and strain.

Nanofibers of poly(lactic-co-glycolic acid)/chitosan/PVA cross-linked with vapors of glutaraldehyde succeeded to increase the tensile strength up to 3.8 MPa and Young's modulus tension up to 106.2 MPa; this is due to the union between the components of the structure [70].

Another cross-linking agent is genipin, which is extracted from the gardenia fruit (*Gardenia jasminoides*). This has been used in hydroxyapatite fibers with chitosan and was found to increase up to four to five times the stiffness with respect to non-cross-linked fibers [15].

5.5. Antimicrobial prophylaxis

Microbiological assays are evaluated through the inhibition provided by a biomaterial in the presence of microorganisms [61]. **Table 3** shows the microorganisms studied in microbiological assays with nanofibers. According to Heunis et al. [72], *Staphylococcus aureus* is the most prevalent microorganism in skin infections.

Infections in wounds are common, because the microorganisms damage the injured tissues causing adverse reactions to the immune system such as inflammation and tissue damage, retarding the healing process [65]. The main pathogenic bacteria that set the wound healing process at risk are *Staphylococcus aureus, Pseudomonas aeruginosa, Streptococcus pyogenes*, some *Proteus, Clostridium*, and coliform species [48].

Study	Microorganism	Inhibitory effect	References
Fabrication of PEO/chitosan/PCL/olive oil nanofibrous scaffolds for wound dressing applications	*Escherichia coli, Staphylococcus aureus*	Superior against *E. coli*	[61]
Evaluation of chitosan nano-dressing for wound healing: characterization, in vitro and in vivo studies	*Escherichia coli, Staphylococcus aureus, Pseudomonas aeruginosa, Bacillus subtilis, Aspergillus niger*	Superior against *B. subtilis*	[63]
The effect of increasing honey concentration on the properties of the honey/poly(vinyl alcohol)/chitosan nanofibers	*Staphylococcus aureus Escherichia coli*	Enhanced against *S. aureus* and *E. coli*	[68]
Chitosan-/polyurethane-blended fiber sheets containing silver sulfadiazine for use as an antimicrobial wound dressing	*Pseudomonas aeruginosa, Staphylococcus aureus*, methicillin-resistant *Staphylococcus aureus* (MRSA)	Superior against MRSA	[73]
Fabrication of an antibacterial non-woven mat of a poly(lactic acid)/chitosan blend by electrospinning	*Escherichia coli, Staphylococcus aureus*	Enhanced incorporating Ag nanoparticles	[74]

Table 3. Antimicrobial assay that involves chitosan nanofibers.

The antimicrobial activity of chitosan is of broad spectrum and has been shown to be effective against Gram-positive and Gram-negative bacteria and many filamentous fungi and yeasts [40]. According to Kumar et al. [75], the cationic amino groups of chitosan have the ability to bind to the anionic groups of the microorganisms by inhibiting the presence of *Escherichia coli, Fusarium, Alternaria,* and *Helminthosporium*. Sarhan et al. [68] attributed this to the polycationic nature of chitosan, which allows it to interact with bacterial membranes that are negatively charged, favoring the loss of permeability, causing cell disruption and subsequently death.

The inhibitory effect in dressings is evaluated by several techniques. Au et al. [74] used the optical density technique in a chitosan/polylactic acid/silver nanofiber blend, mentioning that bacterial cells are opaque and when they propagate in solutions they become turbid; therefore, lower optical density indicates greater antibacterial activity.

Other authors used the bacterial disk inhibition method. Archana et al. [62] reported that in nanofiber dressings of chitosan/pectin/TiO_2 the antibacterial activity is excellent; in addition large surface areas facilitate microbial adsorption and accelerate the antimicrobial activity rate. Sarhan et al. [68] report efficiency against *E. coli* in nanofibers of chitosan and honey; the inhibitory effect due to the polycationic nature of chitosan is potentiated with honey due to its acidity, high sugar content, and hydrogen peroxide production capacity. Additionally, Zarghami et al. [60] found greater efficiency against *E. coli,* Gram-negative in nanofibers of chitosan with poly(ethylene oxide).

The use of metal nanoparticles had been shown to have high antimicrobial activity against bacteria, viruses, and other microorganisms [76]. Lee et al. [73] reported higher efficiency against *S. aureus* in chitosan/polyurethane nanofibers with silver sulfadiazine. Silver is known for its great antimicrobial properties and its use for pharmaceutical applications. Its antimicrobial activity is due to the interaction between silver particles with the bacterial cell, penetrating its wall and causing damage and cell death to components (DNA) [77].

6. Cross-linked copolymers for electrospinned chitosan biocomposites/nanofibers

6.1. Poly(vinyl alcohol)

Poly(vinyl alcohol) (PVA) is a synthetic polymer soluble in water and with excellent chemical resistance. It is recognized mainly for its non-toxicity, biodegradability, and biocompatibility, and it is used in the biomedical area. Since the 1950s it has been commercialized for the formation of highly hydrophilic fibers [17]. It is known that PVA nanofibers dissolve instantaneously in water. That is the main reason why cross-linking is recommended. Destaye et al. [13] cross-linked PVA nanofibers with glutaraldehyde vapors and reported that the concentration of glutaraldehyde increases water retention capacity and the swelling of nanofibers improves their mechanical properties.

For biomedicine, the chitosan nanofibers with PVA have been elaborated by electrospinning because they are more favorable for cell culture compared with only PVA [17]. Recently, chitosan nanofibers blended with PVA are being studied as dressings for the treatment of diabetic foot ulcers. Ahmed et al. [30] demonstrated that nanosized pores protect damaged tissue from bacteria, while high porosity increases fluid absorption and promoting wound healing.

6.2. Sodium alginate

Sodium alginate is a nontoxic polysaccharide with applications in the food and pharmaceutical products. The natural sources are all species of brown seaweed. It is a water-soluble salt of alginic acid. It consists of two uronic acids, β-D-mannuronic acid (M) and α-L-guluronic acid (G), in β-(1–4) union. These acids form homopolymeric blocks M-M or G-G and blocks with an alternating sequence of M-G blocks [78].

The alginate is used in medicine to prepare wound dressings, Tehrani et al. [79]. Fan et al. [80] reported that calcium alginate fibers interact with wound exudates forming a gel, resulting on ion exchange between fiber calcium and sodium ions from exudates. Furthermore, this polymer has been used for the treatment of different wounds due to its biocompatibility properties [81].

Pure sodium alginate solutions cannot be processed by electrospinning because of their high viscosity. But it is possible using organic solvents or with water-soluble synthetic polymer blend such as poly(vinyl alcohol) (PVA) and poly(ethylene oxide) [82]. Islam et al. [16] reported that blend of sodium alginate with PVA can be processed by electrospinning and forms ultrafine nanofibers with uniform surface.

Yu et al. [83] elaborated wound dressings with chitosan/alginate/collagen and hydroxyapatite with potential application in bone tissue engineering due to a suitable structure for cell development. Similarly, Jeong et al. [84] from a polyelectrolyte blend with chitosan/PEO obtained nanofibers for wound dressings, demonstrating their cell promotion and potential use as a dressing.

6.3. Carboxymethylcellulose

Carboxymethylcellulose is a semisynthetic natural polymer obtained by carboxymethylation of cellulose with properties such as biocompatibility, low toxicity, and low degradation rate [85]. Likewise, Ninan et al. [65] attribute their water solubility to their composition from the β-(1 → 4) glucopyranose residues. Its use in biomedicine as a biocompatible material has been proven [86]. Besides, Chen and Fan [87] studied their efficiency in clinical studies with humans and animals, finding that postoperative damages such as abdominal adhesions are reduced.

Chitosan blends with carboxymethylcellulose are possible because it is an anionic polymer with similar structure to chitosan allowing strong ionic cross-linking between them [67].

Fouda et al. [56] elaborated antimicrobial dressings for biological use from chitosan and carboxymethylcellulose blend with silver nanoparticles. Also, Ninan et al. [65] prepared wound dressings from a mixture of pectin, carboxymethylcellulose, and microfibrillated cellulose for skin wound treatment.

6.4. Collagen

Collagen is one of the proteins of the extracellular matrix more abundant in mammals. It is recognized for being reabsorbable with excellent biocompatibility and the ability to promote tissue regeneration [83]. The extracellular matrices in tissues are nanofiber structures that act as wound dressing to attach cells in the tissue, control tissue structure, and regulate the cell phenotype [88].

The main challenge for tissue engineering dressings is to design and create biodegradable matrices that can mimic the composition and structure of extracellular matrices [89]. Cross-linking and blending with biomolecules or synthetic materials have been used to improve the stability of collagen dressings [83]. According to Chen et al. [87], collagen is widely used as a biomaterial in the medical and pharmaceutical field. Collagen and chitosan are blended mimic the components of the extracellular matrix. Chen et al. [90] mentioned that electrospinning using a suitable solvent such as the mixture of 1,1,1,3,3,3 hexafluoro-2-propanol with trifluoroacetic acid to make dressings with application in tissue engineering is possible. Also, Yin et al. [91] used these solvents in electrospinning of collagen/chitosan/poly(L-lactide-co-ε-caprolactone) blend to produce dressings with application in vascular graft.

7. In vivo and in vitro assays

In vivo and in vitro assays evaluate the response of a biomaterial in experimental models or simulations. Biodegradable biomaterials have the characteristic of being completely degraded by the body's enzymes when the support is no longer necessary [55].

Table 4 shows the different in vitro and in vivo tests performed with biomaterials. Archana et al. [62] indicated that the dressings of chitosan/pectin/TiO_2 induce 1.14% of hemolysis of erythrocytes; for that reason they are highly hemocompatible when accepting values up to 5%.

Gautam et al. [92] reported high proliferation, adhesion, and cell morphology in polycaprolactone/elastin/chitosan dressings. This is due to the synergistic effect between the carboxyl group of polycaprolactone, gelatin amino group, and hydroxyl group of chitosan. In addition, in vitro degradation was almost complete after 8 weeks.

Likewise, Kumar et al. [75] in chitosan/pectin/calcium carbonate nanofibers reported that the dressings are biodegradable, because the pectin-chitin matrix gradually degraded up to 60% in 21 days.

Duan et al. [70] reported high cellular viability, because the poly(lactic acid-co-glycolic acid)/chitosan/PVA nanofibers promote attachment and proliferation of fibroblasts; in addition their morphology changed from round to elongated with little cellular activity. Likewise, Coimbra et al. [64] reported cellular growth and proliferation with a typical cellular form of pectin/chitosan dressings, thus demonstrating their biocompatibility and non-toxicity. On the other hand, Mendes et al. [93] evaluated the cellular metabolism and integrity of the membrane in fibroblasts in chitosan/phospholipid nanofibers by reporting biocompatible dressings.

Study	Assessment	Results	References
Application of chitosan/PVA nanofiber as a potential wound dressing for streptozotocin-induced diabetic rats	In vivo (adult Wistar rats)	Good adherence and promote tissue bonding	[14]
Electrospinning of carboxymethyl chitin/poly(vinyl alcohol) nanofibrous scaffolds for tissue engineering applications	In vitro (human mesenchymal stem cells)	High cell adhesion and proliferation	[20]
Evaluation of chitosan nano-dressing for wound healing: characterization, in vitro and in vivo studies	In vitro (mouse fibroblast (NIH 3 T3 and L929)) In vivo (adult male albino rats)	Induces blood clotting Fast wound healing Surface of the wound was covered with new epithelium	[62]
Preparation and chemical and biological characterization of a pectin/chitosan polyelectrolyte complex scaffold for possible bone tissue engineering applications	In vitro (human osteoblast cells (CRL-11372))	Cells adhered and proliferated	[64]
Drug delivery and tissue engineering applications of biocompatible pectin-chitin/nano-$CaCO_3$ composite scaffolds	In vitro (NIH3T3, MG63, and L929 cells)	Efficient cell adhesion and proliferation	[66]
A nanofibrous composite membrane of PLGA-chitosan/PVA prepared by electrospinning	In vitro (rabbit dermal fibroblast from rabbit back skin)	High fibroblast viability	[70]
Fabrication and characterization of PCL/gelatin/chitosan ternary nanofibrous composite scaffold for tissue engineering applications	In vitro (mouse fibroblast cells (L929))	High fibroblast growth and proliferation	[92]
Hybrid electrospun chitosan-phospholipids nanofibers for transdermal drug delivery	In vitro (L929 cells)	Biocompatible nanofibers	[93]

Table 4. In vitro and in vivo assays with chitosan nanofibers.

In vivo assays with animal experimental models on chitosan/PVA nanofibers reported cases of completely healed wounds. This has been attributed to protection of the injured tissue against bacteria that can infect damaged scar tissue by the pores of dressing; in addition the process of fluid absorption and healing was increased after 10 days [14]. Archana et al. [62] achieved wound closure after 16 days with chitosan/pectin/TiO_2 dressings, which were adhered, did not dissolve upon contact with the wound, and had easy removal.

8. Conclusion

Many technological advances continue to exploit the properties offered by polymeric materials as biocomposites. One of the more innovative advances is electrospinning. This technique is simple, low cost, and effective for the development of nanofibers. These polymeric nanocomposites are used as biomaterials for the preparation of dressings.

The surface properties of the nanofibers show the influence of the electrospinning parameters, from their diameter, pore size, porosity, ability to absorb the liquids of the wounds, and their ability to allow gas exchange in the wound, while the mechanical properties are influenced by the composition of the polymer solution. All this provides information about the ability of dressings to adapt to wounds and their ease to manipulate. In addition, it is well known that blending polymers to produce biocomposites is an effective method to improve the performance of materials.

Chitosan is a polycationic polymer widely studied because it has biological properties that make it a good candidate to be used for wound dressings. Thus, the microbiological assays show the capacity of the chitosan dressings against the main microorganisms present in the skin and wounds, demonstrating their efficiency. In vitro tests, through the simulation of wound conditions, reveal their behavior in the body. In contrast, in vivo tests expose the response of the dressing in animal experimental models. However, clinical study reports in human models are required to learn the precise behavior of the biomaterial and their response in the wound.

This chapter exhaustively reviews the literature that discusses the preparation of chitosan dressings by electrospinning and its blending with other polymers, as well as the biological properties that determine its potential medical use in the healing of cutaneous wounds.

Acknowledgements

The first author is grateful to CONACYT (477730). This research was financed under Project No. 248160 from CONACYT-PN2014 and by Project PROFAPI No. 2017-0010 from the Instituto Tecnológico de Sonora.

Conflicts of interests

The authors declare no conflicts of interest.

Summary

As a summary, electrospinned chitosan nanocomposites are promising biomedical wound dressings.

Author details

Claudia A. Vega-Cázarez, Dalia I. Sánchez-Machado and Jaime López-Cervantes*

*Address all correspondence to: jaime.lopez@itson.edu.mx

Instituto Tecnológico de Sonora, Ciudad Obregón, Sonora, México

References

[1] Khor E, Lee YL. Implantable applications of chitin and chitosan. Biomaterials. 2003; **24**:2339-2349. DOI: 10.1016/S0142-9612(03)00026-7

[2] Dash M, Chiellini F, Ottenbrite RM, Chiellini E. Chitosan—A versatile semi-synthetic polymer in biomedical applications. Progress in Polymer Science. 2011;**36**:981-1014. DOI: 10.1016/j.progpolymsci.2011.02.001

[3] Geng X, Kwon O, Jang J. Electrospinning of chitosan dissolved in concentrated acetic acid solution. Biomaterials. 2005;**26**:5427-5432. DOI: 10.1016/j.biomaterials.2005.01.066

[4] Guyal R, Macri LK, Kaplan HM, Kohn J. Nanoparticles and nanofibers for topical drug delivery. Journal of Controlled Release. 2016;**240**:77-92. DOI: 10.1016/j.jconrel.2015.10.049

[5] Abdelrahman T, Newton H. Wound dressings: Principles and practice. Surgery (Oxford). 2011;**29**(10):491-495. DOI: 10.1016/j.mpsur.2011.06.007

[6] Nawalakhe R, Shi Q, Vitchuli N, Noar J, Caldwell JM, Breidt F, Bourham MA, Zhang X, McCord AG. Novel atmospheric plasma enhanced chitosan nanofiber/gauze composite wound dressings. Journal of Applied Polymer Science. 2013;**129**(2):916-923. DOI: 10.1002/APP.38804

[7] Smith IO, Liu XH, Smith LA, Ma PX. Nanostructured polymer scaffolds for tissue engineering and regenerative medicine. WIREs Nanomedicine and Nanobiotechnology. 2009;**1**:226-236. DOI: 10.1002/wnan.026

[8] Bhardwaj N, Kundu SC. Electrospinning: A fascinating fiber fabrication technique. Biotechnology Advances;**201, 28**:325-347. DOI: 10.1016/j.biotechadv.2010.01.004

[9] Li M, Mondrinos MJ, Ghandi MR, Ko FK, Weiss AS, Lelkes PI. Electrospun protein fibers as matrices for tissue engineering. Biomaterials. 2005;**26**:5999-6008. DOI: 10.1016/j.biomaterials.2005.03.030

[10] Hamori M, Nagano K, Kakimoto S, Naruhashi K, Kiriyama A, Nishimura A, Shibata N. Preparation and pharmaceutical evaluation of acetaminophen nano-fiber tablets: Application of a solvent-based electrospinning method for tableting. Biomedicine & Pharmacotherapy. 2016;**78**:14-22. DOI: 10.1016/j.biopha.2015.12.023

[11] Sill TJ, von Recum HA. Electrospinning: Applications in drug delivery and tissue engineering. Biomaterials. 2008;**29**(13):1989-2006. DOI: 10.1016/j.biomaterials.2008.01.011

[12] Kai D, Liow SS, Loh XJ. Biodegradable polymers for electrospinning: Towards biomedical applications. Materials Science and Engineering: C. 2014;**45**:659-670. DOI: 10.1016/j.msec.2014.04.051

[13] Destaye AG, Lin CK, Lee CK. Glutaraldehyde vapor cross-linked nanofibrous PVA mat with in situ formed silver nanoparticles. ACS Applied Materials & Interfaces. 2013;**5**(11):4745-4752. DOI: 10.1021/am401730x

[14] Majd SA, Khorasgani MR, Moshtaghian SJ, Talebi A, Khezri M. Application of chitosan/PVA nano fiber as a potential wound dressing for streptozotocin-induced diabetic rats.

International Journal of Biological Macromolecules. 2016;**92**:1162-1168. DOI: 10.1016/j.ijbiomac.2016.06.035

[15] Frohbergh ME, Katsman A, Botta GP, Lazarovici P, Schauer CL, Wegst GK, Lelkes IP. Electrospun hydroxyapatite-containing chitosan nanofibers crosslinked with genipin for bone tissue engineering. Biomaterials. 2012;**33**:9167-9178. DOI: 10.1016/j.biomaterials.2012.09.009

[16] Islam MS, Karim MR. Fabrication and characterization of poly(vinyl alcohol)/alginate blend nanofibers by electrospinning method. Colloids and Surfaces, A: Physicochemical and Engineering Aspects. 2010;**366**(1):135-140. DOI: 10.1016/j.colsurfa.2010.05.038

[17] Jia YT, Gong J, Gu XH, Kim HY, Dong J, Shen XY. Fabrication and characterization of poly(vinyl alcohol)/chitosan blend nanofibers produced by electrospinning method. Carbohydrate Research. 2007;**67**(3):403-409. DOI: 10.1016/j.carbpol.2006.06.010

[18] Leung V, Hartwell R, Elizei SS, Yang H, Ghahary A, Ko F. Postelectrospinning modifications for alginate nanofiber-based wound dressings. Journal of Biomedical Materials Research Part B. 2014;**102**(3):508-515. DOI: 0.1002/jbm.b.33028

[19] Pakravan M, Heuzey MC, Ajji A. A fundamental study of chitosan/PEO electrospinning. Polymer. 2011;**52**(21):4813-4824. DOI: 10.1016/j.polymer.2011.08.034

[20] Shalumon KT, Annulekha KH, Nair SV, Nair SV, Chennazhi KP, Jayakumar R. Sodium alginate/poly(vinyl alcohol)/nano ZnO composite nanofibers for antibacterial wound dressings. International Journal of Biological Macromolecules. 2011;**49**:247-254. DOI: 10.1016/j.ijbiomac.2011.04.005

[21] Stone SA, Gosavi P, Athauda TJ, Ozer RR. In situ citric acid crosslinking of alginate/polyvinyl alcohol electrospun nanofibers. Materials Letters. 2013;**112**:32-35. DOI: 10.1016/j.matlet.2013.08.100

[22] Tonsonmboon K, Oyen M. Composite electrospun gelatin fiber-alginate gel scaffolds for mechanically robust tissue engineered cornea. Journal of the Mechanical Behavior of Biomedical Materials. 2013;**21**:185-194. DOI: 10.1016/j.jmbbm.2013.03.001

[23] Van der Schueren L, Steyaert I, De Schoenmaker B, De Clerck K. Polycaprolactone/chitosan blend nanofibres electrospun from an acetic acid/formic acid solvent system. Carbohydrate Polymers. 2012;**88**(4):1221-1226. DOI: 10.1016/j.carbpol.2012.01.085

[24] Cui S, Yao B, Sun X, Hu J, Zhou Y, Liu Y. Reducing the content of carrier polymer in pectin nanofibers by electrospinning at low loading followed with selective washing. Materials Science & Engineering. C, Materials for Biological Applications. 2016;**59**:885-893. DOI: 10.1016/j.msec.2015.10.086

[25] Haider A, Haider S, Kang IK. A comprehensive review summarizing the effect of electrospinning parameters and potential applications of nanofibers in biomedical and biotechnology. Arabian Journal of Chemistry. 2015. DOI: 10.1016/j.arabjc.201

[26] Rutledge GC, Fridrikh SV. Formation of fibers by electrospinning. Advanced Drug Delivery Reviews. 2007;**59**(14):1384-1391. DOI: 10.1016/j.addr.2007.04.020

[27] Huang ZM, Zhang YZ, Kotaki M, Ramakrishna S. A review on polymer nanofibers by electrospinning and their applications in nanocomposites. Composites Science and Technology. 2003;**63**(15):2223-2253. DOI: 10.1016/S0266-3538(03)00178-7

[28] Dabirian F, Ravandi SH, Sanatgar RH, Hinestroza JP. Manufacturing of twisted continuous PAN nanofiber yarn by electrospinning process. Fibers and Polymers. 2011;**12**(5):610. DOI: 10.1007/s12221-011-0610-6

[29] Charernsriwilaiwat N, Opanasopit P, Rojanarata T, Ngawhirunpat T, Supaphol P. Preparation and characterization of chitosan-hydroxybenzotriazole/polyvinyl alcohol blend nanofibers by the electrospinning technique. Carbohydrate Polymers. 2010;**81**(3):675-680. DOI: 10.1016/j.carbpol.2010.03.031

[30] Ahmed S, Ikram S. Chitosan based scaffolds and their applications in wound healing. Achievements in the Life Sciences. 2016;**10**(1):27-37. DOI: 10.1016/j.als.2016.04.001

[31] Croisier F, Jérôme C. Chitosan-based biomaterials for tissue engineering. European Polymer Journal. 2013;**49**(4):780-792. DOI: 10.1016/j.eurpolymj.2012.12.009

[32] Berger J, Reist M, Mayer JM, Felt O, Peppas NA, Gurny R. Structure and interactions in covalently and ionically crosslinked chitosan hydrogels for biomedical applications. European Journal of Pharmaceutics and Biopharmaceutics. 2004;**57**(1):19-34. DOI: 10.1016/S0939-6411(03)00161-9

[33] Baldrick P. The safety of chitosan as a pharmaceutical excipient. Regulatory Toxicology and Pharmacology. 2010;**56**(3):290-299. DOI: 10.1016/j.yrtph.2009.09.015

[34] Mansouri S, Lavigne P, Corsi K, Benderdour M, Beaumont E, Fernandes JC. Chitosan-DNA nanoparticles as non-viral vectors in gene therapy: Strategies to improve transfection efficacy. European Journal of Pharmaceutics and Biopharmaceutics. 2004;**57**(1):1-8. DOI: 10.1016/S0939-6411(03)00155-3

[35] Kean T, Thanou M. Biodegradation, biodistribution and toxicity of chitosan. Advanced Drug Delivery Reviews. 2010;**62**(1):3-11. DOI: 10.1016/j.addr.2009.09.004

[36] Waibel KH, Haney B, Moore M, Whisman B, Gomez R. Safety of chitosan bandages in shellfish allergic patients. Military Medicine. 2011;**176**(10):1153-1156

[37] Ng WL, Yeong WY, Naing MW. Polyelectrolyte gelatin-chitosan hydrogel optimized for 3D bioprinting in skin tissue engineering. International Journal of Bioprinting. 2016;**2**(1):53-62. DOI: 10.18063/IJB.2016.01.009

[38] Sharp RG. A review of the applications of chitin and its derivatives in agriculture to modify plant-microbial interactions and improve crop yields. Agronomy. 2013;**3**(4):757-793

[39] Rao SB, Sharma CP. Use of chitosan as a biomaterial: Studies on its safety and hemostatic potential. Journal of Biomedical Materials Research. 1997;**34**(1):21-28. DOI: 0.1002/(SICI)1097-4636(199701)34:1<21::AID-JBM4>3.0.CO;2-P

[40] Kong M, Chen XG, Xing K, Park HJ. Antimicrobial properties of chitosan and mode of action: A state of the art review. International Journal of Food Microbiology. 2010;**144**(1):51-63. DOI: 10.1016/j.ijfoodmicro.2010.09.012

[41] Xia W, Liu P, Zhang J, Chen J. Biological activities of chitosan and chitooligosaccharides. Food Hydrocolloids. 2011;**25**(2):170-179. DOI: 10.1016/j.foodhyd.2010.03.003

[42] De Vrieze S, Westbroek P, Van Camp T, Van Langenhove L. Electrospinning of chitosan nanofibrous structures: Feasibility study. Journal of Materials Science. 2007;**42**(19):8029-8034. DOI: 10.1007/s10853-006-1485-6

[43] Kriegel C, Kit KM, McClements DJ, Weiss J. Influence of surfactant type and concentration on electrospinning of chitosan-poly(ethylene oxide) blend nanofibers. Food Biophysics. 2009;**4**(3):213-228. DOI: 10.1007/s11483-009-9119-6

[44] Lee SH, Wang S. Biodegradable polymers/bamboo fiber biocomposite with bio-based coupling agent. Composites. Part A, Applied Science and Manufacturing. 2006;**37**(1):80-91. DOI: 10.1016/j.compositesa.2005.04.015

[45] Vert M. Polymeric biomaterials: Strategies of the past vs. strategies of the future. Progress in Polymer Science. 2007;**32**(8):755-761. DOI: 10.1016/j.progpolymsci.2007.05.006

[46] Binyamin G, Shafi BM, Mery CM. Biomaterials: A primer for surgeons. Seminars in Pediatric Surgery. 2006;**15**(4):276-283. DOI: 10.1053/j.sempedsurg.2006.07.007

[47] Zhang Y, Lim CT, Ramakrishna S, Huang ZM. Recent development of polymer nanofibers for biomedical and biotechnological applications. Journal of Materials Science. Materials in Medicine. 2005;**16**(10):933-946. DOI: 10.1007/s10856-005-4428-x

[48] Boateng JS, Matthews KH, Stevens HN, Eccleston GM. Wound healing dressings and drug delivery systems: A review. Journal of Pharmaceutical Sciences. 2008;**97**(8):2892-2923. DOI: 10.1002/jps.21210

[49] Newman GR, Walker M, Hobot JA, Bowler PG. Visualisation of bacterial sequestration and bactericidal activity within hydrating Hydrofiber® wound dressings. Biomaterials. 2006;**27**(7):1129-1139. DOI: 0.1016/j.biomaterials.2005.07.046

[50] Caló E, Khutoryanskiy VV. Biomedical applications of hydrogels: A review of patents and commercial products. European Polymer Journal. 2015;**65**:252-267. DOI: 10.1016/j.eurpolymj.2014.11.024

[51] Zahedi P, Rezaeian I, Ranaei-Siadat SO, Jafari SH, Supaphol P. A review on wound dressings with an emphasis on electrospun nanofibrous polymeric bandages. Polymers for Advanced Technologies. 2010;**21**(2):77-95. DOI: 10.1002/pat.1625

[52] Newsom EC, Connolly KL, Nehal KS. Facilitating healing of granulating wounds: Dressings, dermal substitutes, and other methods. Current Dermatology Reports. 2015;**4**(3):125-133. DOI: 10.1007/s13671-015-0108-3

[53] Bhattarai N, Li Z, Edmondson D, Zhang M. Alginate-based nanofibrous scaffolds: Structural, mechanical, and biological properties. Advanced Materials. 2006;**18**(11):1463-1467. DOI: 10.1002/adma.200502537

[54] Haider S, Al-Zeghayer Y, Ali FAA, Haider A, Mahmood A, Al-Masry WA, Aijaz MO. Highly aligned narrow diameter chitosan electrospun nanofibers. Journal of Polymer Research. 2013;**20**(4):105. DOI: 10.1007/s10965-013-0105-9

[55] Lu JW, Zhu YL, Guo ZX, Hu P, Yu J. Electrospinning of sodium alginate with poly(ethylene oxide). Polymer Journal. 2006;**47**(23):8026-8031. DOI: 10.1016/j.polymer.2006.09.027

[56] Fouda MM, El-Aassar MR, Al-Deyab SS. Antimicrobial activity of carboxymethyl chitosan/polyethylene oxide nanofibers embedded silver nanoparticles. Carbohydrate Polymers. 2013;**92**(2):1012-1017. DOI: 10.1016/j.carbpol.2012.10.047

[57] Chen ZG, Wang PW, Wei B, Mo XM, Cui FZ. Electrospun collagen–chitosan nanofiber: A biomimetic extracellular matrix for endothelial cell and smooth muscle cell. Acta Biomaterialia. 2010;**6**(2):372-382. DOI: 10.1016/j.actbio.2009.07.024

[58] Garai S, Sinha A. Biomimetic nanocomposites of carboxymethyl cellulose–hydroxyapatite: Novel three dimensional load bearing bone grafts. Colloids and Surfaces, B: Biointerfaces. 2014;**115**:182-190. DOI: 10.1016/j.colsurfb.2013.11.042

[59] Tarun K, Gobi N. Calcium alginate/PVA blended nano fibre matrix for wound dressing. Indian Journal of Fibre & Textile Research. 2012;**37**:127-132

[60] Zarghami A, Iran M, Mostafazadeh A, Golpour M, Heidarinasab A, Haririan I. Fabrication of PEO/chitosan/PCL/olive oil nanofibrous scaffolds for wound dressing applications. Fibers and Polymers. 2015;**16**(6):1201-1212. DOI 10.1007/s12221-015-1201-8

[61] Chan BP, Leong KW. Scaffolding in tissue engineering: General approaches and tissue-specific considerations. European Spine Journal. 2008;**17**(4):467-479. DOI: 10.1007/s00586-008-0745-3

[62] Archana D, Dutta J, Dutta PK. Evaluation of chitosan nano dressing for wound healing: Characterization, in vitro and in vivo studies. International Journal of Biological Macromolecules. 2013;**57**:193-203. DOI: 10.1016/j.ijbiomac.2013.03.002

[63] Ziabari M, Mottaghitalab V, Haghi AK. Evaluation of electrospun nanofiber pore structure parameters. Korean Journal of Chemical Engineering. 2008;**25**(4):923-932. DOI: 10.1007/s11814-008-0151-x

[64] Coimbra P, Ferreira P, De Sousa HC, Batista P, Rodrigues MA, Correia IJ, Gil MH. Preparation and chemical and biological characterization of a pectin/chitosan polyelectrolyte complex scaffold for possible bone tissue engineering applications. International Journal of Biological Macromolecules. 2011;**48**(1):112-118. DOI: 10.1016/j.ijbiomac.2010.10.006

[65] Ninan N, Muthiah M, Park IK, Kalarikkal N, Elain A, Wong TW, Grohens Y. Wound healing analysis of pectin/carboxymethyl cellulose/microfibrillated cellulose based composite scaffolds. Materials Letters. 2014;**132**:34-37. DOI: 10.1016/j.matlet.2014.06.056

[66] Kumar PS, Ramya C, Jayakumar R, Lakshmanan VK. Drug delivery and tissue engineering applications of biocompatible pectin–chitin/nano $CaCO_3$ composite scaffolds. Colloids and Surfaces, B: Biointerfaces. 2013;**106**:109-116. DOI: 10.1016/j.colsurfb.2013.01.048

[67] Liuyun J, Yubao L, Chengdong X. Preparation and biological properties of a novel composite scaffold of nano-hydroxyapatite/chitosan/carboxymethyl cellulose for bone tissue engineering. Journal of Biomedical Science. 2009;**16**(1):65. DOI: 10.1186/1423-0127-16-65

[68] Sarhan WA, Azzazy HM, El-Sherbiny IM. The effect of increasing honey concentration on the properties of the honey/polyvinyl alcohol/chitosan nanofibers. Materials Science and Engineering: C. 2016;**67**:276-284. DOI: 10.1016/j.msec.2016.05.006

[69] Choi C, Nam JP, Nah JW. Application of chitosan and chitosan derivatives as biomaterials. Industrial and Engineering Chemistry Research. 2016;**33**:1-10. DOI: 10.1016/j.jiec.2015.10.028

[70] Duan B, Yuan X, Zhu Y, Zhang Y, Li X, Zhang Y, Yao K. A nanofibrous composite membrane of PLGA–chitosan/PVA prepared by electrospinning. European Polymer Journal. 2006;**42**(9):2013-2022. DOI: 10.1016/j.eurpolymj.2006.04.021

[71] Chen JP, Chang GY, Chen JK. Electrospun collagen/chitosan nanofibrous membrane as wound dressing. Colloids and Surfaces, A: Physicochemical and Engineering Aspects. 2008;**313**:183-188. DOI: 10.1016/j.colsurfa.2007.04.129

[72] Heunis TDJ, Dicks LMT. Nanofibers offer alternative ways to the treatment of skin infections. BioMed Research International. 2010;**2010**:1-10. DOI: 10.1155/2010/510682

[73] Lee SJ, Heo DN, Moon JH, Park HN, Ko WK, Bae MS, Jung BY. Chitosan/polyurethane blended fiber sheets containing silver sulfadiazine for use as an antimicrobial wound dressing. Journal of Nanoscience and Nanotechnology. 2014;**14**(10):7488-7494. DOI: 10.1166/jnn.2014.9581

[74] Au HT, Pham LN, Vu THT, Park JS. Fabrication of an antibacterial non-woven mat of a poly(lactic acid)/chitosan blend by electrospinning. Macromolecular Research. 2012;**20**(1):51-58. DOI: 10.1007/s13233-012-0010-9

[75] Kumar MNR. A review of chitin and chitosan applications. Reactive and Functional Polymers. 2000;**46**(1):1-27. DOI: 10.1016/S1381-5148(00)00038-9

[76] Rai M, Yadav A, Gade A. Silver nanoparticles as a new generation of antimicrobials. Biotechnology Advances. 2009;**27**(1):76-83. DOI: 10.1016/j.biotechadv.2008.09.002; 10.1016/j.biotechadv.2008.09.002

[77] Sharma VK, Yngard RA, Lin Y. Silver nanoparticles: green synthesis and their antimicrobial activities. Advances in colloid and interface science. 2009;**145**(1-2):83-96. DOI: 10.1016/j.cis.2008.09.002

[78] Wang Q, Hu X, Du Y, Kennedy JF. Alginate/starch blend fibers and their properties for drug controlled release. Carbohydrate Polymers. 2010;**82**(3):842-847. DOI: 10.1016/j.carbpol.2010.06.004

[79] Tehrani NS, Faridi-Majidi R, Atar H, Rezayat M. In: 4th International Conference on Nanostructures (ICNS4). Kish Island, I.R. Iran. 12-14 March; 2012. p. 820-822

[80] Fan L, Du Y, Zhang B, Yang J, Zhou J, Kennedy JF. Preparation and properties of alginate/carboxymethyl chitosan blend fibers. Carbohydrate Polymers. 2006;**65**(4):447-452. DOI: 10.1016/j.carbpol.2006.01.031

[81] Pereira R, Mendes A, Bártolo P. Alginate/Aloe vera hydrogel films for biomedical applications. Procedia CIRP. 2013;**5**:210-215. DOI: 10.1016/j.procir.2013.01.042

[82] Shen W, Hsieh YL. Biocompatible sodium alginate fibers by aqueous processing and physical crosslinking. Carbohydrate Polymers. 2014;**102**:893-900. DOI: 10.1016/j.carbpol.2013.10.066

[83] Yu CC, Chang JJ, Lee YH, Lin YC, Wu MH, Yang MC, Chien CT. Electrospun scaffolds composing of alginate, chitosan, collagen and hydroxyapatite for applying in bone tissue engineering. Materials Letters. 2013;**93**:133-136. DOI: 10.1016/j.matlet.2012.11.040

[84] Jeong SI, Krebs MD, Bonino CA, Samorezov JE, Khan SA, Alsberg E. Electrospun chitosan–alginate nanofibers with in situ polyelectrolyte complexation for use as tissue engineering scaffolds. Tissue Engineering. 2010;**17**(1-2):59-70. DOI: 10.1089/ten.tea.2010.0086

[85] Pasqui D, Torricelli P, Cagna M, Fini M, Barbucci R. Carboxymethyl cellulose—Hydroxyapatite hybrid hydrogel as a composite material for bone tissue engineering applications. Journal of Biomedical Materials Research. Part A. 2014;**102**(5):1568-1579. DOI: 10.1002/jbm.a.34810

[86] Maver T, Kurečič M, Smrke DM, Kleinschek KS, Maver U. Electrospun nanofibrous CMC/PEO as a part of an effective pain-relieving wound dressing. Journal of Sol-Gel Science and Technology. 2016;**9**(3):475-486. DOI: 10.1007/s10971-015-3888-9

[87] Chen H, Fan M. Chitosan/carboxymethyl cellulose polyelectrolyte complex scaffolds for pulp cells regeneration. Journal of Bioactive and Compatible Polymers. 2007;**22**(5):475-491. DOI: 10.1177/0883911507081329

[88] Mo X, Chen Z, Weber HJ. Electrospun nanofibers of collagen-chitosan and P (LLA-CL) for tissue engineering. Front. Mater. Sci. China. 2007;**1**(1):20-23. DOI: 10.1007/s11706-007-0004-2

[89] Huang C, Chen R, Ke Q, Morsi Y, Zhang K, Mo X. Electrospun collagen–chitosan–TPU nanofibrous scaffolds for tissue engineered tubular grafts. Colloids and Surfaces, B: Biointerfaces. 2011;**82**(2):307-315. DOI: 10.1016/j.colsurfb.2010.09.002

[90] Chen Z, Mo X, Qing F. Electrospinning of collagen–chitosan complex. Materials Letters. 2007;**61**(16):3490-3494. DOI: 10.1016/j.matlet.2006.11.104

[91] Yin A, Zhang K, McClure MJ, Huang C, Wu J, Fang J, El-Newehy M. Electrospinning collagen/chitosan/poly(L-lactic acid-co-ϵ-caprolactone) to form a vascular graft: Mechanical and biological characterization. Journal of Biomedical Materials Research. Part A. 2013;**101(**(5):1292-1301. DOI: 10.1002/jbm.a.34434

[92] Gautam S, Chou CF, Dinda AK, Potdar PD, Mishra NC. Fabrication and characterization of PCL/gelatin/chitosan ternary nanofibrous composite scaffold for tissue engineering applications. Journal of Materials Science. 2014;**49**(3):1076-1089

[93] Mendes AC, Gorzelanny C, Halter N, Schneider SW, Chronakis IS. Hybrid electrospun chitosan-phospholipids nanofibers for transdermal drug delivery. International Journal of Pharmaceutics. 2016;**510**(1):48-56. DOI: 10.1016/j.ijpharm.2016.06.016

5

Antifungal Activity of Chitosan against Postharvest Fungi of Tropical and Subtropical Fruits

Porfirio Gutierrez-Martinez, Aide Ledezma-Morales, Luz del Carmen Romero-Islas, Anelsy Ramos-Guerrero, Jovita Romero-Islas, Carolina Rodríguez-Pereida, Paloma Casas-Junco, Leonardo Coronado-Partida and Ramsés González-Estrada

Abstract

In the present chapter, results about the efficacy of chitosan (Chi) on sporulation, mycelial growth, germination, as well as quality parameters on fruits are shown. The results demonstrate that chitosan can control various phytopathogen isolates from diverse fruits. The pathogens in the genera *Colletotrichum, Fusarium,* and *Rhizopus* are involved in important postharvest disease losses throughout the world. In Nayarit, producers had reported high postharvest losses not only at field but also during the commercial chain with their products, besides the resistance of several pathogens to fungicides, which traditionally are applied for controlling diseases. In this sense, the aim of this research group is focused on the research of alternative and effective methods for controlling postharvest diseases. In vivo results are promising due to a good control in important tropical fruits like banana, avocado, mango, and jackfruit. An enhancement in the chitosan antimicrobial activity is reported with the combination with GRAS substances, as well as the use of nanotechnology. Chitosan can be an environment-friendly alternative to the use of chemical fungicides for controlling postharvest diseases in fruits.

Keywords: chitosan, fruits, fungal growth, vegetables

1. Introduction

Mexico is an important exporter of fruits worldwide [1]. However, important postharvest losses have been reported. In this sense, postharvest diseases represent a major factor of losses during storage and shelf life of produce, due to the deterioration of quality and microbial contamination [2]. In many countries traditionally, postharvest decay control is obtained using chemical fungicides, but nowadays consumers are concerned about food safety and environmental issues [3]. The use of antimicrobial packaging can be effective during the storage period, handling, or transport of fruits [4]. In recent years, various investigations have reported the efficacy of the application of chitosan (Chi) in the control of postharvest pathogenic microorganisms, due to the diverse properties like the ability to form films, biodegradability, antimicrobial properties, and the elicitor function [5]. Chitosan has become a useful compound due to its fungicidal effect and its induction of plant defense mechanisms for controlling postharvest diseases of fruit and vegetables [6]. In a previous study, chitosan was applied successfully on strawberry (*Fragaria x ananassa*); the coating decreased the respiratory rate, reduced water losses, as well as preserved the firmness during storage time on treated fruit [7]. Besides, previous studies reported resistance-inducing properties of chitosan in the form of defense responses (enzymes POD, PPO, and PAL) in fruits [8]. The objective of this chapter article was to summarize information about the application of chitosan with other alternative methods, including GRAS substances and the use of nanotechnology, against important fungi that affect tropical and subtropical fruits.

1.1. Fruits and vegetables

1.1.1. Health issues

Nowadays, an interest in the health benefits of fruit and vegetable consumption is increasing. The easy consumption, the good taste, and the nutritional value of fresh fruits and vegetables are important characteristics that have allowed consumers to be more aware about the benefits of a healthy diet. The consumption of fruits and vegetables contributes to the wellness and nutritional health of consumers, due to their high content of phytochemicals as well as other components that may act synergistically with phytochemicals (ascorbic acid, carotenoids, and phenolic compounds) [9].

1.1.2. Production

Worldwide, the main tropical fruit producers and export countries are the Far East, Latin America, and the Caribbean, most of which are developing countries, while a high percentage of developed countries are importers of these fruits. The main tropical fruits for exportation are mango, pineapple, papaya, and avocado, which represent approximately 75% of the exportation of fresh tropical products [10]. The postharvest losses of fruits and vegetables caused by microorganisms worldwide are of the order of 5–25% in developed countries and 20–50% in developing countries; in developed countries they have technologies that allow to diminish or avoid the attack of microorganisms.

1.1.3. Postharvest losses

The main causes of quantitative postharvest fruit losses are classified as crop and harvest practices, availability and conditions of transport, pests and infections, climatic conditions, consumer preferences and attitudes, infrastructure, as well as financial availability of the markets [11]. Fruits can be infected at field or during the postharvest management [12]. Diseases are the principal cause of postharvest losses in tropical and subtropical fruits; anthracnose is the main postharvest disease in various tropical fruits caused by *Colletotrichum gloeosporioides* [13].

1.2. Chitosan: origin, structure, and antimicrobial properties

Chitosan is a polysaccharide derived from chitin, which is the second most abundant polysaccharide in the world, after cellulose [14]. Chitin is a polysaccharide of animal origin and is the main constituent of the outer skeleton of insects and crustaceans like shrimp, crabs, and lobster [15]. Chitosan is the *N*-deacetylated derivative of chitin [16]. The molecular weight of chitosan ranged between 300 and 1000 kDa depending on the source of chitin. Chitosan is a copolymer of N-acetyl-D-glucose amine and D-glucose amine as shown in **Figure 1**.

Important chemical properties of chitosan are as follows: linear polyamine, reactive amino groups, reactive hydroxyl groups available, and chelates metal ions, specially transition metals. Between the biological properties of the chitosan, the most important one is the biocompatibility (natural, biodegradable, safe, and nontoxic) [17]. Various mechanisms of action have been proposed; however, this process is not fully understood. It is important to mention that the antimicrobial activity of the chitosan on pathogenic microorganisms depends on different factors like the strain, molecular weight, concentration, degree of deacetylation, and type of chitosan, among others [5]. The interaction of chitosan with the microorganism

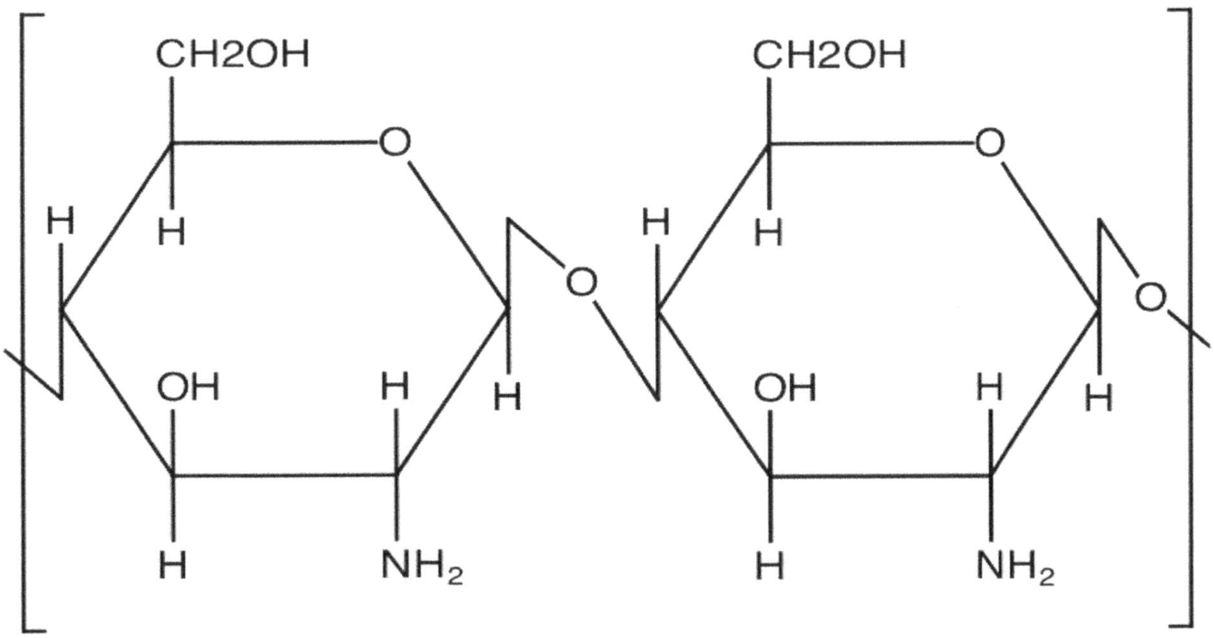

Figure 1. Chemical structure of chitosan.

results in different changes: (a) changes on cell permeability, due to the polycationic nature of the chitosan amino group and the electronegative charges in the outer surface of the fungal or bacteria membrane [18]; (b) affectation on homeostasis (K^+, Ca^{2+}), leading to the efflux of small molecules affecting fungal respiration [19]; (c) microbial starvation, when chitosan acts as chelating agent of metals and essential nutrients affecting microbial development [20]; and (d) inhibition on synthesis of mRNA and proteins, related to their ability to pass through the cell membrane of a microorganism and subsequently bind to DNA [21].

1.2.1. Effects of the antifungal activity of chitosan in the control of postharvest pathogens isolated from various fruits

Chitosan is considered one of the most promising products for the control of several important postharvest fungi in fruits and vegetables [22–25]. In vitro tests used a completely randomized block design. Data were subjected to analysis of variance (ANOVA), and a Tukey test ($p \leq 0.05$) was used for the comparison of means. In our research, the molecular weight of chitosan as well as the concentration used plays an important role in antifungal efficacy against the pathogens tested. Several factors influence the antimicrobial activity of chitosan; among them are the type of chitosan and the concentration [21]. It is reported that the antimicrobial activity of chitosan also depends of the molecular weight, is better when chitosan of low molecular weight and oligochitosan instead of high molecular weight chitosan is applied. In this sense, high molecular weight chitosan cannot pass through the microbial membrane and acts against microbial development [21]. Concerning to the efficacy of chitosan at different concentrations is reported that at lower concentrations, chitosan binds cell surface of microorganisms (negatively charged), disturbing the cell membrane, and causes death of microbial cell by leakage of the intracellular components; however, at higher concentrations, chitosan may coat the microbial surface and prevent the leakage of intracellular components [21].

1.2.1.1. Mycelial growth

The application of chitosan (Chi) at 1.0, 1.5, and 2.0% was effective against the pathogen *Colletotrichum* sp. isolated from banana (*Musa paradisiaca*), 100% of mycelial growth inhibition was observed ($p \leq 0.05$). Conversely, for the pathogen of *Fusarium* sp. isolated from banana, higher concentrations of the chitosan (1.5 and 2%) were applied to obtain a good inhibition (93.2 and 100%, respectively ($p \leq 0.05$)) (**Figure 2**) [26]. In a study, against *Colletotrichum* sp. isolated from banana (*Musa sapientum*), different concentrations where applied (1.0, 1.5 and 2.0%) obtaining a 78.94, 92.1, and 98.68%, respectively ($p \leq 0.05$). Conversely, with the application of chitosan at 0.5%, only 49% of growth inhibition of *Colletotrichum* sp. was obtained ($p \leq 0.05$) [27].

As shown on **Figure 2**, *Colletotrichum gloeosporioides* T147 isolated from avocado (*Persea americana* mill.) c.v. Hass was successfully inhibited as the concentration of chitosan increased, and the percentage of inhibition ranged from 88.85 to 92.97% ($p \leq 0.05$) [28]. In a recent study, the ability of chitosan-pepper tree (*Schinus molle*) essential oil biocomposites against *Colletotrichum gloeosporioides* was evaluated (data not shown) [29]. In this study, an important

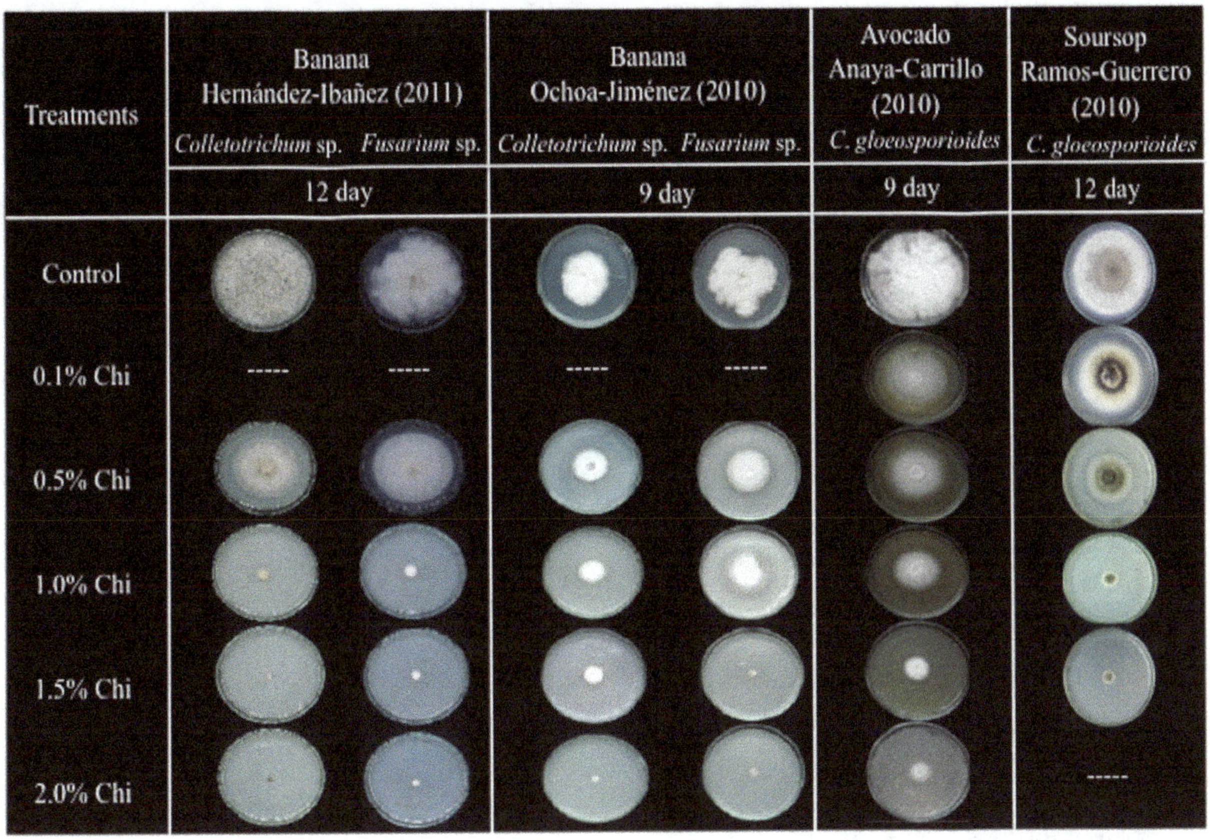

Figure 2. Inhibition of the mycelial growth of pathogens isolated from various fruits by applying chitosan at different concentrations. Control, sterile distilled water; chi, chitosan; -----, concentration not analyzed.

inhibitory effect on the viability of *Colletotrichum gloeosporioides* increased proportionally to the used concentration; the authors concluded that the antifungal activity of the biocomposites against the pathogen can be associated to a synergic effect between the chitosan and pepper tree essential oil. The principal mechanism of pepper tree (*Schinus molle*) essential oil acts against the cytoplasmic membrane, causing it to lose its integrity by assisting the chitosan to enter the interior of the cell. This leads to dissipation of the proton-motive forces and the inhibition of the respiratory enzymes responsible for the cell wall synthesis; this in turn inhibits spore germination and germ tube elongation [30]. Good results were obtained with the application of chitosan at 1.0 and 1.5% against *Colletotrichum gloeosporioides* LSC-120 isolated from soursop (*Annona muricata* L.); up to 90% was achieved using 1.5% of the chitosan ($p \leq 0.05$), whereas this strain was not totally controlled in vitro at low concentrations tested (0.1 or 0.5%) [31]. Similar results were obtained in a study with *Alternaria alternata* isolated from mango (*Mangifera indica* L.) c.v. Tommy Atkins; chitosan treatments at 0.05, 0.1, 0.5, and 1.0% inhibited the mycelial growth of the pathogen by 11.5, 23.1, 55.0, and 70.0%, respectively ($p \leq 0.05$) [32]. *Rhizopus* sp. isolated from jackfruit (*Artocarpus heterophyllus* L.) was controlled (up to 98% ($p \leq 0.05$)) with the application of chitosan in combination with H_2O_2 (peroxide hydrogen) in vitro tests (**Figure 3A**). Synergistic effects were reported with the combinations of chitosan with potassium sorbate or H_2O_2. The pathogen only was totally inhibited with the combinations of potassium sorbate and chitosan at high concentrations of the treatments (**Figure 3B**) [33].

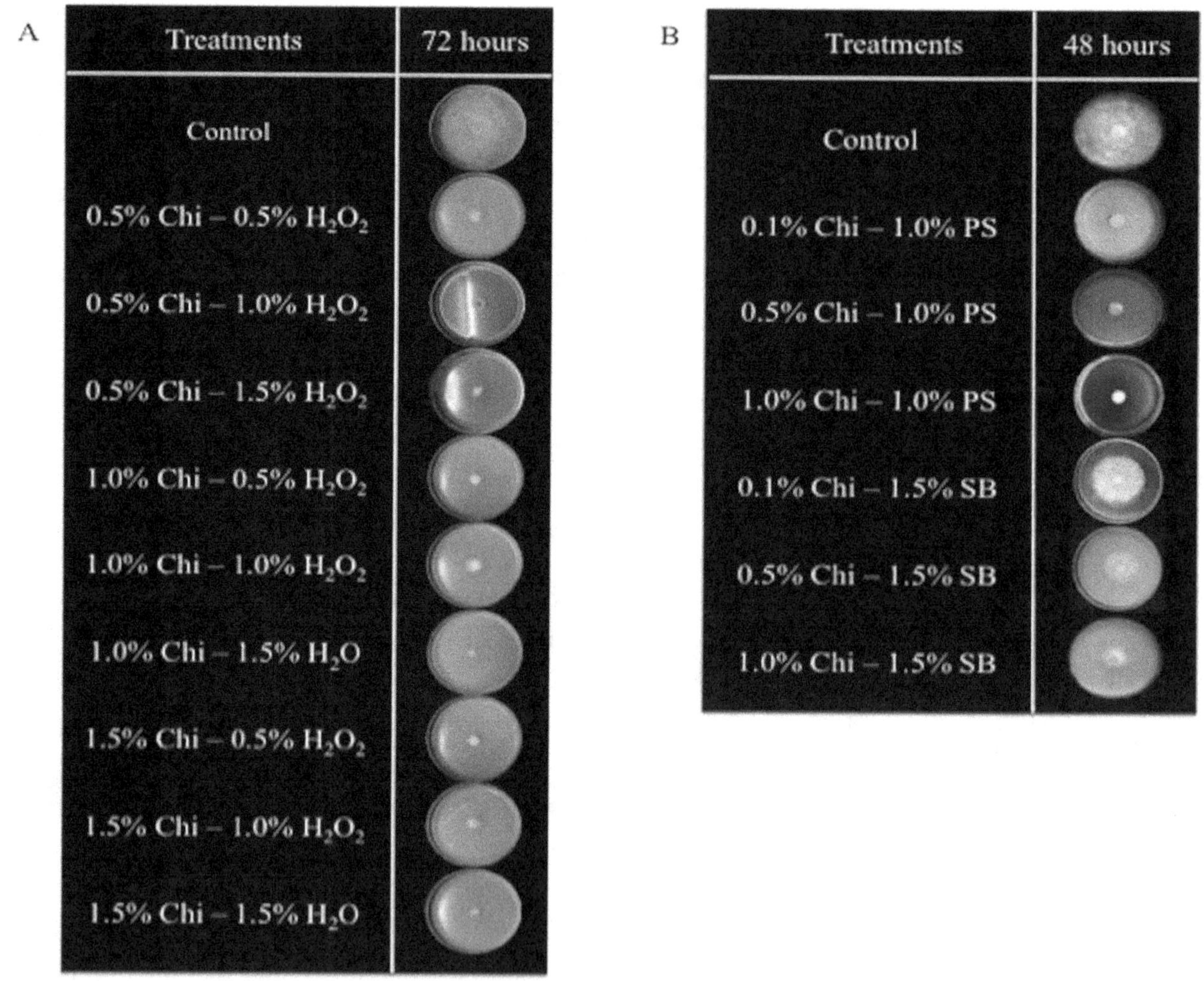

Figure 3. Effect of the interactions of chitosan, hydrogen peroxide, potassium sorbate, and/or sodium bicarbonate at different concentrations in the inhibition of mycelial growth of *Rhizopus* sp. isolated from jackfruit (*Artocarpus heterophyllus* L.). (A) Arciniega-Castro (2014) and (B) Coronado-Partida (2015). Control, sterile distilled water; chi, chitosan; H_2O_2, hydrogen peroxide; PS, potassium sorbate; SB, sodium bicarbonate.

1.2.1.2. Sporulation

In a study with fungus isolated from banana (*Musa paradisiaca*), only the application of chitosan at concentrations of 1.0% the strains of *Colletotrichum* sp. and *Fusarium* sp. showed a decrease in the final concentration of spores [26]. For the fungus of *Fusarium* sp. isolated from banana (*Musa sapientum*), chitosan treatments of 0.5 and 1.0% showed a decrease on the final concentration of the spores, and this process was totally inhibited applying concentrations of 1.5 and 2.0% of chitosan (**Table 1**) [27]. In the same way, *Colletotrichum gloeosporioides* T147 isolated from avocado (*Persea americana* mill) c.v. Hass and *Annona muricata* L. was successfully inhibited using concentrations of chitosan at 1.5 and 2.0% and 1.0 and 1.5%, respectively [28, 31]. Chitosan at concentrations of 0.5, 1.0, and 1.5% in combination with H_20_2 at 0.5, 1.0, and 1.5% affected the development of the spores; this may be due to a synergistic effect between the mechanisms of actions of both compounds that affect the sporulation conditions of the pathogens. The application of combinations of chitosan and H_20_2 at concentrations greater than 0.5% decreased the final concentration of the spores of *Rhizopus* sp. isolated from

Treatments	SoursopRamos-Guerrero (2012)	AvocadoAnaya-Carrillo (2013)	BananaHernández-Ibañez (2011)		BananaOchoa-Jiménez (2010)	
	C. gloeosporioides	*C. gloeosporioides*	*Colletotrichum sp.*	*Fusarium sp.*	*Colletotrichum sp.*	*Fusarium sp.*
Control	3.4×10^6 a	4.0×10^7 a	2.5×10^7 a	9.9×10^7 a	3.8×10^7 a	2.8×10^7 a
0.5% Chi	2.2×10^6 b	3.6×10^7 b	1.0×10^7 b	7.7×10^7 b	1.5×10^7 b	8.7×10^6 b
1.0% Chi	1.4×10^6 c	2.0×10^7 c	6.4×10^6 c	5.0×10^6 c	1.1×10^7 b	2.7×10^6 c
1.5% Chi	1.2×10^6 c	2.2×10^7 c	3.8×10^6 d	2.9×10^6 d	6.0×10^6 d	0 d
2.0% Chi	---	2.2×10^7 c	1.8×10^5 e	3.3×10^5 e	3.7×10^6 c	0 d

The values with different letters in columns are significantly different (Tukey's honestly significant difference; $p \leq 0.05$). Control = Sterile distilled water; Chi = Chitosan; --- = not determined.

Table 1. Effect of chitosan at different concentrations on the sporulation of different pathogens isolated from different fruits.

Treatments	Jack fruitArciniega-Castro (2014)	Treatments	Jack fruitCoronado-Partida (2015)
	Rhizopus sp.		*Rhizopus* sp.
Control	4.75×10^6 a	Control	1.4×10^8 a
0.5% Chi—0.5% H_2O_2	4.3×10^4 f	0.1% Chi—1.0% PS	0 e
0.5% Chi—1.0% H_2O_2	2.83×10^5 b	0.5% Chi—1.0% PS	0 e
0.5% Chi—1.5% H_2O_2	0 g	1.0% Chi—1.0% PS	0 e
1.0% Chi—0.5% H_2O_2	8.5×10^4 d	0.1% Chi—1.5% SB	3.6×10^7 b
1.0% Chi—1.0% H_2O_2	1.0×10^5 c	0.5% Chi—1.5% SB	2.0×10^7 c
1.0% Chi—1.5% H_2O_2	0 g	1.0% Chi—1.5% SB	1.2×10^7 d
1.5% Chi—0.5% H_2O_2	6.6×10^4 e	---	---
1.5% Chi—1.0% H_2O_2	1.0×10^5 c	---	---
1.5% Chi—1.5% H_2O_2	0 g	---	---

The values with different letters in columns are significantly different (Tukey's honestly significant difference; $p \leq 0.05$). Control = Sterile distilled water; Chi = Chitosan; H_2O_2 = Hydrogen peroxide; PS = Potassium sorbate; SB = Sodium bicarbonate.

Table 2. Synergistic effect between chitosan, hydrogen peroxide, potassium sorbate and/or sodium bicarbonate at different concentrations on the sporulation of ***Rhizopus*** sp isolated from jackfruit (***Artocarpus heterophyllus*** L.).

jackfruit (*Artocarpus heterophyllus* L.) [33]. The use of chitosan in combination with H_2O_2 was effective to reduce the spore's production of *Rhizopus* sp. isolated from jackfruit (*Artocarpus heterophyllus* L.) (**Table 2**) [34]. In the same sense, the synergistic effect was also evidenced with the use of organic salts (potassium sorbate and sodium bicarbonate) applied at different concentrations with chitosan, obtaining good results on sporulation as shown in **Table 2** [33].

1.2.1.3. Spore germination

Promising results have been obtained in different investigations on *Colletotrichum* sp. (banana) and *Fusarium* sp. (banana), with the total inhibition on germination applying chitosan at different concentrations (0.5, 1.0, 1.5, and 2.0%) [26, 27]. Conversely, at concentrations of 0.5 and 1.0%, only 50% of the inhibition was obtained ($p \leq 0.05$) against *Colletotrichum gloeosporioides* T147 isolated from avocado (*Persea americana* mill.) c.v. Hass [28]. Good results were reported by Chávez-Magdaleno and Luque-Alcaraz [29] with the application of biocomposites of chitosan loaded with essential oils against *Colletotrichum gloeosporioides* T147, reporting up to 95% of inhibition on spore development ($p \leq 0.05$). Phytopathogens isolated from soursop (*Colletotrichum gloeosporioides* LSC-120), mango (*Alternaria alternata*), and jackfruit (*Rhizopus* sp.) were successfully controlled by the application of chitosan alone or in combination with organic salts [31–34].

1.2.2. Effects of the application of chitosan on postharvest disease control in fruits

The use of chitosan as a coating on fresh fruits is a real alternative on the control of postharvest diseases. An important biological function of the chitosan is like an inducer of defense mechanisms in fruit and vegetable products, causing a reduction and/or inhibition of the development of diseases [35–37]. Besides, the application of chitosan with other natural methods of biological control and nanoparticles is another promising alternative for controlling postharvest diseases [38]. Chitosan nanoparticles can improve the antimicrobial activity, which is associated with the position of the amino groups favoring the binding to the cell surface and an alteration with the normal functions of the membrane, thus inhibiting the growth of the pathogen [16]. There are several studies on the application of chitosan alone or in combination with natural methods such as resistance inducer or as a disease control agent [33, 34, 39, 40]. The analysis of disease incidence, severity, and quality parameters used a completely randomized block design. Data were subjected to analysis of variance (ANOVA), and a Tukey test ($p \leq 0.05$) was used for a means of comparison.

1.2.2.1. Disease incidence and severity

In a study on banana fruits *Musa paradisiaca* and *Musa sapientum*, a total inhibition on infected fruits with *Colletotrichum* sp. and *Fusarium* sp. were reported with the application of chitosan at 1.5% compared to control (80% of incidence) [26, 27]. Related to severity, control fruits presented a damage around the crown on bananas; conversely, fruits treated do not present visible damage (**Figure 4**A, B). As shown in **Figure 4**, the absence was reported on avocado fruits c.v. Hass treated with chitosan (1.5%) and inoculated with *Colletotrichum gloeosporioides* T147. The application of biocomposites of chitosan-pepper tree (*Schinus molle*) essential oil in avocado fruits (*Persea americana* mill.) c.v. Hass infected with *Colletotrichum gloeosporioides* successfully controlled anthracnose disease in a preventive and curative way (**Figure 5**) [39]. In a study on soursop fruits (*Annona muricata* L.) artificially and naturally infected with *Colletotrichum gloeosporioides* LSC-120, a total control was obtained by the application of chitosan at 1.0% (**Figure 6**) [31]. The combination of chitosan with peroxide and GRAS substances in jackfruit (*Artocarpus heterophyllus* L.) was effective in inhibiting the

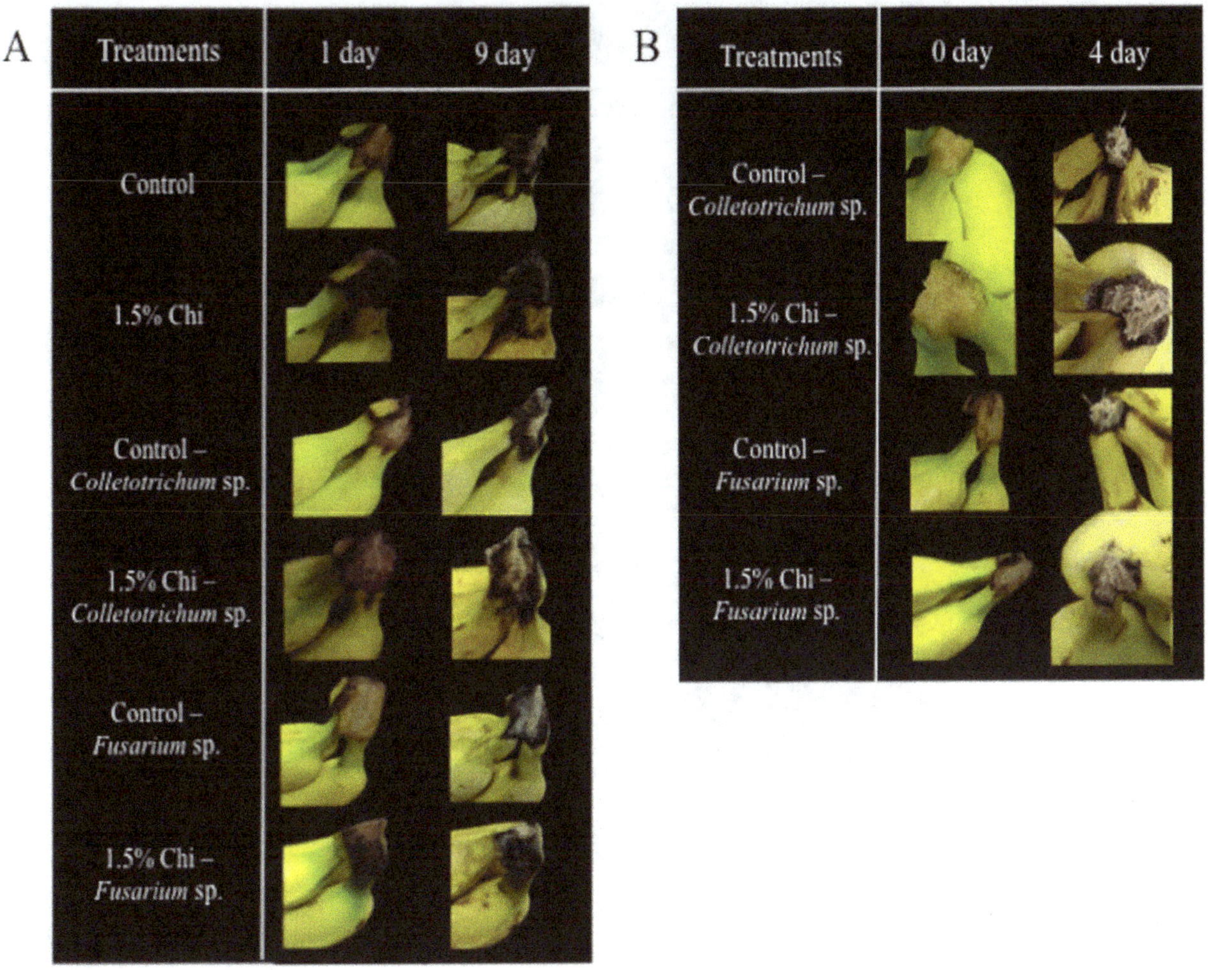

Figure 4. Crowns of banana fruits with 1.5% chitosan with and without inoculation of *Colletotrichum* sp. or *Fusarium* sp., on days 9 and 4 of storage. (A) Banana (*Musa paradisiaca*) stored at 15°C and (B) banana (*Musa sapientum*) stored at 25°C. Control, sterile distilled water; chi, chitosan.

development of soft rot disease by *Rhizopus* sp. The application of 1.0% Chi–1.0% H_2O_2 in jackfruit inoculated with the pathogen was able to totally inhibit the development of soft rot (**Figure** 7A) [34]. In other studies, the application of chitosan with potassium sorbate (SP) or sodium bicarbonate (BS) in jackfruit (*Artocarpus heterophyllus* L.) was effective for controlling the development of *Rhizopus* sp. The combination of 1.0% Chi–1.0% SP with and without inoculation of *Rhizopus* sp. fruits does not present the presence of infection (**Figure** 7B) [33]. Conversely, with the use of BS with chitosan, only 10% of severity reduction was obtained ($p \leq 0.05$). It is concluded that the combination of chitosan with SP can be an alternative to control *Rhizopus* sp. infection in jackfruit. The principal mode of action of the bicarbonate ion is through its buffering capacity, whereby an alkaline environment is sustained, and inhibition of microorganisms occurs due to the use of energy from microbial cells to produce an acidic environment [41]. The antimicrobial activity of potassium sorbate is associated to an alteration of the activity of Krebs cycle enzymes as well as the integrity of cell membranes [42]. The effective conditions for the control of diseases were for bananas *Musa paradisiaca* (15°C) and *Musa sapientum* (25°C), and avocado (*Persea americana* mill.) is 1.5% chitosan with 90–95% of relative humidity. For fruits of soursop (*Annona muricata* L.)

Figure 5. Avocado fruits (*Persea americana* mill.) c.V. Hass coated with chitosan nanoparticles and chitosan-pepper tree essential oil with and without inoculation of *Colletotrichum gloeosporioides* T147 for 10 days at 25°C. Control, sterile distilled water; chi, chitosan, NPs, nanoparticles.

and jackfruit (*Artocarpus heterophyllus* L.), the effective concentration was 1.0% chitosan at 20°C with 90–95% of relative humidity.

The efficacy of chitosan for controlling postharvest diseases depends not only on its antimicrobial properties. Romanazzi and Sanzani [43] reported that the arrays of defense mechanisms are activated in fruits exposed to biotic or abiotic stress, including chitosan application with or without inoculation of the pathogen. Chitosan induces the synthesis of phenolic compounds (chlorogenic acid, caffeic acid) and hydrolase antifungal enzymes (chitinases and β-(1,3)-glucanases) that hydrolyze the main components of the cell wall of fungi causing inhibition of their growth [44]. On the other hand, chitosan coating can serve not only as a protective barrier to fungal infection but also as a barrier to gaseous exchange affecting the fungal development on fruits.

1.2.2.2. Quality parameters

1.2.2.2.1. Weight loss

The ability of chitosan to form coatings is well documented; this property is useful to preserve the quality of fruits and vegetables. Chitosan applied at 1.5% on banana fruits was useful to avoid water losses compared to control [26]. The same concentration of chitosan was effective on bananas (*Musa sapientum*) with lower water losses on fruits compared to control (**Table 3**) [27].

Figure 6. Anthracnose severity in soursop fruits treated with chitosan, inoculated or not with *C. gloeosporioides* LSC-120, and stored 20°C for 9 days. Control, sterile distilled water; chi, chitosan.

The ability to maintain the quality of fruits at different temperatures has been reported also for mango fruits (*Mangifera indica* L.) c.v. Tommy Atkins treated with chitosan at 1.0% and stored at 12°C. In the same sense, fruits treated with 1.0% chitosan and stored at 25°C showed the highest weight loss (10.5%) [32]. Novel formulations of chitosan like the use of nanoparticles have been applied with good results with the application of treatments (2.4%) compared to control (12.3%) [39]. The decreased weight loss in fruits is due to the presence of chitosan coating on the surface of the fruit, acting as a physical barrier to moisture loss and therefore delaying dehydration [45].

1.2.2.2.2. Firmness, pH, and total soluble solids

The loss of firmness is a factor that affects the quality of the fruits. In fruits of soursop (*Annona muricata* L.) treated with chitosan, at the end of the evaluation, fruits had a greater firmness (25 N) compared to control fruits (6.4 N) ($p \leq 0.05$). The total soluble solids (TSS) of the soursop fruits treated with 1.0% of chitosan and control showed a continuous increase (10–18% ($p \leq 0.05$)), and in the case of pH, in fruits treated with 1.0% of chitosan, an increase during storage was obtained, with a pH ranging from 3.49 to 4.10 ($p \leq 0.05$) [31]. The coating formed in banana fruits (*Musa paradisiaca*) by chitosan (1.5%) maintains the losses of firmness compared to control (data not shown). The concentration used on fruits does not affect the TSS, pH, and titratable acidity [26]. Mango fruits treated with 1.0% of chitosan no significant changes on TSS, pH, and titrable acidity were reported. The use of chitosan nanoparticles resulted effectively to

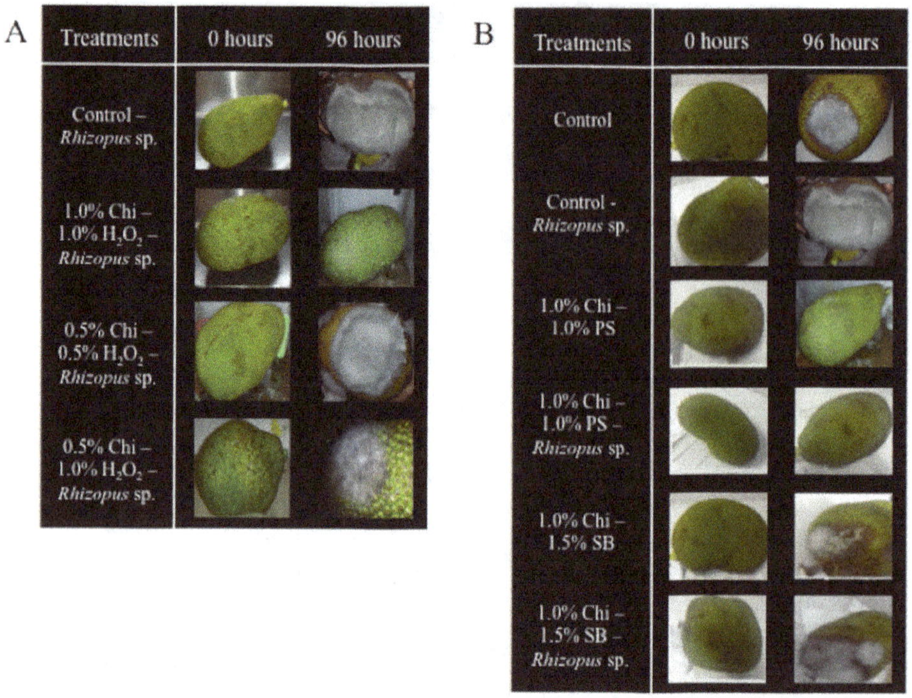

Figure 7. Severity of soft rot infection in jackfruit (*Artocarpus heterophyllus* L.) treated with the combination of chitosan, potassium sorbate, and sodium bicarbonate inoculated with *Rhizopus* sp. at 25°C. (A) Arciniega-Castro (2014) and (B) Coronado-Partida (2015). Control, sterile distilled water; chi, chitosan; H_2O_2, hydrogen peroxide; PS, potassium sorbate; SB, sodium bicarbonate.

Fruits	Temperature	Treatments	% Weight loss
Banana (*Musa paradisiaca*)	15°C	Control	8 ± 0.23[a]
		1.5% Chi	2 ± 0.12[b]
Banana (*Musa sapientum*)	25°C	Control	9.8 ± 0.34[a]
		1.5% Chi	8 ± 0.47[a]
Mango (*Mangifera indica* L.) c.v. Tommy Atkins	12°C	Control	5 ± 0.25[a]
		1.0% Chi	3 ± 0.39[b]
	25°C	Control	15 ± 0.5[a]
		1.0% Chi	10.5 ± 0.46[b]
Soursop (*Annona muricata* L.)	20°C	Control	18 ± 0.7[a]
		1.0% Chi	13.3 ± 0.55[b]
Avocado (*Persea americana* mill.) c.v. Hass	25°C	Control	12.3 ± 0.5[a]
		NPs Chi	2.4 ± 0.43[b]
		NPs – Chi - P	1.5 ± 0.51[b]

Data are means ± standard deviation. The values with different superscripts are significantly different (Tukey's honestly significant difference; $p \leq 0.05$). Control = Sterile distilled water; NPs = Nanoparticles; Chi = Chitosan, P: Pepper tree essential oil.

Table 3. Weight loss of fruits at different temperatures treated with chitosan.

maintain firmness (29 N) on avocado fruits (*Persea americana* mill.) c.v. Hass compared to control (15 N) ($p \leq 0.05$) [39]. In terms of firmness, it decreases as maturation progresses due to changes occurring at the level of the cell wall, where there is hydrolysis of the pectic compounds due to the action of the enzymes cellulase, pectin methylesterase, and polygalacturonase, which in turn degrade high molecular weight polymers such as cellulose and hemicellulose [46]. The coating of chitosan at different concentrations on fruits does not change the quality parameters (TSS, pH, and titratable acidity). This may be due to the fact that chitosan does not interfere in the metabolism cycles (synthesis of sugars, synthesis of organic molecules) [47].

2. Conclusions

The use of chitosan in agriculture commodities can be a suitable alternative to the use of fungicides for controlling postharvest diseases, as well as to preserve the quality of fruits.

Acknowledgements

The authors are pleased to thank SAGARPA-CONACYT for funding the project (number 291472 -2017-2102).

Conflict of interest

Replace the entirety of this text with the "conflict of interest" declaration.

Author details

Porfirio Gutierrez-Martinez[1,2]*, Aide Ledezma-Morales[1,2], Luz del Carmen Romero-Islas[1,2], Anelsy Ramos-Guerrero[1,2], Jovita Romero-Islas[1,2], Carolina Rodríguez-Pereida[1,2], Paloma Casas-Junco[1,2], Leonardo Coronado-Partida[1,2] and Ramsés González-Estrada[1,2]

*Address all correspondence to: pgutierrez@ittepic.edu.mx

1 Tecnológico Nacional de México, Mexico City, Mexico

2 Laboratorio Integral de Investigación en Alimentos, Biochemical Engineering Department, Instituto Tecnológico de Tepic, Tepic, Mexico

References

[1] SAGARPA. Secretaría de agricultura, ganadería, desarrollo rural, pesca y alimentación; 2015

[2] Mari M, Bautista-Baños S, Sivakumar D. Decay control in the postharvest system: Role of microbial and plant volatile organic compounds. Postharvest Biology and Technology. 2016;**122**:70-81

[3] González-Estrada RR et al. Antimicrobial soy protein based coatings: Application to Persian lime (Citrus latifolia Tanaka) for protection and preservation. Postharvest Biology and Technology. 2017;**132**:138-144

[4] Van Long NN, Joly C, Dantigny P. Active packaging with antifungal activities. International Journal of Food Microbiology. 2016;**220**:73-90

[5] Bautista-Baños S et al. Chitosan: A versatile antimicrobial polysaccharide for fruit and vegetables in postharvest–A review. Revista Chapingo Serie Horticultura. 2017; **23**(2):103-121

[6] Romanazzi G et al. Shelf life extension of fresh fruit and vegetables by chitosan treatment. Critical Reviews in Food Science and Nutrition. 2017;**57**(3):579-601

[7] Gol NB, Patel PR, Rao TR. Improvement of quality and shelf-life of strawberries with edible coatings enriched with chitosan. Postharvest Biology and Technology. 2013;**85**:185-195

[8] Al-Qurashi AD, Awad MA. Postharvest chitosan treatment affects quality, antioxidant capacity, antioxidant compounds and enzymes activities of 'El-Bayadi' table grapes after storage. Scientia Horticulturae. 2015;**197**:392-398

[9] Yahia EM. Fruit and Vegetable Phytochemicals: Chemistry and Human Health. Vol. 2. Newark: John Wiley & Sons; 2017

[10] FAO. Food and Agriculture Organization of the United Nations. Tropical fruits. 2017. http://www.fao.org/docrep/006/y5143e/y5143e1a.htm. [Last accessed: 1/15/2018]

[11] Aulakh J, Regmi A. Post-Harvest Food Losses Estimation-Development of Consistent Methodology. Rome: FAO; 2013

[12] El-Ghaouth A et al. Biological control of postharvest diseases of fruits and vegetables. In: Khachatourians GG, Arora DK, editors. Applied Mycology and Biotechnology. Vol 2. Agriculture and Food Production. Elsevier Science B.V. 2002. pp. 219-238

[13] Droby S et al. Postharvest pathology of tropical and subtropical fruit and strategies for decay control. Postharvest Biology and Technology of Tropical and Subtropical Fruits. 2011;**1**:194-223

[14] Ibrahim H, El-Zairy E. Chitosan as a biomaterial—structure, properties, and electrospun nanofibers. In: Bobbarala V, editor. Concepts, Compounds and the Alternatives of Antibacterials. Rijeka: Intech; 2015

[15] Goy RC, Britto DD, Assis OB. A review of the antimicrobial activity of chitosan. Polímeros. 2009;**19**(3):241-247

[16] Kumar MNR. A review of chitin and chitosan applications. Reactive and Functional Polymers. 2000;**46**(1):1-27

[17] Rinaudo M. Chitin and chitosan: Properties and applications. Progress in Polymer Science. 2006;**31**(7):603-632

[18] Palma-Guerrero J et al. Membrane fluidity determines sensitivity of filamentous fungi to chitosan. Molecular Microbiology. 2010;**75**(4):1021-1032

[19] Peña A, Sánchez NS, Calahorra M. Effects of chitosan on Candida albicans: Conditions for its antifungal activity. BioMed Research International. 2013;**2013**:1-15

[20] Wang X et al. Chitosan-metal complexes as antimicrobial agent: Synthesis, characterization and structure-activity study. Polymer Bulletin. 2005;**55**(1):105-113

[21] Hosseinnejad M, Jafari SM. Evaluation of different factors affecting antimicrobial properties of chitosan. International Journal of Biological Macromolecules. 2016;**85**:467-475

[22] El Ghaouth A et al. Antifungal activity of chitosan on post-harvest pathogens: Induction of morphological and cytological alterations in Rhizopus stolonifer. Mycological Research. 1992;**96**(9):769-779

[23] Reddy MB et al. Effect of chitosan on growth and toxin production by Alternaria alternata f. Sp. lycopersici. Hortscience. 1997;**32**(3):467-468

[24] Ben-Shalom N et al. Controlling gray mould caused by Botrytis cinerea in cucumber plants by means of chitosan. Crop Protection. 2003;**22**(2):285-290

[25] Hirano S, Nagao N. Effects of chitosan, pectic acid, lysozyme, and chitinase on the growth of several phytopathogens. Agricultural and Biological Chemistry. 1989;**53**(11):3065-3066

[26] Hernández–Ibáñez A. Efecto del quitosano sobre el control de la pudrición de la corona e inducción de genes de defensa en plátano (Musa paradisiaca) C.V. enano gigante. Tepic, México: Instituto Tecnológico de Tepic; 2011

[27] Ochoa-Jiménez VA. Efecto del quitosano en la inducción de Resistencia a Fusarium sp. y Colletotrichum sp. en frutos de plátano (musa sapientum). Nayarit, México, Tepic: Instituto Tecnológico de Tepic; 2010

[28] Anaya-Carrillo M. Efecto del quitosano en el control de Colletotrichum sp. de frutos de aguacate (Persea americana mill) c.v. Hass del estado de Nayarit. Nayarit, México, Tepic: Instituto Tecnológico de Tepic; 2013

[29] Chávez-Magdaleno ME et al. Effect of chitosan-pepper tree (Schinus molle) essential oil biocomposites on the growth kinetics, viability and membrane integrity of Colletotrichum gloeosporioides. Revista Mexicana de Ingeniería Química. 2017;**17**(1):29-45

[30] Mari M, Bautista-Baños S, Sivakumar D. Decay control in the postharvest system: Role of microbial and plant volatile organic compounds. Postharvest Biology and Technology. 2016;**122**:70-81

[31] Ramos-Guerrero A. Evaluación del quitosano como un sistema alternativo en el control postcosecha de Colletotrichum sp. en frutos de guanábana (Annona muricata). Nayarit, México, Tepic: Instituto Tecnológico de Tepic; 2012

[32] López-Mora LI et al. Evaluación de la actividad antifúngica del quitosano en Alternaria alternata y en la calidad del mango 'Tommy Atkins' durante el almacenamiento. Revista Chapingo Serie horticultura. 2013;**19**(3):315-331

[33] Coronado-Partida LD. Efecto del Quitosano y sustancias GRAS en el control postcosecha de Rhizopus sp. y en la inducción de la actividad enzimática de la Peroxidasa (POD) y Polifenoloxidasa (PFO) en frutos de jaca (Artocarpus heterophyllus L.). Nayarit, México, Tepic: Instituto Tecnológico de Tepic; 2015

[34] Arciniega-Castro PR. Análisis de la interacción de dos inductores de resistencia en frutos de Jaca (Artocarpus heterophyllus L.) ante la enfermedad por Rhizopus sp. Nayarit, México, Tepic: Instituto Tecnológico de Tepic; 2014

[35] Bautista-Baños S, Barrera-Necha L. Tecnologías empleadas en el control de las enfermedades postcosecha de los productos hortofrutícolas. Memoria de investigación; 2001. pp. 111-118

[36] Meng X et al. Physiological responses and quality attributes of table grape fruit to chitosan preharvest spray and postharvest coating during storage. Food Chemistry. 2008;**106**(2):501-508

[37] Chien P-J, Sheu F, Lin H-R. Coating citrus (Murcott tangor) fruit with low molecular weight chitosan increases postharvest quality and shelf life. Food Chemistry. 2007;**100**(3):1160-1164

[38] Bautista-Baños S et al. Quitosano: Una alternativa natural para reducir microorganismos postcosecha y mantener la vida de anaquel de productos hortofrutícolas. Revista Iberoamericana de tecnología postcosecha. 2005;**7**(1):1-6

[39] Chávez-Magdaleno ME. Síntesis y caracterización de nanopartículas de quitosano, complejo quitosano-pirul y su efecto sobre Colletotrichum gloeosporioides en frutos de aguacate Hass. Tepic, Nayarit, México: Instituto Tecnológico de Tepic; 2016

[40] Chen J et al. Combination effect of chitosan and methyl jasmonate on controlling Alternaria alternata and enhancing activity of cherry tomato fruit defense mechanisms. Crop Protection. 2014;**56**:31-36

[41] Mills A, Platt H, Hurta RA. Effect of salt compounds on mycelial growth, sporulation and spore germination of various potato pathogens. Postharvest Biology and Technology. 2004;**34**(3):341-350

[42] Hervieux V et al. Effect of organic and inorganic salts on the development of Helminthosporium solani, the causal agent of potato silver scurf. Plant Disease. 2002; **86**(9):1014-1018

[43] Romanazzi G et al. Induced resistance to control postharvest decay of fruit and vegetables. Postharvest Biology and Technology. 2016;**122**:82-94

[44] Palma-Guerrero J et al. Effect of chitosan on hyphal growth and spore germination of plant pathogenic and biocontrol fungi. Journal of Applied Microbiology. 2008; **104**(2):541-553

[45] Zhu X et al. Effects of chitosan coating on postharvest quality of mango (Mangifera indica L. cv. Tainong) fruits. Journal of Food Processing and Preservation. 2008;**32**(5):770-784

[46] Silveira A. Fisiología y bioquímica de los productos MPF. In: V Congreso Iberoamericano de Postcosecha y Agroexportaciones. Universidad de Cartagena. Colombia: Cartagena. p. 2007

[47] Park Y, Jung S, Gorinstein S. Ethylene treatment of 'Hayward' kiwifruits (Actinidia deliciosa) during ripening and its influence on ethylene biosynthesis and antioxidant activity. Scientia Horticulturae. 2006;**108**(1):22-28

6

Chitoneous Materials for Control of Foodborne Pathogens and Mycotoxins in Poultry

Daniel Hernandez-Patlan, Bruno Solis-Cruz,
Billy M. Hargis and Guillermo Tellez

Abstract

Public concern with the incidence of antibiotic-resistant bacteria, particularly among foodborne pathogens has been challenging the poultry industry to find alternative means of control. Chitosan is a modified, natural biopolymer derived by deacetylation of chitin. The antimicrobial activity and film-forming property of chitosan makes it a potential source of food preservative or coating material of natural origin for improvement of quality and shelf life of various foods of agriculture, poultry, beef and seafood origin. In addition to its use as an antimicrobial, it has been shown that it has good properties as a mycotoxin adsorbent. The purposes of the present chapter is to summarize our experience using chitin-chitosan from Deacetylated 95% food grade chitosan (Paragon Specialty Products LLC Rainsville, AL) or *Aspergillus oryzae* meal (Fermacto®, PetAg Inc., Hampshire IL) to control foodborne pathogens, improve performance, biological sanitizer and mycotoxin binder in commercial poultry.

Keywords: chitosan, Fermacto®, *Salmonella*, mycotoxins, gut health

1. Introduction

Chitin ($C_8H_{13}O_5N$)n) is a long-chain polymer of a N-acetylglucosamine (**Figure 1(a)**), a derivative of glucose, and is found in many places globally. It is the main component of the cell walls of fungi, the exoskeletons of arthropods such as crustaceans (e.g., crabs, lobsters and shrimps) and insects, the radula of mollusks, and the beaks of cephalopods, including squid and octopi [1]. In terms of structure, chitin may be compared to the polysaccharide cellulose and, in terms of function, to the protein keratin [2]. Depending on its source, two types of chitin allomorphs

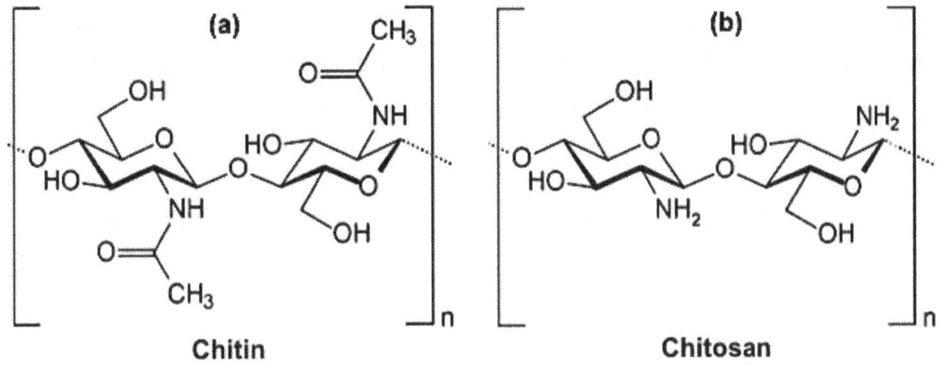

Figure 1. Chemical structure (a) of chitin poly(N-acetyl-β-d-glucosamine) and (b) of chitosan (poly(d-glucosamine) repeat units.

can occur, the α and β forms, which can be differentiated by infrared and solid-state NMR spectroscopy together with X-ray diffraction [3]. Chitosan is a high molecular weight polysaccharide linked by a β-1,4 glycoside and is composed of N-acetyl-glucosamine and glucosamine (**Figure 1(b)**). It is a natural biopolymer derived by deacetylation of chitin and the most widespread polycationic biopolymer [3]. However, although chitosan is obtained from chitin, the applications of the latter compared to chitosan are limited because it is chemically inert [4] and because of its poor solubility [5]. Unlike chitin, chitosan is soluble but in an acidic media since at neutral or alkaline pH it is insoluble. The properties of chitosan can be modified by changing the degree of deacetylation, pH and ionic strength. At neutral pH, most chitosan molecules will lose their charge and precipitate when it is in solution [6].

The application of chitin is focused on obtaining soluble derivatives in aqueous media such as chitosan [3]. Chitosan has several applications in fields such as waste and water treatment, agriculture, fabric and textiles, cosmetics, nutritional enhancement, and food processing. Given its low toxicity and allergenicity, and its biocompatibility, biodegradability and bioactivity, it is a very attractive substance for diverse applications in the pharmaceutical and medical fields, since it has been used for systemic and local delivery of drugs and vaccines [7]. However, one of the most important application is its antimicrobial activity against bacteria, filamentous fungi and yeasts. Chitosan has wide spectrum of activity against Gram-positive and Gram-negative bacteria but it is more effective against Gram-negative bacteria [8, 9]. Furthermore, it has been reported that the antimicrobial activity and film-forming property of chitosan makes it a potential source of food preservative or coating material of natural origin for improvement of quality and shelf life of various foods of agriculture, poultry, beef and seafood origin [3, 10]. The mechanism of the antimicrobial activity of chitosan has not yet been fully elucidated, but several hypotheses have been proposed. The most feasible hypothesis is a change in cell permeability due to interactions between the positively charged chitosan molecules and the negatively charged microbial cell membranes [11, 12]. Other mechanisms include the interaction of diffused hydrolysis products with microbial DNA, which leads to the inhibition of mRNA and protein synthesis and chelation of metals, spore elements, and essential nutrients [5, 13].

The antimicrobial activity of chitosan depends on both intrinsic and extrinsic factors. Among the intrinsic factors are the molecular weight and degree of deacetylation of chitosan. While the extrinsic factors include pH, temperature and reactive time [14]. Moreover, it has been observed that when the chitosan is in nanoparticle form, it has better antimicrobial properties since its small particle size gives it a greater surface area and high reactivity which could enhance the charge interaction with the bacterial surface and of this way to produce a superior antimicrobial effect [15].

2. Antimicrobial effect of chitosan on *Salmonella* in broiler chickens

Salmonella enterica serovars continue to be among the most important foodborne pathogens worldwide due to the considerable human rates of illness reported, the wide hosts species that are colonized by members of this remarkable pathogen genus, which serve as vectors and reservoirs for spreading these agents to animal and human populations. **Figure 2** shows the distribution of the major serotypes of Salmonella with importance in public health. Furthermore, the public concern for the appearance of resistant strains to many antibiotics, particularly among zoonotic pathogens such as common *Salmonella* isolates, is also challenging the poultry industry to find alternative means of control [16]. For these reasons, continued research on sustainable alternatives to antibiotic growth promoters for animal production is needed.

Interest in chitosan, a biodegradable, nontoxic, non-sensitizing, and biocompatible polymer isolated from shellfish, arises from the fact that chitosan is reported to exhibit numerous beneficial effects, including strong antimicrobial and antioxidant activities in foods [18]. Its application in agriculture, horticulture, environmental science, industry, microbiology, and medicine are well reported. A significant interest in the antimicrobial activities of chitosan either as solution, or as powders, edible films and coating against foodborne pathogens, spoilage bacteria, and pathogenic viruses and fungi in several food categories has been extensively investigated [19]. We have evaluated the effect *in vitro* and *in vivo* of chitosan on *Salmonella typhimurium* in broiler chickens [20]. In our *in vitro* crop assay experiments (**Table 1**), 0.2% chitosan significantly

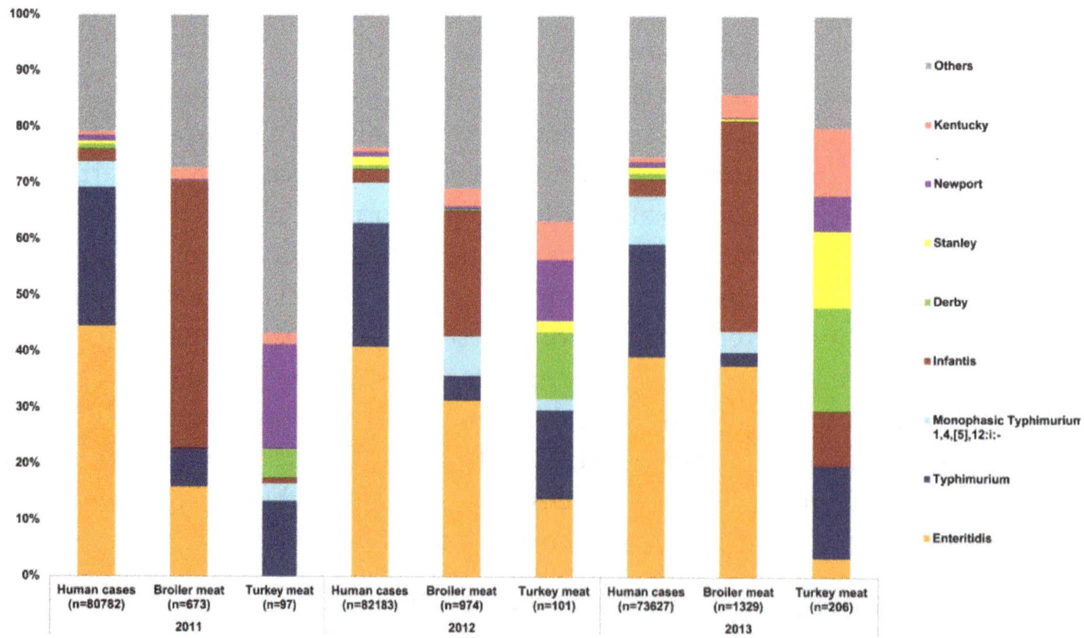

Figure 2. Distribution of the major serotypes of non-typhoidal *Salmonella* associated with human cases (salmonellosis) and poultry meat in EU, 2011 to 2013 [17].

Treatment	Trial 1		Trial 2		Trial 3	
	30 min	6 h	30 min	6 h	30 min	6 h
Control	5.22 ± 0.15^a	7.62 ± 0.01^a	5.19 ± 0.11^a	6.99 ± 0.03^a	6.05 ± 0.18^a	7.95 ± 0.31^a
Chitosan (0.2%)	3.94 ± 0.20^b	3.04 ± 0.20^b	3.49 ± 0.24^b	4.40 ± 0.19^b	5.05 ± 0.19^b	5.31 ± 0.26^b

$^{a\text{-}b}$Values within columns with different lowercase superscripts differ significantly (P < 0.05).

Table 1. Antimicrobial activity of chitosan on *Salmonella typhimurium* in an in vitro crop assay.

Treatment	Trial 1		Trial 2	
	Cecal tonsils (CT)	Log_{10} ST/g of CT	Cecal tonsils (CT)	Log_{10} ST/g of CT
Control	15/20 (75%)	4.20 ± 0.82^a	15/20 (75%)	5.00 ± 0.62^a
Chitosan (0.2%)	9/20 (45%)	2.28 ± 0.75^b	12/20 (60%)	3.34 ± 0.72^b

$^{a\text{-}b}$Values within columns with different lowercase superscripts differ significantly (P < 0.05).

Table 2. Effect of chitosan on *Salmonella enteritidis* cecal tonsils colonization in 7-days-old broiler chickens.

reduced the cfu of ST at 30 min or 6 h compared with control (P < 0.05). In the *in vivo* experiments with 40 day-of-hatch broiler chicks and challenged with 2 × 10^5 cfu ST, dietary 0.2% chitosan significantly reduce the cfu/g of ST in the ceca in both experiments (**Table 2**). However, no significant reduction in the incidence of ST in cecal tonsils colonization was observed, suggesting that 0.2% chitosan significantly reduced the cfu of ST/gram *in vitro* and *in vivo*.

3. Effect of chitosan as a biological sanitizer on chicken skin

Chickens contain large numbers of microorganisms in their gastrointestinal tract and on their feathers and feet; therefore, storage quality of fresh chicken is partially dependent on the bacteria present on the integument prior to slaughter. Pathogenic microorganisms present in chicken carcasses after processing and throughout scalding and picking can contaminate equipment and other carcasses [21]. Pathogenic bacteria such as *Salmonella spp.* and *Campylobacter* spp. are able to attach to skin and penetrate in skin layers or feather follicles, facilitating their presence on chicken skin and carcass during poultry processing [22]. Critical control point determination at broiler processing has become very important, especially because of the recent attention on

Treatment	Trial 1		Trial 2	
	1 h	24 h	1 h	24 h
Control	6.57 ± 0.11^a	6.03 ± 0.02^a	6.78 ± 0.06^a	7.36 ± 0.06^a
Chitosan (0.5%)	6.23 ± 0.03^a	5.81 ± 0.06^b	7.06 ± 0.08^a	6.60 ± 0.17^b

$^{a\text{-}b}$Values within columns with different lowercase superscripts differ significantly (P < 0.05).

Table 3. *Salmonella typhimurium* (log10 cfu ± standard error)/square cm of chicken skin treated with 0.5% chitosan solution.

Hazard Analysis and Critical Control Points (HACCP) for reduction of microbial contamination of meat and poultry [23]. For all these reasons, strategies to reduce bacterial contamination on poultry carcasses are important. However, most of the bacterial reduction strategies for poultry comprise the use of antimicrobial chemicals in rinses or washes and their efficacy is reduced by the presence of organic matter. Therefore, it grows the need of biological sanitizers in the processing plant to prevent carcass to carcass cross-contamination by pathogenic bacteria and to lower the potential of foodborne diseases. Interest in chitosan, a biocompatible polymer derived from shellfish, as a biological sanitizer arises from reports showing several beneficial effects such as antimicrobial and antioxidative activities in foods [2]. The use of chitosan in industry, agriculture, and medicine is well described [13]. The antimicrobial activities of chitosan against foodborne pathogens has been broadly investigated in the food industry [24]. Research conducted in our laboratory on the effect of chitosan as a biological sanitizer in chicken skin contaminated with *Salmonella* Typhimurium and aerobic Gram-negative spoilage bacteria present on chicken skin, have revealed that 0.5% chitosan for 30 s dipping ST contaminated skin samples in a solution of 0.5% chitosan reduced ($P < 0.05$) the recovery of ST by 24 h as well as the presence of spoilage-causing psychrotrophic bacteria below detectable levels [19], (**Table 3**). The antimicrobial activity and film-forming characteristic of chitosan makes it a potential source of food preservative, increasing quality and shelf life of different types of foods [10]. The mechanism of the antimicrobial activity of chitosan has not yet been fully elucidated; nevertheless, different hypotheses have been proposed. The most realistic hypothesis is that chitosan is able to change cell permeability due to interactions between the positive charges of its molecules and the negative charges of the bacterial cell membranes [1]. Other hypotheses include the chelation of metals and essential nutrients, inhibiting bacterial growth had also suggested that high molecular weight chitosan could be able to form a polymer membrane around the bacterial cell, preventing it from receiving nutrients [25].

4. Prebiotic properties of *Aspergillus oryzae* to control foodborne pathogens improve performance and bone mineralization in poultry

Prebiotics are non-digestible food ingredients that are selectively fermented by gut bacteria and are known to have positive effects on gastrointestinal (GI) physiology. Some prebiotics have been shown to selectively stimulate the growth of endogenous lactic acid bacteria in the gut thereby improving the health of the host [26]. Prebiotics selectively modify the colonic microflora and can potentially influence gut metabolism [27]. The commercially available mycelium product of *Aspergillus oryzae*, Fermacto® (PetAg Inc. Hampshire, IL 60140 USA), referred to as *Aspergillus* meal (AM), has no live cells or spores and is proven to enhance the digestive efficiency of the GI tract [28]. *Aspergillus* fiber contains beta-glucans [29], fructooligosaccharides (FOS) [30], chitosan [31], and mannanoligosaccharides (MOS) [32]. Beta-glucan is considered as a powerful immune-enhancing nutritional supplement that affects the intestinal villi and primes the innate immune system to help the body defend itself against viral and bacterial invaders [33]. MOS protect the GI tract from invading toxins and pathogens by binding toxin active sites [34]. FOS and chitosan refer to a class of non-digestible carbohydrates that are readily fermented by beneficial bacteria in the intestine [30]. A healthy population of these beneficial bacteria in the digestive tract enhances the digestion and absorption of nutrients, detoxification and elimination processes, and helps boost the immune system [35]. With an increase in the dependence on livestock as an important food source, it becomes crucial to achieve good health in order to make rearing of animal food sources safe and beneficial to both animals and humans.

Several studies have demonstrated that prevention of *Salmonella* colonization in chickens can be achieved by feeding prebiotics [36]. According to Lowry et al. [37], dietary beta-glucan reduces SE colonization significantly in chickens. In their experiment, SE from L/S was recovered from 76% of non-treated birds, while only 7% of the birds were positive for SE in the treated group, (**Figure 3**). Moreover, in the same study, heterophils isolated from birds treated with dietary beta-glucan contained 40% ($p < 0.05$) more SE than heterophils isolated from untreated birds, (**Table 4**). Heterophils form the first line of defense and killing of *Salmonella* by heterophils is well-described. This corroborates the immunostimulatory effect of beta-glucans and FOS are widely used as prebiotics in a broad range of animal species, and these carbohydrates have been tested with success for protection against *Salmonella* infections in chickens and other avian pathogens [35, 38, 39]. Kim et al. conducted a study where dietary MOS (0.05%) and FOS (0.25%) had an effect on intestinal microflora of broiler chickens, suggesting the use of these prebiotics as an alternative to the use of growth-promoting antibiotics [40]. Finally, chitosan is a modified, natural biopolymer derived by deacetylation of chitin, the main component of the cell walls of fungi and exoskeletons of arthropods. As mention before, chitosan exhibits numerous beneficial effects, including strong antimicrobial and antioxidative activities. Its application in agriculture, horticulture, environmental science, industry, microbiology, and medicine are well reported [10]. According to Huang et al., the use of 0.01 or 0.015% of oligo-chitosan in the diet increased serum levels of immunoglobulins in broiler chickens, suggesting a potential immunomodulatory effect [41]. There have been numerous studies that report the use of chitosan as a mucosal adjuvant, by enhancing IgA levels. It is well known that IgA is active across mucosal surfaces and is the predominant class of antibody against enteric pathogens [42, 43]. The commercial prebiotic supplement derived from *Aspergillus* sp. mycelium is

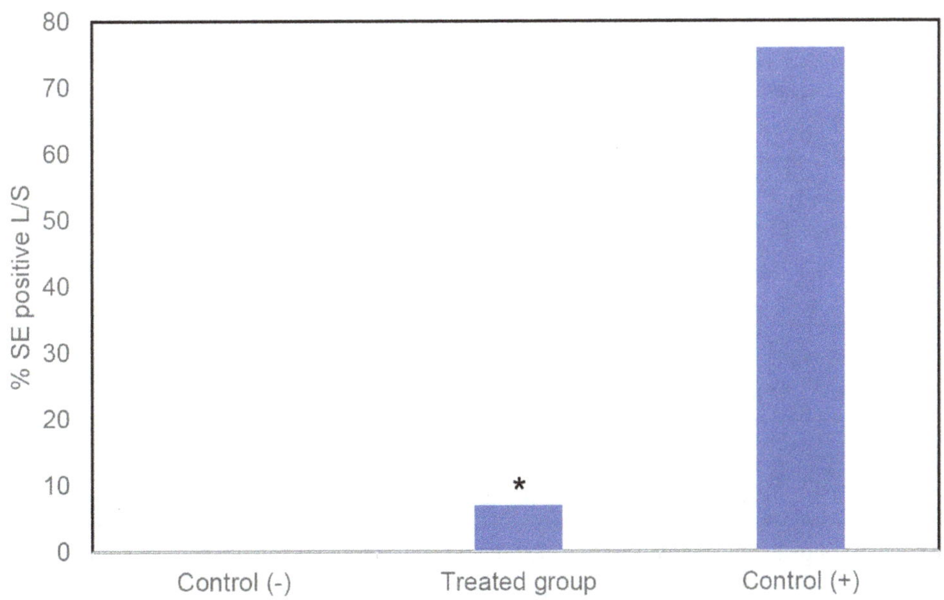

Figure 3. Effects of dietary β-glucan on SE organ invasion in immature chickens. β-glucan fed as a nutritional supplement to neonatal chickens 3 days prior to SE challenge. (*indicates statistically significant differences, $P < 0.05$).

Treatments	Percent heterophils + SE	Mean #SE/heterophil	Phagocytic index (PI)
Control feed	38.54 ± 0.05^b	4.38 ± 1.08^b	175.54 ± 44.92^b
β-glucan feed	78.84 ± 0.03^a	8.20 ± 0.76^a	644.10 ± 57.07^a

$^{a-b}$Values within columns with different lowercase superscripts differ significantly ($P < 0.05$).

Table 4. Effects of feeding β-glucan on chicken heterophil phagocytosis.

unique because it contains all of the above mentioned prebiotic ingredients. Additionally, AM contains 16% protein and 45% fiber and may be used with low levels of protein and amino acid diets to improve performance in commercial poultry [28, 44]. Even though the exact mechanisms of action for prebiotics have not been defined, it may be speculated that the effect is due to changing intestinal flora that promotes the growth of beneficial bacteria. This product has also been shown to benefit poultry through stimulation of growth, most probably by increasing absorption of feed ingredients and improving digestibility [45, 46].

In a recent study conducted in our laboratory we evaluated the effect of 0.2% dietary *Aspergillus* meal (AM) against horizontal transmission of *Salmonella* spp. in turkeys and chickens [36]. The results of this study showed that dietary supplementation with 0.2% *Aspergillus* Meal was able to reduce *Salmonella enteritidis* horizontal transmission in turkeys, (**Table 5**) and *Salmonella* Typhimurium horizontal transmission in broiler chickens, by reducing the overall colonization levels in birds, (**Table 6**). Although the mechanism of action is not totally understood, the reduction in *Salmonella* colonization may be related to a synergistic effect between beta-glucan, MOS, chitosan, and FOS present in the *Aspergillus oryzae* mycelium. In a previous work, we showed that dietary AM induces important changes on intestinal morphometry in turkey poults such as increased number of acid mucin cells in the duodenum, neutral mucin cells in the ileum, and

Groups	Day 10 cecal tonsils	Day 20 cecal tonsils	Day 30 cecal tonsils
Control—No AM	20/20 (100%)	18/20 (90%)	15/20 (75%)
AM	15/20 (75%)*	12/20 (60%)*	8/20 (40%)*

*(a) in the Control-No AM group and (b) in the AM group in both tables.
$^{a-b}$Values within columns with different lowercase superscripts differ significantly ($P < 0.05$).

Table 5. Effect of dietary *Aspergillus* meal against horizontal transmission of *Salmonella enteritidis* at 10, 20 and 30 days of age in turkeys.

Groups	Trial 1		Trial 2	
	Liver/spleen	Cecal tonsils	Liver/spleen	Cecal tonsils
Control—No AM	18/20 (90%)	20/20 (100%)	19/20 (95%)	18/20 (90%)
AM	6/20 (30%)*	5/20 (25%)*	8/20 (40%)*	6/20 (30%)*

*(a) in the Control-No AM group and (b) in the AM group in both tables.
$^{a-b}$Values within columns with different lowercase superscripts differ significantly ($P < 0.05$).

Table 6. Effect of dietary *Aspergillus* meal against horizontal transmission of *Salmonella typhimurium* at 10 days of age in chickens.

	Control	Aspergillus meal
Body weight (Kg)	600.32 ± 52.26[b]	720.87 ± 63.82[a]
FC (feed: gain)	1.34 ± 0.03[a]	1.23 ± 0.02[b]
Mortality (%)	2.00%[a]	2.50%[a]

[a-b]Values within columns with different lowercase superscripts differ significantly ($P < 0.05$).

Table 7. Effect of *Aspergillus* meal on productive parameters in turkey poults at 30 days of ages.

sulphomucin cells in the duodenum and ileum, as well as increased villi height and villi surface area of both duodenum and ileum when compared to control, suggesting that AM prebiotic has an impact on the mucosal architecture and goblet cells proliferation in the duodenum and ileum of neonate poults [45]. Our extended studies using dietary AM prebiotic supplemented for 30 days, have shown significantly increased the body weight of neonate poults and improved feed conversion when compared with poults that received the control basal diet, and interestingly, energy and protein content in the ileum was significantly lower in poults that received dietary AM prebiotic compared with control poults suggesting better digestibility, absorption of those nutrients and bone mineralization [46]. FOS have been shown to stimulate calcium (Ca) and magnesium (Mg) absorption in the intestine and increase bone mineral concentrations in humans and rats as well as stimulate net Ca transport from the epithelium of the small and large intestine [30, 38].

The gastrointestinal tract serves as the interface between diet and the metabolic events that sustain life. Intestinal villi, which play a crucial role in digestion and absorption of nutrients, are underdeveloped at hatch and maximum absorption capacity is attained by 10 days of age [47]. Understanding and optimizing the maturation and development of the intestine in poultry will improve feed efficiency, growth and overall health of the bird. Studies on nutrition and metabolism during the early phase of growth in poults may, therefore, help in optimizing nutritional management for maximum growth. By dietary means it is possible to affect the development of the gut and the competitiveness of both beneficial and harmful bacteria, which can alter not only gut dynamics, but also many physiologic processes due to the end products metabolized by symbiotic gut microflora [48]. Additives such as enzymes, probiotics and prebiotics are now extensively used throughout the world [49–51]. Our studies suggest that the increase in performance and bone parameters in neonatal poults fed with 0.2%AM (**Table 7**), may be related to a synergistic effect between beta-glucan, MOS, chitosan and FOS from *Aspergillus niger* mycelium [45, 46].

5. Evaluation of chitosan as binding adsorbent material to prevent mycotoxicosis poultry

Mycotoxins are secondary toxic metabolites produced by filamentous fungi which, even at low concentrations, represent an important danger for both animal and human health [52, 53]. Currently, over 300 mycotoxins have been identified worldwide, being aflatoxins, ochratoxins, zearalenone, trichothecenes, and fumonisins, the most frequently found with synergistic toxic effects reported when more than one of these mycotoxins are present in the feed [54, 55]. Mycotoxins are chemically and structurally different, representing serious public health risk factors since mycotoxins have been shown to have carcinogenic, teratogenic, nephrotoxic, and hepatotoxic effects after the consumption of contaminated grains or animal food products [56]. On the other hand, mycotoxins are equally important in the animal food industry, causing significant economic losses due to diminished performance and productivity, decreased reproductive parameters, and an increased mortality rate associated with the toxicological effects in liver, kidneys, and immune system [52, 57, 58]. Researchers have developed some methods in order to reduce the harmful effects of grains contaminated with mycotoxins. These include physical (thermal and irradiation inactivation); chemical (ozonation and ammoniation); and, biological (bacterial degradation or adsorption [57, 59, 60]. Nevertheless, toxin sequestering agents are the most common and reliable products used for the feed industry due to its economic practicality and aptness for nutritional insight [61, 62]. Several studies have demonstrated that cellulosic materials have adsorption capacities for heavy metal ions and other pollutants [63, 64]. Similarly, some researchers have evaluated the binding activity of chitosan (CS) against several mycotoxins [2, 65]. As a biological polymer, chitosan has been shown to have promising uses as an adsorbent for the removal of various mycotoxins, heavy metal ions, and dyes [65]. Furthermore, it has been tested in the removal of OTA from contaminated drinks, demonstrating that chitosan can reduce the levels of this mycotoxin [1, 66]. On the other hand, some in vitro methods have been developed to evaluate the adsorbent capacity of mycotoxin sequestering products [67, 68]. However, these methods may not be directly applicable to poultry diets because they do not use the successive incubation at different pH and enzyme activity conditions similar to the different gastrointestinal compartments of poultry. Recently, we evaluated the adsorption capacity of CS on Aflatoxin B1 (AFB_1); Fumonisin B1 (FUB_1); Ochratoxin (OTA); Trichothecene (T-2); Deoxynivalenol (DON); and, Zearalenone (ZEA), using an *in vitro* digestive model that simulates three gastrointestinal compartment of poultry [69]. In that study, deacetylated 95%, high molecular weight (350 kDa) Chitosan (CS, Paragon Specialty Products, LLC, Rainsville, AL, USA) was tested and acetylated with an aqueous solution of acetic acid 1% (v/v). Then, this solution was dropped into NaOH

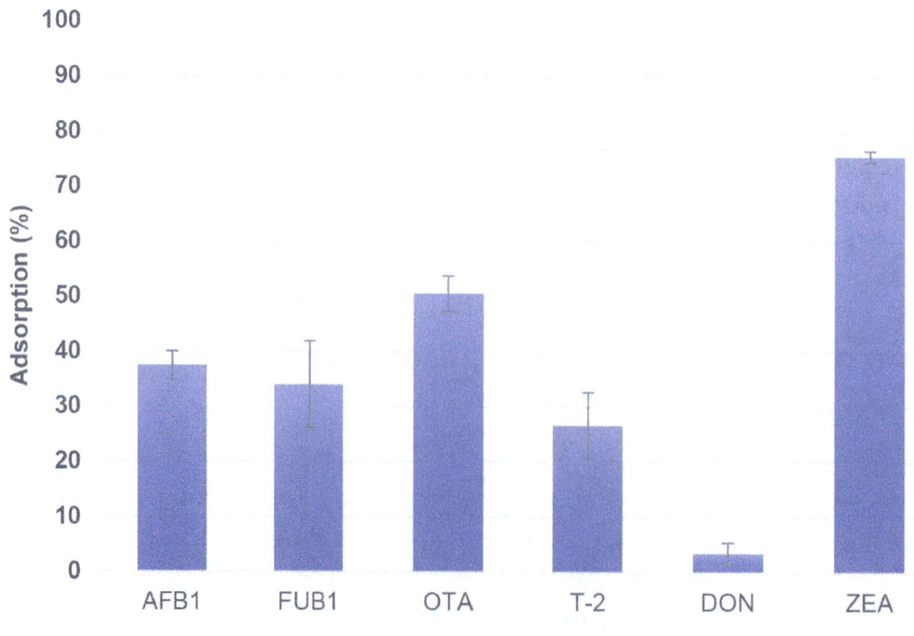

Figure 4. Percentage of adsorption of different mycotoxins on chitosan. Bars are the mean values. Error bars displays an interval around each mean, which are based on Fisher's least significant difference (LSD) procedure.

0.5 M solution and the formed chitosan particles were rinsed three times with pure water and dried [69, 70]. The results showed a moderate adsorbent capacity of CS against five of the six mycotoxins evaluated, except for DON since only 3.5% was adsorbed, (**Figure 4**). Similar results were obtained in another study using non-crosslinked chitosan against different mycotoxins but it is a fact that cross-linking is related to a higher adsorption capacity and pH can affect it [70]. The mycotoxins adsorption capacity of CS is due to the electrostatic interactions. At alkaline pH, the CS is positively charged, while mycotoxins such as AFB1, FUB1, OTA and ZEA are negatively charged [70–72]. In the case of DON and T-2, the interactions appeared to be minor, causing poor adsorption. These results are very similar to those obtained in other studies [70]. Therefore, it could be said that ionic interactions are the main mechanism of mycotoxin adsorption of chitosan.

6. Chitosan nanocarriers: a strategy to improve solubility, permeability and stability of drugs

Another application of chitosan (CS) is its use in nanotechnology for the development of drug delivery systems such as nanoparticles and nanocapsules. These systems emerge as a strategy to improve the dissolution of drugs with low solubility and increase its permeability, which translates into an increase in bioavailability, a greater specificity and also an increase in the stability of drugs against physiological and environmental conditions [73]. In our laboratory, we have developed two nanocapsular systems capable of loading a phytopharmaceutical named Curcumin. This molecule has also been the subject of study in the poultry industry, given its properties, including its antioxidant action, the immunomodulatory, anticoccidial, anti-inflammatory, antimicrobial and growth promotion effects, the latter as an alternative to antibiotic growth promoters in order to maintain the performance and health of the birds [74–76]. In our laboratory, we have already evaluated the antimicrobial activity of curcumin against Salmonella *enteritidis* in an in vitro model that simulates the three compartments of the chicken gastrointestinal tract. The results obtained show that at a dose of 1%, the concentration of Salmonella *enteritidis* decreases slightly but not significantly with respect to the control [77]. However, one of the problems of curcumin, even when it is administered at high doses (12 g/day) is its low bioavailability due to its poor solubility and therefore poor absorption, as well as its rapid metabolism and systemic elimination [76]. In this sense, the development of nanocapsular systems aimed to increase the solubility, permeability and stability of curcumin. Such systems were named chitosan nanocapsules (NC-CS) and Alginate nanocapsules (NC-ALG) and were composed of an oily core of vitamin E surrounded by a biodegradable polymeric shell of either chitosan or alginate respectively.

Both systems were obtained by the a slightly modification of the solvent displacement technique [78]. However, the formation of NC-CS is based on the electrostatic and hydrophobic interactions as well as the hydrogen bonding and van der Waals forces that take place between the chitosan dissolved in an acidic aqueous phase and the lipid cores of Vitamin E formed in the organic phase, causing the polymer to be adsorbed on the surface of the lipid cores [4, 79–81], (**Figure 5(a)**). While NC-ALG were prepared using the "Single-stage procedure" based on the dipolar ionic interactions between the polymer (ALG), which is dissolved in the aqueous phase and the cationic surfactant (CTAB) present in the organic phase which also contains the oil [82], (**Figure 5(b)**).

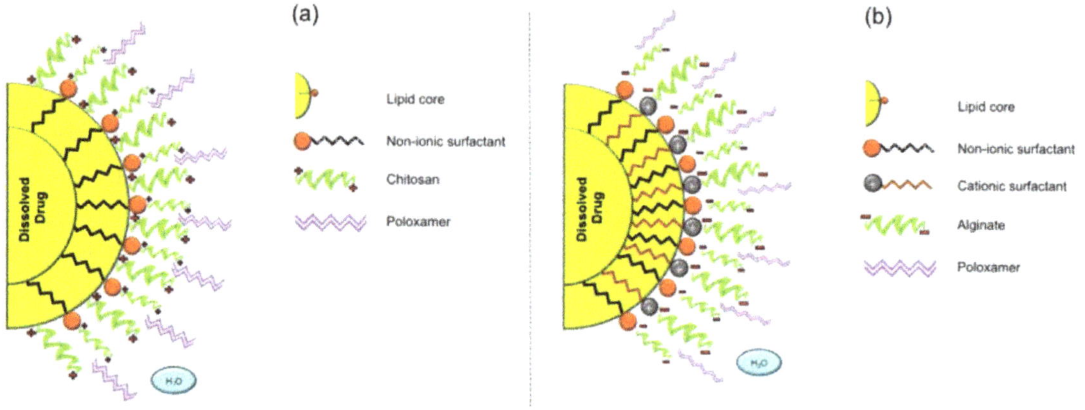

Figure 5. Structural illustration of NC-CS (a), NC-ALG (b) and its components.

Formulation	Size (nm)	PDI	ζ potential (mV)	%E. E.	pH
NC-NC	116.7 ± 3.2	0.107	24.4 ± 2.1	>98	4.67 ± 0.08
NC-ALG	178 ± 7.9	0.149	−49.0 ± 2.3		6.08 ± 0.06

PDI: polydispersity index; EE: encapsulation efficiency. Values are given as mean ± SD; n = 3.

Table 8. Physicochemical characteristics of the nanocapsules obtained.

The systems were characterized physicochemically in terms of particle size, surface charge, polydispersity index (PDI) and curcumin encapsulation efficiency, (**Table 8**). Particle size and (PDI) were determined by dynamic light scattering (DLS) using a Malvern Zetasizer Nanoseries 3600 (Worcestershire, UK). The zeta potential values were calculated from the mean electrophoretic mobility values, as determined by Laser Doppler Velocimetry (LDV) using a Malvern Zetasizer Nanoseries 3600 (Worcestershire, UK). The particle size of NC-CS was round 116.7 nm with a PDI of 0.107 and presented positive surface charge (24.4 mV) while NC-ALG was round 178 nm with a PDI of 0.149 and a negative surface charge (−49.0 mV). Curcumin encapsulation efficiency was determined indirectly by Centrifugation-Filtration. Quantification of curcumin was performed by high performance liquid chromatography (HPLC, Merck-Hitachi, Japan) at 425 nm, using a reverse phase Hypersil® Division C8 column (150 × 3 mm, 5 μm; ThermoQuest, Hemel Hempstead, England). Curcumin encapsulation efficiency of both formulations, was >90%, with a final concentration of curcumin around 750 μg/ml [Unpublished work from our laboratory].

The stability to storage conditions is a parameter that must be evaluated in nanoparticulate systems (**Table 9**). In that study, the storage stability of NC-CS and NC-ALG was around 3 and 2 months respectively. In the case of NC-Cs, after 3 months of storage, the decrease in particle size and the precipitation of CUR were presented with greater magnitude since the chitosan begins to hydrolyze gradually and the viscosity of the formulation based on nanocapsules decreased during the storage period [83]. On the other hand, the results obtained for NC-ALG suggest that the stability of this type of formulation is around 2 months [Unpublished work from our laboratory]. These results are very similar to those reported in other studies, in which they report that the particle size of NC-ALG decreases between month 1 and 5 of storage [84].

Formulation	NC-CS			NC-ALG		
Time (months)	1	2	3	1	2	3
Z-average (nm)	115.2 ± 2.9	101.3 ± 4.1	95.45 ± 3.0	161.9 ± 2.1	157.7 ± 4.77	149.3 ± 2.9
PDI	0.115	0.145	0.196	0.178	0.213	0.245
ζ potential (mV)	23.8 ± 2.1	23.9 ± 3.2	24.5 ± 2.8	−48.7 ± 1.9	−46.8 ± 2.5	−44.3 ± 2.7

Values are given as the mean ± SD; n = 3.

Table 9. Physicochemical characteristics of the nanocapsules during stability studies under storage conditions (4°C).

An important parameter that was take into account in these nanosystems was the cellular toxicity on caco-2 cells. For this, the conditions for the maintenance of the cell cultures were made according to Déat-Lainé et al. [85] with slight modifications. Before starting the study, the formulations were diluted in cell culture medium (DMEM: Dulbecco's Modified Eagle Medium) in order to obtain the treatments with different polymer concentrations. Cell viability was determined by MTT assay [86]. In **Figure 6**, the results showed that even at high polymer concentrations (500 μg/mL) the cell viability is above 80%. However, it is a fact that the toxicity increases as so does the polymer concentration. Other studies in Caco-2 cells have shown similar results to those obtained in our laboratory and agree that the toxicity of chitosan nanoparticles is due to their physicochemical properties such as size and surface charge and also to the molecular weight of the chitosan and the concentration at which the cells are exposed [87, 88]. In the case of NC-ALG, the toxicity was lower since the interactions between the carboxyl groups of alginate and cell membranes are weaker because they are of the electrostatic type. The toxicity in these systems is more related to the particle size [89]. From the toxicity study, the polymer concentration to carry out the permeability studies was selected. It should be mentioned that this concentration did not compromise cellular viability.

Permeability studies were carried out on a monolayer of caco-2 cells and the quantification of curcumin was performed by UPLC-TQ-ESI-MS/MS (Waters ACQUITY UPLC system, Milford, MA, USA). Chromatographic analysis was performed on a Waters ACQUITY BEH Shield RP 18 column (2.1 × 100 mm, 1.7 μm). The polymer concentration used was 500 μg/mL of each polymer. Results show that the permeability of curcumin increased 28.6 and 14.6 times when it was in NC-CS and NC-ALG respectively, compared to the dispersion of curcumin in cell culture medium (DMEM: Dulbecco's Modified Eagle Medium) [Unpublished work from our laboratory]. The increase in the permeability of curcumin in NC-CS is due to the ability of chitosan to temporarily open the tight junctions, which are related to a decrease in the value of transepithelial electrical resistance (TEER, MERSSTX01 electrode, Millicell ERS-2, Millipore, Billerica, MA, USA) (**Table 10**). The mechanism by which chitosan has this capacity is based on the interaction of its protonated amino groups with cell membranes, followed by a reversible structural reorganization of the binding proteins and a specific redistribution of the actin F cytoskeleton and the ZO-1 protein [90, 91]. Furthermore, it has been reported that particles positively charged, with spherical shape and with a monodisperse population have improved cellular uptake through the caveolae-mediated endocytosis and macropinocytosis pathway

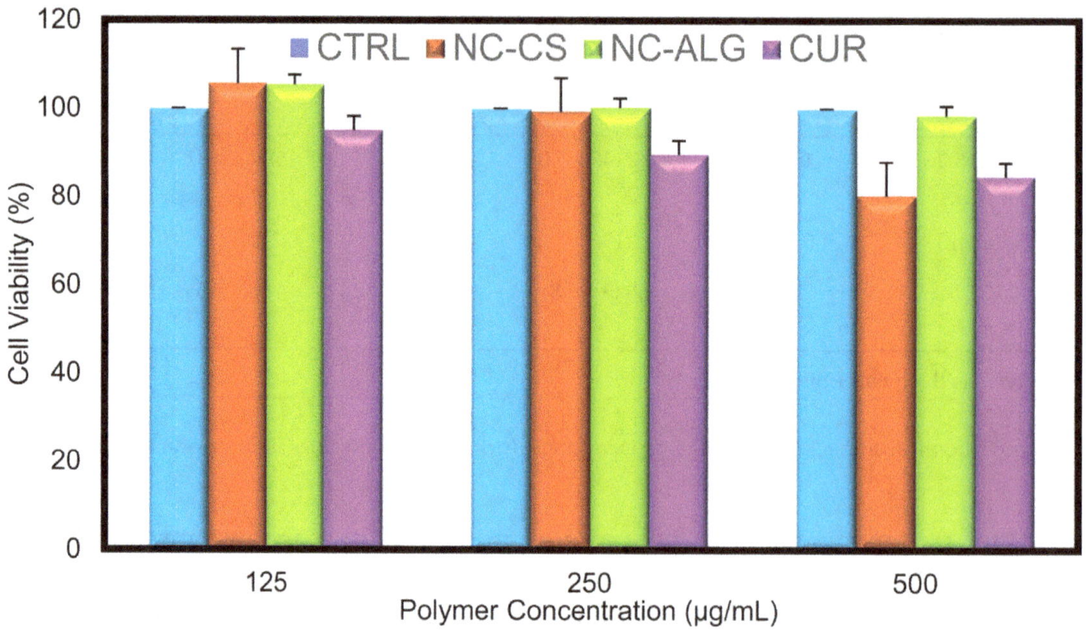

Figure 6. Cell viability by the MTT assay on Caco-2 cells 2 h after the addition of NC-CS, NC-ALG and CUR at different concentrations. Values are given as mean ± SD; n = 3.

Formulation	$P_{APP} \times 10^{-6}$ (cm/s)	R	TEER (%)			
			0 (h)	1 (h)	2(h)	12 (h)
CUR	4.96 ± 0.36	–	100	92 ± 1	93 ± 2	99 ± 4
NC-CS	141.60 ± 37.62	28.6*	100	81 ± 3	82 ± 4	97 ± 6
NC-ALG	72.38 ± 19.33	14.6*	100	90 ± 2	88 ± 6	100 ± 2

Values are given as the mean ± SD;
n = 3.*p < 0.05 significantly different from CUR.

Table 10. Mean apparent permeability (Papp) and the absorption enhancement ratio (R) of NC-CS, NC-ALG and CUR across Caco-2 cells monolayers after 2 h incubation, as well as the values of transepithelial electrical resistance (TEER) determined at different times.

[65, 92]. Meanwhile, the mechanism of passage of NC-ALG through the monolayer of caco-2 cells depends largely on the particle size mainly. So, the main mechanisms are endocytic such as clathrin-mediated endocytosis, caveolae-mediated endocytosis and micropinocytosis [92, 93]. The results suggest that the use of NC-CS and NC-ALG to improve the bioavailability of curcumin is an interesting strategy to enhance the antimicrobial effect. Previous studies using an *in vitro* digestive model that simulates three gastrointestinal compartments of poultry have demonstrated that raw curcumin does not have good antimicrobial activity against *Salmonella enteritidis* [77]. However, when a solid dispersion of curcumin/PVP K30 was used, it decreased the concentration of *Salmonella enteritidis* more than 3 log in the compartment that simulates the intestine [Unpublished work from our laboratory]. Additional *in vivo* studies in 1-day-old chickens challenged with 10^4 CFU of *Salmonella enteritidis*/bird has shown that the solid dispersion of curcumin/PVP K30 administered in the feed at a concentration of 0.1% decreased more than 2 log the concentration of *Salmonella enteritidis* in ceca-cecal tonsil isolates [Unpublished

work from our laboratory]. Since nanocapsules increased the solubility and permeability of curcumin, the antimicrobial activity of nanocapsules loaded with curcumin developed in our laboratory is being carried out both *in vitro* and *in vivo* against *Salmonella enteritidis*.

7. Conclusion

As seen in this chapter, chitin and its derivatives, such as chitosan, are biopolymers with a wide variety of applications in different areas. Chitosan as a functional biopolymer has different properties. Some of these properties are its intrinsic nutritional value, such as antioxidant properties and health-promoting bioactivities against many chronic diseases, including hypercholesterolemia, hypertension, inflammation and immune diseases. In the case of chitin, its application is more limited given its poor solubility in aqueous medium, however, it has been reported that it has practically the same properties as its derivatives.

Every year millions of people are affected and thousands of them die due to infections and intoxication as a result of foodborne outbreaks, which also cause billions of dollars' worth of damage, public health problems and agricultural product loss. A considerable portion of these outbreaks is related to the consumption of contaminated products with foodborne pathogens and mycotoxins. In this sense, one of the main applications of chitosan is its antimicrobial effect against Gram-positive bacteria such as Gram-negative bacteria, having better activity with the latter due to the ionic interaction that takes place between the positively charged chitosan molecules and the negatively charged microbial cell membranes. Studies conducted on chickens and turkeys challenged with Salmonella *enteritidis* and typhimurium show the antimicrobial capacity of chitosan when it is administered in the feed. Furthermore, in vitro studies have demonstrated its properties as an adsorbent, since it can interact ionically with mycotoxins such as AFB1, FUB1, OTA and ZEA given that they are negatively charged, nevertheless, it is a fact that cross-linking is related to a higher adsorption capacity.

Finally, another application of chitosan is its use in nanotechnology for the development of nanoparticles and nanocapsules. These systems are an important strategy to improve the solubility, permeability and stability of molecules that are difficult to formulate. In the case of curcumin, a phytopharmaceutical that has become the subject of study in the poultry industry given its properties, including its antioxidant action, the immunomodulatory, anticoccidial, anti-inflammatory, antimicrobial and growth promotion effects, has problems of solubility and permeability, which causes low bioavailability. However, its association or encapsulation in nanoparticulate systems has shown that the solubility and permeability of this are improved. This suggests that the use of curcumin loaded in chitosan nanocapsules could increase its antimicrobial activity derived from the combination of the effects between chitosan and curcumin on different bacteria.

Acknowledgements

This research was supported by the Arkansas Bioscience Institute under the project: Development of an avian model for evaluation early enteric microbial colonization on the gastrointestinal tract and immune function.

The authors thank the CONACyT for the doctoral scholarship number 447447 and the financial support obtained through the program PAPIIT IN218115 of DGAPA-UNAM.

Author details

Daniel Hernandez-Patlan[1], Bruno Solis-Cruz[1], Billy M. Hargis[2] and Guillermo Tellez[2]*

*Address all correspondence to: gtellez@uark.edu

1 Laboratorio 5: LEDEFAR, Unidad de Investigacion Multidisciplinaria, Facultad de Estudios Superiores (FES) Cuautitlan, Universidad Nacional Autonoma de Mexico (UNAM), Cuautitlan Izcalli, Estado de Mexico, Mexico

2 Department of Poultry Science, University of Arkansas, Fayetteville, Arkansas, USA

References

[1] Bornet A, Teissedre PL. Chitosan, chitin-glucan and chitin effects on minerals (iron, lead, cadmium) and organic (ochratoxin A) contaminants in wines. European Food Research and Technology. 2008;**226**:681-689. DOI: 10.1007/s00217-007-0577-0

[2] Szymczyk P, Filipkowska U, Jóźwiak T, Kuczajowska-Zadrożna M. Phosphate removal from aqueous solutions by chitin and chitosan in flakes. Progress on Chemistry and Application of Chitin and its Derivatives. 2016;**21**:260-272

[3] Rinaudo M. Chitin and chitosan: Properties and applications. Progress in Polymer Science. 2006;**31**:603-632

[4] Ahmed TA, Aljaeid BM. Preparation, characterization, and potential application of chitosan, chitosan derivatives, and chitosan metal nanoparticles in pharmaceutical drug delivery. Drug Design, Development and Therapy. 2016;**10**:483

[5] Kumar MNVR. A review of chitin and chitosan applications. Reactive and Functional Polymers. 2000;**46**:1-27

[6] Şenel S, McClure SJ. Potential applications of chitosan in veterinary medicine. Advanced Drug Delivery Reviews. 2004;**56**:1467-1480

[7] Kong M, Chen XG, Xing K, Park HJ. Antimicrobial properties of chitosan and mode of action: A state of the art review. International Journal of Food Microbiology. 2010;**144**:51-63

[8] Helander IM, Nurmiaho-Lassila E-L, Ahvenainen R, Rhoades J, Roller S. Chitosan disrupts the barrier properties of the outer membrane of Gram-negative bacteria. International Journal of Food Microbiology. 2001;**71**:235-244

[9] Chung Y-C, Su YP, Chen C-C, Jia G, Wang HL, Wu JCG, Lin JG. Relationship between antibacterial activity of chitosan and surface characteristics of cell wall. Acta Pharmacologica Sinica. 2004;**25**:932-936

[10] Filipkowska U, Jóźwiak T, Szymczyk P. Application of cross-linked chitosan for phosphate removal from aqueous solutions. Progress on Chemistry and Application of Chitin and its Derivatives. 2014;**19**:5-14

[11] Luo Y, Wang Q. Recent advances of chitosan and its derivatives for novel applications in food science. Journal of Food Processing & Beverages. 2013;**1**:1-13

[12] Rao SB, Sharma CP. Use of chitosan as a biomaterial: Studies on its safety and hemostatic potential. Journal of Biomedical Materials Research. 1997;**34**:21-28

[13] Muzzarelli RAA. Chitins and chitosans as immunoadjuvants and non-allergenic drug carriers. Marine Drugs. 2010;**8**:292-312

[14] Divya K, Vijayan S, George TK, Jisha MS. Antimicrobial properties of chitosan nanoparticles: Mode of action and factors affecting activity. Fibers and Polymers. 2017;**18**:221-230

[15] Romainor ANB, Chin SF, Pang SC, Bilung LM. Preparation and characterization of chitosan nanoparticles-doped cellulose films with antimicrobial property. Journal of Nanomaterials. 2014;**2014**:130

[16] Boyle EC, Bishop JL, Grassl GA, Finlay BB. Salmonella: From pathogenesis to therapeutics. Journal of Bacteriology. 2007;**189**:1489-1495

[17] Antunes P, Mourão J, Campos J, Peixe L. Salmonellosis: The role of poultry meat. Clinical Microbiology and Infection. 2016;**22**:110-121

[18] Naksuriya O, Okonogi S. Comparison and combination effects on antioxidant power of curcumin with gallic acid, ascorbic acid, and xanthone. Drug Discoveries & Therapeutics. 2015;**9**:136-141

[19] Menconi A, Hernandez-Velasco X, Latorre JD, Kallapura G, Pumford NR, Morgan MJ, Hargis BM, Tellez G. Effect of chitosan as a biological sanitizer for *Salmonella typhimurium* and aerobic Gram negative spoilage bacteria present on chicken skin. International Journal of Poultry Science. 2013;**12**:318

[20] Menconi A, Pumford NR, Morgan MJ, Bielke LR, Kallapura G, Latorre JD, Wolfenden AD, Hernandez-Velasco X, Hargis BM, Tellez G. Effect of chitosan on *Salmonella typhimurium* in broiler chickens. Foodborne Pathogens and Disease. 2014;**11**:165-169

[21] Corrier DE, Byrd JA, Hargis BM, Hume ME, Bailey RH, Stanker LH. Presence of *Salmonella* in the crop and ceca of broiler chickens before and after preslaughter feed withdrawal. Poultry Science. 1999;**78**:45-49

[22] Byrd JA, Hargis BM, Caldwell DJ, Bailey RH, Herron KL, McReynolds JL, Brewer RL, Anderson RC, Bischoff KM, Callaway TR. Effect of lactic acid administration in the drinking water during preslaughter feed withdrawal on *Salmonella* and *Campylobacter* contamination of broilers. Poultry Science. 2001;**80**:278-283

[23] Cox JM, Pavic A. Advances in enteropathogen control in poultry production. Journal of Applied Microbiology. 2010;**108**:745-755

[24] Choi EH, Yang HP, Chun HS. Chitooligosaccharide ameliorates diet-induced obesity in mice and affects adipose gene expression involved in adipogenesis and inflammation. Nutrition Research. 2012;**32**:218-228

[25] Agnihotri SA, Mallikarjuna NN, Aminabhavi TM. Recent advances on chitosan-based micro- and nanoparticles in drug delivery. Journal of Controlled Release. 2004;**100**:5-28

[26] Gibson GR, Roberfroid MB. Dietary modulation of the human colonic microbiota: Introducing the concept of prebiotics. The Journal of Nutrition. 1995;**125**:1401

[27] Liu X, Cao S, Zhang X. Modulation of gut microbiota–brain axis by probiotics, prebiotics, and diet. Journal of Agricultural and Food Chemistry. 2015;**63**:7885-7895

[28] Torres-Rodriguez A, Sartor C, Higgins SE, Wolfenden AD, Bielke LR, Pixley CM, Sutton L, Tellez G, Hargis BM. Effect of *Aspergillus* meal prebiotic (Fermacto) on performance of broiler chickens in the starter phase and fed low protein diets. Journal of Applied Poultry Research. 2005;**14**:665-669

[29] Mizutani O, Shiina M, Yoshimi A, Sano M, Watanabe T, Yamagata Y, Nakajima T, Gomi K, Abe K. Substantial decrease in cell wall α-1, 3-glucan caused by disruption of the kexB gene encoding a subtilisin-like processing protease in *Aspergillus oryzae*. Bioscience, Biotechnology, and Biochemistry. 2016;**80**:1781-1791

[30] Guio F, Rugeles LD, Rojas SE, Palomino MP, Camargo MC, Sánchez OF. Kinetic modeling of fructooligosaccharide production using *Aspergillus oryzae* N74. Applied Biochemistry and Biotechnology. 2012;**167**:142-163

[31] Bays HE, Evans JL, Maki KC, Evans M, Maquet V, Cooper R, Anderson JW. Chitin-glucan fiber effects on oxidized low-density lipoprotein: A randomized controlled trial. European Journal of Clinical Nutrition. 2013;**67**:2-7

[32] Uchima CA, Tokuda G, Watanabe H, Kitamoto K, Arioka M. Heterologous expression and characterization of a glucose-stimulated β-glucosidase from the termite *Neotermes koshunensis* in *Aspergillus oryzae*. Applied Microbiology and Biotechnology. 2011;**89**:1761-1771

[33] Cox E, Verdonck F, Vanrompay D, Goddeeris B. Adjuvants modulating mucosal immune responses or directing systemic responses towards the mucosa. Veterinary Research. 2006;**37**:511-539

[34] Corrigan A, Horgan K, Clipson N, Murphy RA. Effect of dietary supplementation with a *Saccharomyces cerevisiae* mannan oligosaccharide on the bacterial community structure of broiler cecal contents. Applied and Environmental Microbiology. 2011;**77**:6653-6662

[35] Janssens GP j, Millet S, Van Immerseel F, De Buck J, Hesta M. The impact of prebiotics and sal monellosis on apparent nutrient digestibility and *Salmonella typhimurium* var. Copenhagen excretion in adult pigeons (*Columba livia domestica*). Poultry Science. 2004;**83**:1884-1890

[36] Londero A, Menconi A, Reginatto AR, Bacocina AI, Wolfenden A, Shivaramaiah S, Hargis BM, Tellez G. Effect of an *Aspergillus* meal prebiotic on salmonella infection in turkeys and broiler chickens. International Journal of Poultry Science. 2011;**10**:946-951

[37] Lowry VK, Farnell MB, Ferro PJ, Swaggerty CL, Bahl A, Kogut MH. Purified β-glucan as an abiotic feed additive up-regulates the innate immune response in immature chickens against *Salmonella enterica* serovar Enteritidis. International Journal of Food Microbiology. 2005;**98**:309-318

[38] Xu ZR, Hu CH, Xia MS, Zhan XA, Wang MQ. Effects of dietary fructooligosaccharide on digestive enzyme activities, intestinal microflora and morphology of male broilers. Poultry Science. 2003;**82**:1030-1036

[39] Dahiya JP, Wilkie DC, Van Kessel AG, Drew MD. Potential strategies for controlling necrotic enteritis in broiler chickens in post-antibiotic era. Animal Feed Science and Technology. 2006;**129**:60-88

[40] Kim HG, Lee SY, Kim NR, Lee HY, Ko MY, Jung BJ, Kim CM, Lee JM, Park JH, Han SH. *Lactobacillus plantarum* lipoteichoic acid down-regulated *Shigella flexneri* peptidoglycan-induced inflammation. Molecular Immunology. 2011;**48**:382-391

[41] Huang RL, Yin YL, Wu GY, Zhang YG, Li TJ, Li LL, Li MX, Tang ZR, Zhang J, Wang B. Effect of dietary oligochitosan supplementation on ileal digestibility of nutrients and performance in broilers. Poultry Science. 2005;**84**:1383-1388

[42] Merino-Guzmán R, Latorre JD, Delgado R, Hernandez-Velasco X, Wolfenden AD, Teague KD, Graham LE, Mahaffey BD, Baxter MFA, Hargis BM. Comparison of total immunoglobulin A levels in different samples in leghorn and broiler chickens. Asian Pacific Journal of Tropical Biomedicine. 2017;**7**:116-120

[43] Beal RK, Wigley P, Powers C, Hulme SD, Barrow PA, Smith AL. Age at primary infection with *Salmonella enterica* serovar Typhimurium in the chicken influences persistence of infection and subsequent immunity to re-challenge. Veterinary Immunology and Immunopathology. 2004;**100**:151-164

[44] Harms RH, Miles RD. Research note: Influence of Fermacto® on the performance of laying hens when fed diets with different levels of methionine. Poultry Science. 1988;**67**:842-844

[45] Tellez G, Nava GM, Vincente JL, De Franceschi M, Morales EJ, Prado O, Hargis BM. Evaluation of dietary *Aspergillus* meal on intestinal morphometry in turkey poults. International Journal of Poultry Science. 2010;**9**:75-878

[46] Reginatto AR, Menconi A, Londero A, Lovato M, Rosa AP, Shivaramaiah S, Wolfenden AD, Huff WE, Huff GR, Rath NC. Effects of dietary *Aspergillus* meal prebiotic on turkey poults production parameters and bone qualities. International Journal of Poultry Science. 2011;**10**:496-499

[47] Uni Z, Ganot S, Sklan D. Posthatch development of mucosal function in the broiler small intestine. Poultry Science. 1998;**77**:75-82

[48] Tellez G, Higgins SE, Donoghue AM, Hargis BM. Digestive physiology and the role of microorganisms. Journal of Applied Poultry Research. 2006;**15**:136-144

[49] Latorre JD, Hernandez-Velasco X, Kuttappan VA, Wolfenden RE, Vicente JL, Wolfenden AD, Bielke LR, Prado-Rebolledo OF, Morales E, Hargis BM. Selection of *Bacillus* spp. for

cellulase and xylanase production as direct-fed microbials to reduce digesta viscosity and Clostridium perfringens proliferation using an in vitro digestive model in different poultry diets. Frontiers in Veterinary Science. 2015;**2**:25

[50] Latorre JD, Hernandez-Velasco X, Wolfenden RE, Vicente JL, Wolfenden AD, Menconi A, Bielke LR, Hargis BM, Tellez G. Evaluation and selection of *Bacillus* species based on enzyme production, antimicrobial activity, and biofilm synthesis as direct-fed microbial candidates for poultry. Frontiers in Veterinary Science. 2016;**3**:95

[51] Latorre JD, Hernandez-Velasco X, Vicente JL, Wolfenden R, Hargis BM, Tellez G. Effects of the inclusion of a Bacillus direct-fed microbial on performance parameters, bone quality, recovered gut microflora, and intestinal morphology in broilers consuming a grower diet containing corn distillers dried grains with solubles. Poultry Science. 2017;**96**:2728-2735

[52] Hussein HS, Brasel JM. Toxicity, metabolism, and impact of mycotoxins on humans and animals. Toxicology. 2001;**167**:101-134

[53] Zain ME. Impact of mycotoxins on humans and animals. Journal of Saudi Chemical Society. 2011;**15**:129-144

[54] Armando MR, Pizzolitto RP, Dogi CA, Cristofolini A, Merkis C, Poloni V, Dalcero AM, Cavaglieri LR. Adsorption of ochratoxin A and zearalenone by potential probiotic *Saccharomyces cerevisiae* strains and its relation with cell wall thickness. Journal of Applied Microbiology. 2012;**113**:256-264

[55] Streit E, Schatzmayr G, Tassis P, Tzika E, Marin D, Taranu I, Tabuc C, Nicolau A, Aprodu I, Puel O. Current situation of mycotoxin contamination and co-occurrence in animal feed—Focus on Europe. Toxins (Basel). 2012;**4**:788-809

[56] Smith LE, Stoltzfus RJ, Prendergast A. Food chain mycotoxin exposure, gut health, and impaired growth: A conceptual framework. Advances in Nutrition: An International Review. 2012;**3**:526-531

[57] Greco MV, Franchi ML, Rico Golba SL, Pardo AG, Pose GN. Mycotoxins and mycotoxigenic fungi in poultry feed for food-producing animals. Scientific World Journal. 2014;**2014**:1-9

[58] Galarza-Seeber R, Latorre JD, Bielke LR, Kuttappan VA, Wolfenden AD, Hernandez-Velasco X, Merino-Guzman R, Vicente JL, Donoghue A, Cross D. Leaky gut and mycotoxins: Aflatoxin B1 does not increase gut permeability in broiler chickens. Frontiers in Veterinary Science. 2016;**3**:10

[59] Jouany JP. Methods for preventing, decontaminating and minimizing the toxicity of mycotoxins in feeds. Animal Feed Science and Technology. 2007;**137**:342-362

[60] Kolosova A, Stroka J. Evaluation of the effect of mycotoxin binders in animal feed on the analytical performance of standardised methods for the determination of mycotoxins in feed. Food Additives & Contaminants: Part A. 2012;**29**:1959-1971

[61] Avantaggiato G, Solfrizzo M, Visconti A. Recent advances on the use of adsorbent materials for detoxification of *Fusarium* mycotoxins. Food Additives and Contaminants. 2005;**22**:379-388

[62] Di Natale F, Gallo M, Nigro R. Adsorbents selection for aflatoxins removal in bovine milks. Journal of Food Engineering. 2009;**95**:186-191

[63] Hokkanen S, Bhatnagar A, Sillanpää M. A review on modification methods to cellulose-based adsorbents to improve adsorption capacity. Water Research. 2016;**91**:156-173

[64] Tan KB, Abdullah AZ, Horri BA, Salamatinia B. Adsorption mechanism of microcrystalline cellulose as green adsorbent for the removal of cationic methylene blue dye. Journal of the Chemical Society of Pakistan. 2016;**38**:651-664

[65] Zhao L, Yang G, Shi Y, Su C, Chang J. Co-delivery of Gefitinib and chloroquine by chitosan nanoparticles for overcoming the drug acquired resistance. Journal of Nanobiotechnology. 2015;**13**:57. DOI: 10.1186/s12951-015-0121-5

[66] Mine Kurtbay H, Bekçi Z, Merdivan M, Yurdakoç K. Reduction of ochratoxin A levels in red wine by bentonite, modified bentonites, and chitosan. Journal of Agricultural and Food Chemistry. 2008;**56**:2541-2545

[67] Ledoux DR, Rottinghaus GE. In vitro and in vivo testing of adsorbents for detoxifying mycotoxins in contaminated feedstuffs. Biotechnology Feed Industry. Nottingham, UK: Nottingham University Press; 1999. pp. 369-379

[68] Kong C, Shin SY, Kim BG. Evaluation of mycotoxin sequestering agents for aflatoxin and deoxynivalenol: An in vitro approach. Spring. 2014;**3**:346

[69] Solís-Cruz B, Hernández-Patlán D, Beyssac E, Latorre JD, Hernandez-Velasco X, Merino-Guzman R, Tellez G, López-Arellano R. Evaluation of chitosan and cellulosic polymers as binding adsorbent materials to prevent aflatoxin B1, fumonisin B1, ochratoxin, trichothecene, deoxynivalenol, and zearalenone mycotoxicoses through an in vitro gastrointestinal model for poultry. Polymers (Basel). 2017;**9**:529

[70] Zhao Z, Liu N, Yang L, Wang J, Song S, Nie D, Yang X, Hou J, Wu A. Cross-linked chitosan polymers as generic adsorbents for simultaneous adsorption of multiple mycotoxins. Food Control. 2015;**57**:362-369

[71] Daković A, Tomašević-Čanović M, Rottinghaus GE, Matijašević S, Sekulić Ž. Fumonisin B 1 adsorption to octadecyldimethylbenzyl ammonium-modified clinoptilolite-rich zeolitic tuff. Microporous and Mesoporous Materials. 2007;**105**:285-290

[72] Bazin I, Faucet-Marquis V, Monje M-C, El Khoury M, Marty J-L, Pfohl-Leszkowicz A. Impact of pH on the stability and the cross-reactivity of ochratoxin A and citrinin. Toxins (Basel). 2013;**5**:2324-2340

[73] Ghadi A, Mahjoub S, Tabandeh F, Talebnia F. Synthesis and optimization of chitosan nanoparticles: Potential applications in nanomedicine and biomedical engineering. Caspian Journal of Internal Medicine. 2014;**5**:156

[74] Khan RU, Naz S, Javdani M, Nikousefat Z, Selvaggi M, Tufarelli V, Laudadio V. The use of turmeric (*Curcuma longa*) in poultry feed. World's Poultry Science Journal. 2012;**68**:97-103

[75] Khalafalla RE, Müller U, Shahiduzzaman M, Dyachenko V, Desouky AY, Alber G, Daugschies A. Effects of curcumin (diferuloylmethane) on *Eimeria tenella* sporozoites in vitro. Parasitology Research. 2011;**108**:879-886

[76] Anand P, Kunnumakkara AB, Newman RA, Aggarwal BB. Bioavailability of curcumin: Problems and promises. Molecular Pharmaceutics. 2007;**4**:807-818

[77] Hernandez-Patlan D, Solis-Cruz B, Méndez-Albores A, Latorre JD, Hernandez-Velasco X, Tellez G, López-Arellano R. Comparison of PrestoBlue® and plating method to evaluate antimicrobial activity of ascorbic acid, boric acid and curcumin in an in vitro gastrointestinal model. Journal of Applied Microbiology. 2017;**124**:423-430. DOI: 10.1111/jam.13659

[78] Rivera-Rodriguez GR, Lollo G, Montier T, Benoit JP, Passirani C, Alonso MJ, Torres D. In vivo evaluation of poly-l-asparagine nanocapsules as carriers for anti-cancer drug delivery. International Journal of Pharmaceutics. 2013;**458**:83-89. DOI: 10.1016/j.ijpharm.2013.09.038

[79] Alishahi A, Mirvaghefi A, Tehrani MR, Farahmand H, Shojaosadati SA, Dorkoosh FA, Elsabee MZ. Shelf life and delivery enhancement of vitamin C using chitosan nanoparticles. Food Chemistry. 2011;**126**:935-940

[80] Luo Y, Teng Z, Wang Q. Development of zein nanoparticles coated with carboxymethyl chitosan for encapsulation and controlled release of vitamin D3. Journal of Agricultural and Food Chemistry. 2012;**60**:836-843

[81] Dimzon IKD, Ebert J, Knepper TP. The interaction of chitosan and olive oil: Effects of degree of deacetylation and degree of polymerization. Carbohydrate Polymers. 2013;**92**:564-570

[82] Oyarzun-Ampuero FA, Rivera-Rodríguez GR, Alonso MJ, Torres D. Hyaluronan nanocapsules as a new vehicle for intracellular drug delivery. European Journal of Pharmaceutical Sciences. 2013;**49**:483-490

[83] Rokhati N, Widjajanti P, Pramudono B, Susanto H. Performance comparison of α- and β-amylases on chitosan hydrolysis. ISRN Chemical Engineering. 2013;**2013**:186159

[84] Ma H, Qi X, Maitani Y, Nagai T. Preparation and characterization of superparamagnetic iron oxide nanoparticles stabilized by alginate. International Journal of Pharmaceutics. 2007;**333**:177-186. DOI: 10.1016/j.ijpharm.2006.10.006

[85] Déat-Lainé E, Hoffart V, Garrait G, Beyssac E. Whey protein and alginate hydrogel microparticles for insulin intestinal absorption: Evaluation of permeability enhancement properties on Caco-2 cells. International Journal of Pharmaceutics. 2013;**453**:336-342

[86] Vázquez-Durán A, Díaz-Torres R, Ramírez-Noguera P, Moreno-Martínez E, Méndez-Albores A. Cytotoxic and genotoxic evaluation of tortillas produced by microwave heating during alkaline-cooking of aflatoxin-contaminated maize. Journal of Food Science. 2014;**79**:T1024-T1029

[87] Schipper NG, Varum KM, Artursson P. Chitosans as absorption enhancers for poorly absorbable drugs. 1: Influence of molecular weight and degree of acetylation on drug transport across human intestinal epithelial (Caco-2) cells. Pharmaceutical Research. 1996;**13**:1686-1692

[88] Prego C, Torres D, Fernandez-Megia E, Novoa-Carballal R, Quiñoá E, Alonso MJ. Chitosan–PEG nanocapsules as new carriers for oral peptide delivery: Effect of chitosan pegylation degree. Journal of Controlled Release. 2006;**111**:299-308

[89] Xiang Y, Liu Y, Mi B, Leng Y. Hydrated polyamide membrane and its interaction with alginate: A molecular dynamics study. Langmuir. 2013;**29**:11600-11608. DOI: 10.1021/la401442r

[90] Van der Lubben IM, Verhoef JC, Borchard G, Junginger HE. Chitosan and its derivatives in mucosal drug and vaccine delivery. European Journal of Pharmaceutical Sciences. 2001;**14**:201-207. DOI: 10.1016/S0928-0987(01)00172-5

[91] Amidi M, Mastrobattista E, Jiskoot W, Hennink WE. Chitosan-based delivery systems for protein therapeutics and antigens. Advanced Drug Delivery Reviews. 2010;**62**:59-82. DOI: 10.1016/j.addr.2009.11.009

[92] Salatin S, Yari Khosroushahi A. Overviews on the cellular uptake mechanism of polysaccharide colloidal nanoparticles. Journal of Cellular and Molecular Medicine. 2017;**21**:1668-1686. DOI: 10.1111/jcmm.13110

[93] Li Q, Liu C-G, Yu Y. Separation of monodisperse alginate nanoparticles and effect of particle size on transport of vitamin E. Carbohydrate Polymers. 2015;**124**:274-279. DOI: 10.1016/j.carbpol.2015.02.007

7

Ampicillin-Loaded Chitosan Nanoparticles for In Vitro Antimicrobial Screening on *Escherichia coli*

Marilyn Porras-Gómez, Jose Vega-Baudrit and Santiago Núñez-Corrales

Abstract

Purpose: To develop ampicillin-loaded chitosan nanoparticles by modified ionic gelation method for evaluating their antimicrobial activity onto *Escherichia coli*.

Methods: Ampicillin-loaded chitosan nanoparticles (CHT-NPs) prepared by ionic gelation method with sodium tripolyphosphate as cross-linking agent. Drug release parameters (zeta potential, particle size, entrapment efficiency, and *in vitro* drug release) were assessed in relation to CHT-NP antimicrobial profile on *E. coli*. Antibacterial properties of CHT-NP formulation with ampicillin were found better than mere ampicillin without CHT-NPs.

Results: SEM, AFM images revealed dimensions of CHT-NPs with irregularity in shape/size. Optimized concentrations of chitosan 0.5% w/v with three different ratios (0.05% to 0.3% w/v) of TPP proved optimal for the evaluation of antibacterial profile of CHT-NPs. *In vitro* ampicillin-loaded CHT-NP delivery studies revealed an initial burst followed by slow sustained drug release, demonstrating superior antimicrobial activity than plain ampicillin, due to the synergistic effect of chitosan and ampicillin.

Conclusion: Chitosan content and cross-linking agent concentrations are control factors in synthesis of the optimized CHT-NP formulation. CHT-NPs with ampicillin cargo capably sustained ampicillin delivery due to NP size and increased surface charge, resulting in efficient growth inhibition in assays with *E. coli*.

Keywords: ampicillin, chitosan nanoparticles, drug delivery, *Escherichia coli*, ionic gelation

1. Introduction

Drug delivery systems are effective for implementing sustained release of many kinds of drugs. Chitosan (CHT), in particular chitosan nanoparticles (CHT-NPs), have been frequently used in drug delivery applications [1]. CHT is the generic name for a family of strongly polycationic derivatives of poly-N-acetyl-D-glucosamine (chitin) found in the exoskeletons of crustaceans such as crabs and shrimps. It can also be found in the cell wall of fungi and bacteria. Structurally, it is a linear polymer of cationic character formed by units of 2-amino-2-deoxy-D-glucose and 2-acetamido-2-deoxy-D-glucose linked by 1–4 bonds [2]. Having a positive charge, CHT is ideal for many drug delivery applications [3, 4]. CHT is biodegradable, non-toxic, non-immunogenic and biocompatible as well as the only naturally occurring polycationic polymer. Along with its derivatives, CHT has received a great deal of attention from the pharmaceutical industry as antimicrobial and antifungal agent [3, 4]. An extensive review of biocompatibility, hydrophilicity, biodegradability and broad spectrum gram-negative/positive antibacterial and anti-fungal effects of chitosan can be found [2].

CHT-NPs exhibit great drug encapsulation efficiency (*EE*) and diverse release profiles, adequate for transporting different types of drugs in many environments [5]. By controlling synthesis parameters conveniently (i.e. reagent concentrations, stirring speed), their diameter, surface charge and other properties can be easily modified thereby leading to a versatile delivery vehicle. CHT-NPs can be synthesized using the ion gelation method with tripolyphosphate (TPP) as cross-linking agent [6, 7]. Ionic cross-linking of CHT is a typical non-covalent interaction, which can be realized by association with negatively charged multivalent ions of TPP. There is a considerable body of literature on the production of CHT-TPP using this method, with many variations concerning the concentrations and ratios of the initial components, of which [1, 7] are seminal works in these matter. Ion gelation allows encapsulation of various compounds, including β-lactams.

Ampicillin is a beta-lactam antibiotic with amino-penicillin skeleton that easily penetrates outer membrane gram-positive/negative bacteria and irreversibly inhibits the transpeptidase enzyme—needed for synthesis of the bacterial cell wall [8]—in the third and final stage of synthesis in binary fission leading cell lysis (bacteriolytic) [9]. Bacterial resistance to antibiotics, including ampicillin, has been observed and investigated for three decades as a serious threat and health crisis [10]. Several mechanisms have been found liking transpeptidase synthesis and activation pathways to bacterial resistance to ampicillin in *E. coli*. Given that ampicillin drug acts as a broad-spectrum penicillin/antibiotic in treatment of infections caused by gram-positive/negative bacteria, and the broad-spectrum β-lactamases among *Enterobacteriaceae* [11] that is evolutionary conserved, finding novel synergistic ways to curb such general and versatile resistance mechanisms is essential for ensuring continued effectiveness of antimicrobial agents [12].

In this work, ampicillin-loaded CHT-NPs were synthesized and the effectiveness of their antimicrobial activity was evaluated against ampicillin and CHT-NPs on *E. coli*. A set of synthesis conditions was investigated in terms of their effect upon particle size, morphology and surface charge. After selection of the most viable delivery system from the later analysis, the β-lactamic antibiotic (ampicillin) was encapsulated in order to evaluate the encapsulation efficiency of the CHT-NPs. A release profile was obtained in order to assess their applicability as antimicrobial agents against gram-negative bacteria. The aim of the present study is the development of a simple method of synthesis of nano-carriers capable of transporting and releasing a drug such as ampicillin, ultimately inhibiting growth on *E. coli*.

2. Materials and methods

2.1. Reagents

CHT with molecular weight between 100,000 and 300,000 (ACROS Organics™), sodium tripolyphosphate and glacial acetic acid were used. Ultra-pure water was obtained using the Milli-Q A10 system (Millipore). Ampicillin sodium salt (Fisher BioReagents) was used as the antimicrobial agent.

2.2. Bacterial strain

E. coli ATCC 25922 was used for this study.

2.3. Synthesis of chitosan nanoparticles

CHT-NPs were produced using a modified ion gelation method. Briefly, CHT was dissolved at 0.5% w/v in 2% v/v acid acetic solution. Sodium tripolyphosphate was dissolved in ultra-pure water to obtain a 0.1% w/v concentration. An ampicillin volume of 1 ml was then added and later 1 ml of the TPP solution was added drop-wise to 1.5 ml of the CHT solution and magnetically stirred for 1 h. The final suspension was centrifuged at 11,000 rpm for 20 min [6, 13]. Three concentrations of CHT (0.1, 0.3 and 0.5% w/v) and five concentrations of TPP (0.005, 0.01, 0.05, 0.1 and 0.3% w/v) were employed in order to determine their optimal ratio in terms of particle size and surface charge.

2.4. Statistical analysis

In order to evaluate the role of synthesis parameters on particle size and surface charge, including random variations in CHT and then TPP concentrations, selection of linear mixed effects models was performed using standard the Akaike information criterion (AIC), the Bayes information criterion (BIC) and the negative 2 log likelihood criterion (−2LL) [14]. The generalized form (Ω_0^2) of the usual coefficient of determination (R_2) was used [15]. A total of five models (including interactions) were evaluated for each response variable, that is, particle size and ζ potential. Calculations were performed using the *R* Statistical Computing package [16] with the lmer4 package for linear mixed effects models [17]. The Staterthwaite approximation of p values was computed for the corresponding observable (**Table 1**) [18].

Model	Response variable	Fixed effects
1	Size	1
2	Size	CHT concentration
3	Size	TPP concentration
4	Size	CHT + TPP concentrations
5	Size	CHT + TPP + CHT*TPP concentrations
6	ζ potential	1
7	ζ potential	CHT concentration
8	ζ potential	TPP concentration
9	ζ potential	CHT + TPP concentrations
10	ζ potential	CHT + TPP + CHT*TPP concentrations

The value of 1 in effects represents the case of a purely random model used as the baseline.

Table 1. Statistical models for evaluating formulations.

2.5. Determination of particle size and ζ potential

The determination of particle size (apparent hydrodynamic diameter) was performed by dynamic light scattering (DLS) and surface electric charge using a Zetasizer Malvern Nano SZ-90 particle analyser, reported as either ζ potential or electrophoretic mobility [7].

2.6. Morphology analysis of CHT-NPs

In order to study the morphology of nanoparticles, topographic images of CHT-NPs were taken on a multi- mode atomic force microscope (AFM) Asylum Research MFP-3D. The AFM probes used for this study were rectangular silicon probes with a nominal spring constant of 40 nN/nm. Similarly, image visualization was carried out in a scanning electron microscope (SEM) Hitachi S-3700 with a 15 nm gold coating on the diluted samples (1/10) using an aluminum base at an acceleration voltage of 15 kV [7].

2.7. Determination of encapsulation efficiency of loaded CHT-NPs

The encapsulation efficiency (*EE*) of the CHT-NPs was determined according to the method described in the previous studies [19]. In brief, the nanoparticle suspensions were separated by centrifuging at 11,000 rpm for 20 min and the contents of ampicillin in the supernatants were measured by HPLC-DAD Perkim Elmer. A blank sample of CHT-NPs without ampicillin was obtained but treated similarly as the drug-loaded CHT-NPs. All samples were measured in triplicate. The *EE* were calculated using.

$$EE = \frac{F}{T} \times 100\% \tag{1}$$

where F is the free amount of ampicillin in the supernatant, T is total amount of ampicillin.

2.8. Release profile of loaded CHT-NP

Release studies were carried out in PBS (pH 7.4) as follows: 1.5 ml ampicillin-loaded CHT-NPs and 1.5 ml PBS were incubated at 37°C and shaken at 200 rpm. Triplicate samples were analyzed at each time step, between 0 and 24 h. The samples were centrifuged and the concentrations of ampicillin released in the supernatant were determined by HPLC-DAD.

2.9. Determination of antimicrobial activity of loaded CHT-NPs

The spectrophotometrically adjusted inoculum (100 μl) of 10^4 bacterial cells was added to each well in the sterile flat-bottomed microtiter plate containing the test CHT-NPs. The design of experiments includes duplicated wells of ampicillin-loaded nanoparticles with three different concentrations of ampicillin (5, 10, 20 mg/ml), two wells with ampicillin as growth inhibition control, two wells containing bacterial suspension with CHT-NPs (growth control) and two wells containing only media (background control) were included in this plate. For the case of wells with ampicillin and CHT-NPs, dilutions were halved at each consecutive level in the gradient. Optical densities were measured for 24 h at 37°C using a multi-detection microplate reader Biotek Synergy HT at 600 nm and automatically recorded for each well every 30 min. Turbidimetric growth curves were obtained depending on the changes in the optical density of bacterial growth for each CHT NP sample and the drug-free growth control.

3. Results and discussion

3.1. Particle size and surface charge of loaded CHT-NPs

Figure 1 corresponds to measures of nanoparticle size (a) and ζ potential (b) according to the concentrations indicated above for both CHT and TPP. As TPP concentration increases, diameter decreases at 0.3 and 0.5% w/v CHT concentrations (a). The opposite occurs at a CHT concentration of 0.1% w/v. In the case of ζ potential (b), all CHT concentrations yield decreasing values as TPP concentration increases.

Statistical models revealed that the best alternative for explaining nanoparticle size in terms of CHT and TPP concentrations is model 5, which includes both factors as well as their interaction (**Table 2**). Both AIC and −2LL have a larger distance from the baseline than in model 3 (size only depending on TPP concentration), being also more significant and with a better fit than it. **Figure 1(a)** shows that nanoparticle size decreases as TPP concentration increases and, simultaneously, CHT concentration decreases. In that sense, model 5 also indicates that while TPP is the largest driver of nanoparticles diameter ($\chi^2 = 3.536$, $df = 4$, $p(\chi^2) < 0.001$), the interaction between CHT and TPP is responsible for the observed change in slope (dotted lines for each line). CHT concentration seems not to have a large role by itself.

A similar analysis was carried out for ζ potential (**Figure 1(b)**). By the same criteria, model 10 was the best criteria for all AIC, BIC and −2LL simultaneously. While all factors are statistically significant, TPP concentration.

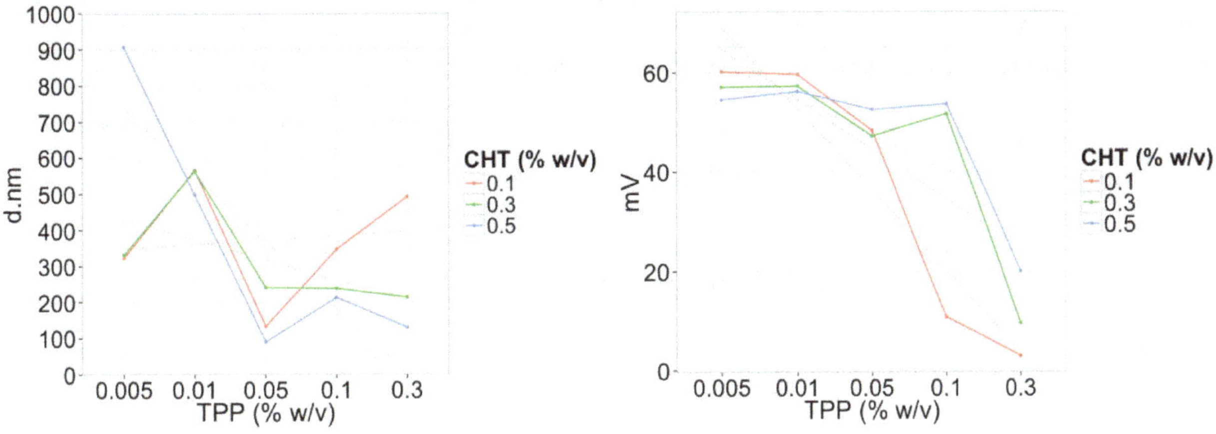

(a) (b)

Figure 1. Nanoparticle diameter (nm) and ζ potential (mV) at different concentrations of CHT and TPP. CHT concentration has a marked inverse effect to that of TPP on nanoparticle size, evidence of interference between both factors. A similar trend is identifiable for ζ potential between the lowest and highest TPP concentrations. (**a**) Nanoparticle diameter. (**b**) ζ potential.

Model	*df*	AIC	BIC	−2LL	χ^2	*df* (χ^2)	*p*(χ^2)	Ω_0^2
2	6	597.35	608.19	−292.67	0.206	2	0.902	0.853
3	8	590.54	604.99	−287.27	10.805	2	0.005*	0.844
4	10	594.11	612.17	−287.05	0.433	2	0.805	0.843
5	18	566.67	599.19	−265.33	43.440	8	<0.001*	0.852

Model 1 is not included (comparison baseline). *Significant *p* values.

Table 2. Results for selection of linear mixed effects models using maximum likelihood for nanoparticles size as the response variable.

central ($\chi^2 = 6604.98$, $df = 4$, $p(\chi^2) < 0.001$): as it increases, ζ potential decreases thanks to a lower amount of available binding sites in the CHT matrix (**Table 3**).

Finally, the most significant levels for both size (diameter in nm) and ζ potential that include CHT and TPP concentrations were identified ($p < 0.001$) as candidates (**Table 4**). In agreement with existing literature, nanoparticles with a large ζ potential ($\zeta \geq 40$ mV) are desirable due to their good stability [20, 21] having sizes in between 200 and 580 nm [5, 22] for the particular case of CHT-NPs. From **Figure 2**, it is clear that levels 1 and 2 reported in **Table 4** (matching Samples M and N in **Table 5**) comply with these requirements. The choice of sample N maximizes the value for the ζ potential while preserving a small nanoparticle size within the range mentioned above.

3.2. Morphology analysis of loaded CHT-NPs

A batch of CHT-NPs without cargo was synthesized and visualized using AFM imaging according to sample preparation N (**Figure 3**). Scan areas are (A) 5 μm × 5 μm and (B) 1 μm × 1 μm respectively for 3A and 3B. The distribution of nanoparticle diameters is reported in **Figure 4** after post-processing of the AFM image.

Model	*df*	AIC	BIC	−2LL	χ^2	$df(\chi^2)$	$p(\chi^2)$	Ω_0^2
7	6	278.95	289.79	−133.477	0.862	2	0.650	0.995
8	8	261.73	276.18	−122.866	21.222	2	<0.001*	0.995
9	10	261.64	279.71	−120.822	4.087	2	0.13	0.995
10	18	196.41	228.93	−80.204	81.2370	8	<0.001*	0.995

Model 6 is not included (comparison baseline). *Significant *p* values.

Table 3. Results for selection of linear mixed effects models using maximum likelihood for ζ potential as the response variable.

Level	Concentrations
1	CHT 0.5%w/v, TPP 0.05%w/v
2	CHT 0.5%w/v, TPP 0.1%w/v
3	CHT 0.5%w/v, TPP 0.3%w/v

Only levels with $p < 0.001$ are shown here.

Table 4. Significant levels for CHT and TPP concentrations simultaneously reported in model 5 for nanoparticle size and model 10 for ζ potential.

Additionally, SEM images were performed upon samples with and without ampicillin cargo (**Figure 5**). Images 5A, 5B and 5C are from a CHT-NPs sample. It verifies that CHT-NPs have a dispersed, corrugated and spherical morphology with a diameter between 100 and 200 nm. Complementary, images 5D, 5E and 5F belong to 10 mg/ml ampicillin-loaded CHT-NPs sample with an *EE* more than 40% (**Figure 6**) these samples have a smooth spherical morphology with a diameter between 500 and 1000 nm, showing an aggregation effect between nanoparticles.

3.3. Encapsulation efficiency of loaded CHT-NPs

For encapsulation efficiency (**Figure 6**) the initial ampicillin concentration was compared against final encapsulated concentration (**Figure 6(a)**) and later transformed into *EE* (**Figure 6(b)**). *EE* lies within a range of 15–41% with a peak value at a concentration of ampicillin of 4 mg/ml in the final CHT-NPs solution volume.

3.4. Release profile of loaded CHT-NP

A release profile was obtained for the ampicillin-loaded CHT-NPs (**Figure 7**). Released percentage was calculated in relation to the encapsulated ampicillin concentration. Release percentage oscillates between 5 and 20% across 24 h. The burst effect is clearly observable (between 0 and 2 h), to be later succeeded by a more stable behavior (from 2 to 18 h) and rising finally in the last stage (from 18 to 24 h). The observed pattern suggests that swelling of the first layer in the polymeric matrix releases a large amount of ampicillin in the medium (0–1 h), and becomes more stable until the innermost layers are reached, where the remaining

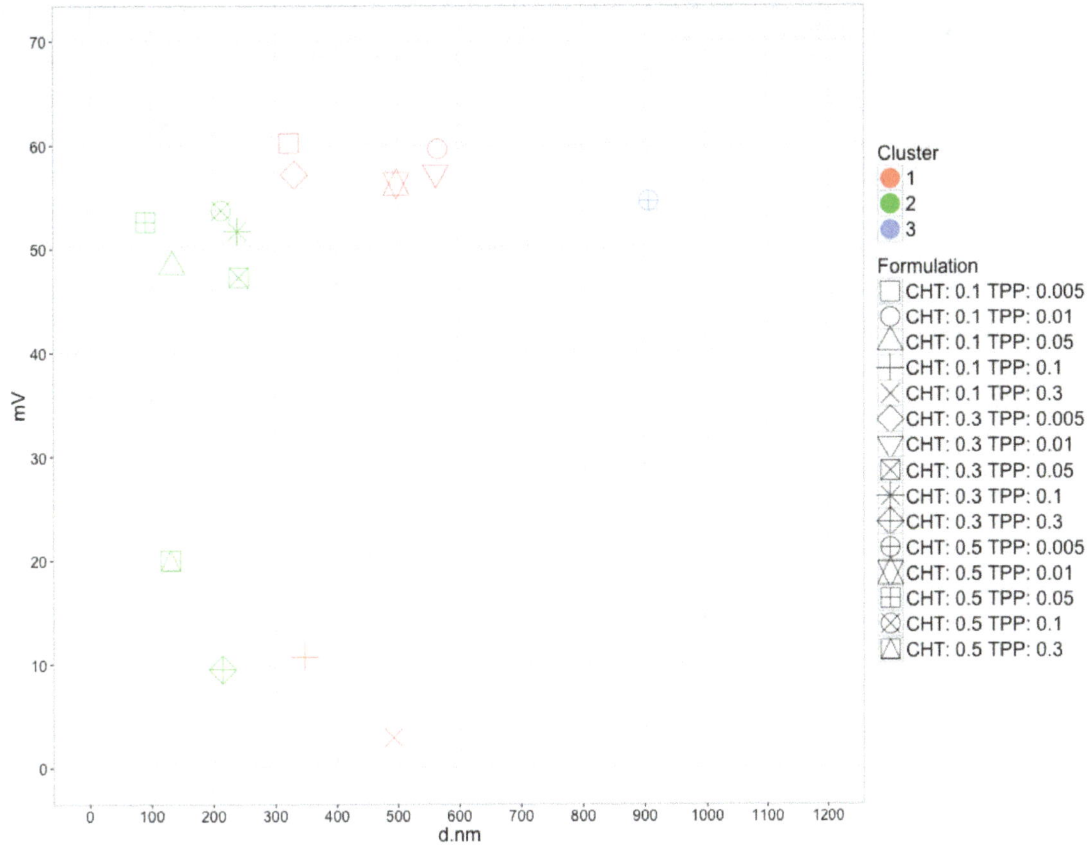

Figure 2. Scatter plot for diameter and surface charge of different CHT-NPs. The classification (clusters) was computed using the Hartigan-Wong *k*-means clustering algorithm [23]. Formulations indicate both CHT and TPP concentrations used for each nanoparticle type. The top left points in the green cluster (above 45 mV and below 300 nm) are the best delivery targets with respect to size and ζ potential, which appear at medium to low concentrations of TPP (0.05–0.1% w/v) and mostly at medium and high concentrations of CHT (except for C with 0.1% w/v). Most experimentally found diameters are below 600 nm. In the case of surface charge, data are partitioned in two groups: above 40 mV and below 20 mV with no nanoparticle in the middle.

Sample	CHT	TPP	*d*	ζ
A	0.1	0.005	323.57 ± 186.84	60.23 ± 2.85
B	0.1	0.01	565.60 ± 97.49	59.67 ± 0.92
C	0.1	0.05	133.60 ± 18.34	48.37 ± 0.31
D	0.1	0.1	347.43 ± 17.60	10.73 ± 0.59
E	0.1	0.3	492.63 ± 16.35	2.92 ± 0.16
F	0.3	0.005	331.73 ± 168.06	57.17 ± 2.28
G	0.3	0.01	562.73 ± 155.08	57.33 ± 2.06
H	0.3	0.05	241.90 ± 34.51	47.30 ± 0.53
I	0.3	0.1	239.30 ± 27.14	51.73 ± 1.25
J	0.3	0.3	214.93 ± 11.40	9.53 ± 0.40
K	0.5	0.005	906.97 ± 264.60	54.63 ± 2.26
L	0.5	0.01	498.30 ± 49.73	56.27 ± 0.95

Sample	CHT	TPP	d	ζ
M	0.5	0.05	90.88 ± 30.90	52.63 ± 2.53
N	0.5	0.1	213.50 ± 29.76	53.73 ± 3.33
O	0.5	0.3	130.56 ± 15.65	20.03 ± 1.46

CHT and TPP concentrations are given in terms of w/v percentage. Diameter d is given in nm. ζ potential is given in mV. Standard deviation is included in both cases.

Table 5. CHT and TPP concentrations for synthesis of nanoparticles in **Figure 2**.

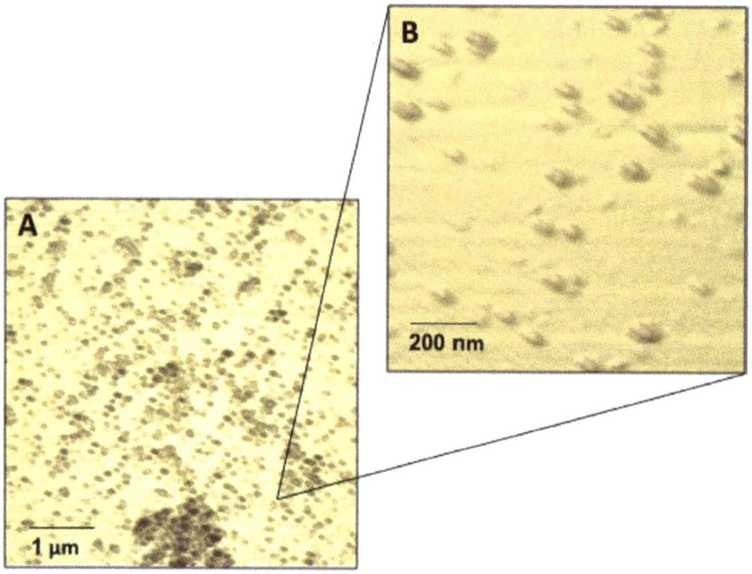

Figure 3. AFM image of CHT-NPs without ampicillin cargo. AFM images were performed as a first assessment of CHT-NP synthesis process. **Figure 5** provides the contrast between these nanoparticles and ampicillin loaded CHT-NPs. **A** corresponds to a scan area of 5 μm × 5 μm and **B** to 1 μm × 1 μm.

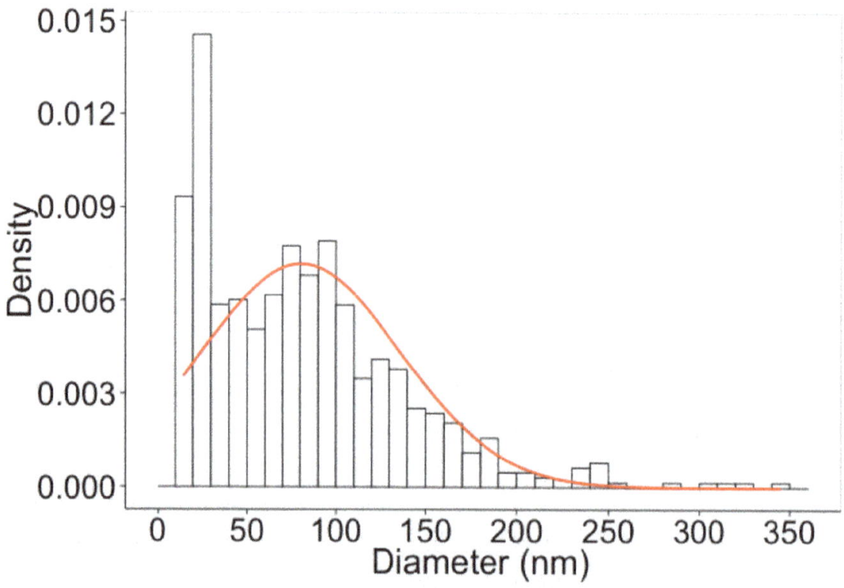

Figure 4. Histogram of CHT-NPs diameters for the AFM image in **Figure 3**. $N = 632$, $\bar{d} = 80.5$ nm, $\sigma_d = 55.6$ nm.

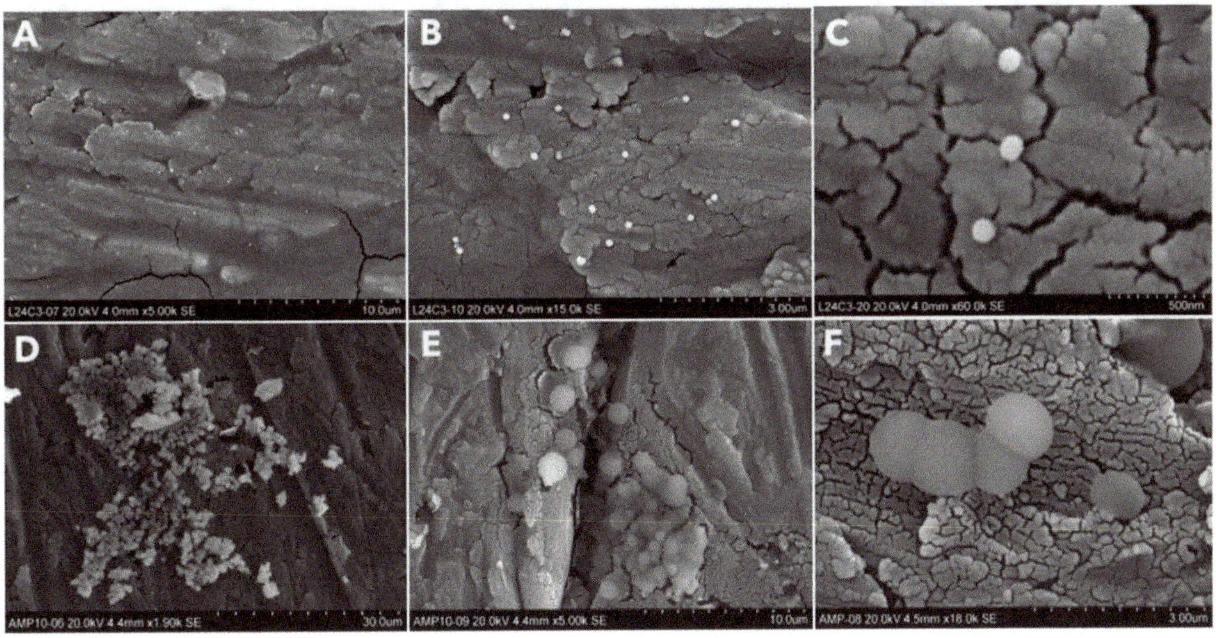

Figure 5. SEM image of CHT-NPs, without and with antibiotic cargo (ampicillin). **A**, **B** and **C** are images of nanoparticles without cargo and **D**, **E** and **F** encapsulate ampicillin. Nanoparticle diameter and size appears to increase in order to gain stability in ampicillin-loaded nanoparticles, which also has an additional aggregation effect.

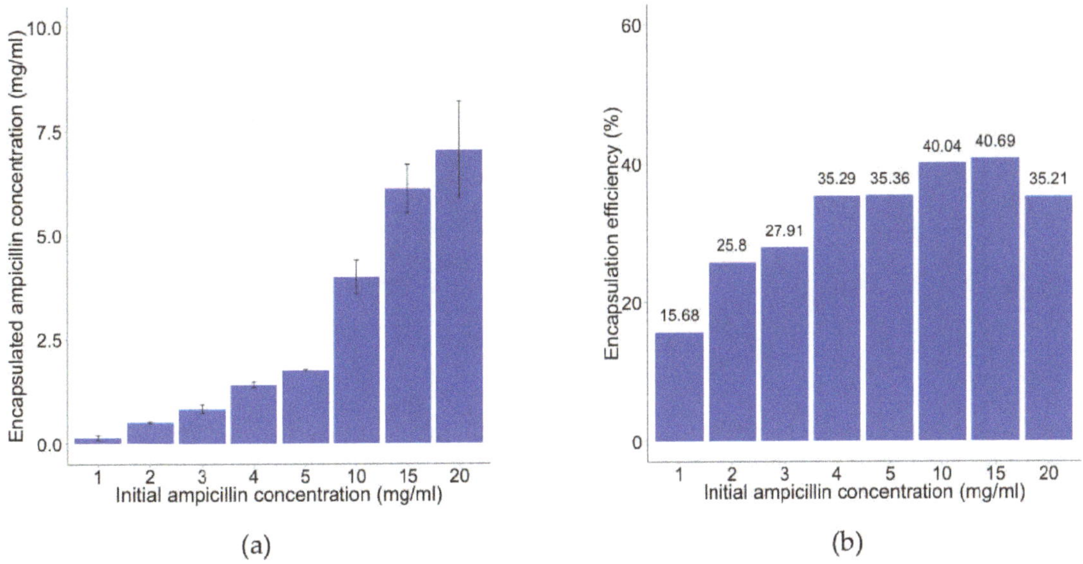

Figure 6. Cargo efficiency of the CHT-NPs from initial concentrations, obtained as described in [19]. (**a**) Measured encapsulated ampicillin concentration through HPLC. (**b**) Corresponding encapsulation efficiency, calculated as *EE* in Eq. (1).

contents are finally released. The latter is consistent with what is described by Carbinato et al. [24] for cross-linked pectin/high-amylose starch matrices.

3.5. Growth inhibition assays for *E. coli* ATCC 25922

Figure 8 shows the growth inhibition assays for *E. coli* ATCC 25922, an ampicillin–susceptible reference strain. Under the conditions of this assay, this strain shows a minimal inhibitory

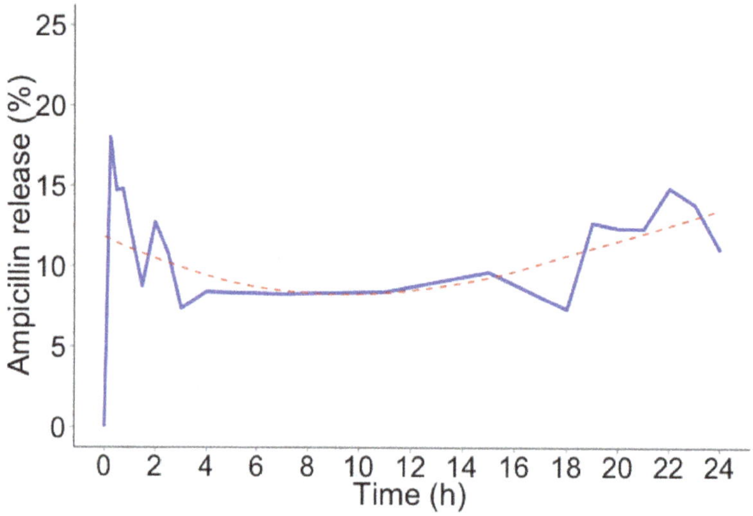

Figure 7. Release profile of ampicillin-loaded CHT-NPs. A burst effect was observed for the first hour, followed by a stabilization phase (1–4 h), a steady release phase (4–15 h) and posterior relative increase (15–24 h).

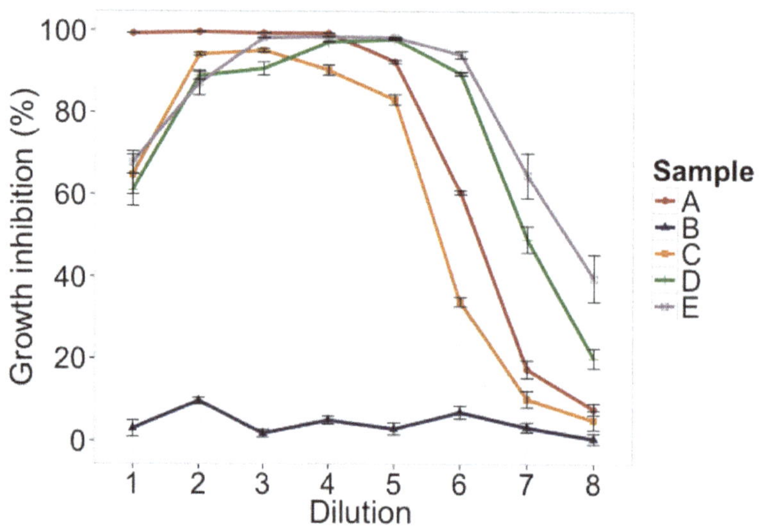

Figure 8. Growth inhibition assays for *E. coli*. Samples are as follows: (**A**) ampicillin 20 mg/ml, (**B**) CHT-NPs, ampicillin-loaded NPs at (**C**) 5 mg/ml, (**D**) 10 mg/ml and (**E**) 20 mg/ml. As expected, sample **B** produces no inhibition, while **A–C** remain effective until the fifth dilution and **D–E** to the sixth dilution.

concentration (MIC) of 2 μg/ml, which is in agreement with the range values (2–8 μg/ml) previously reported [25]. A clinical *E. coli* isolate is considered as non-susceptible when its MIC value is >8 μg/ml. The highest ampicillin concentration used as control was 500 μg/ml (dilution 1, **Figure 8**), generating a concentration gradient by double dilutions through 3.91 μg/ml (dilution 8, **Figure 8**). The three ampicillin-loaded CHT-NPs systems (C, D and E) show a similar growth inhibition pattern with respect to the positive control (ampicillin), indicating that under these conditions ampicillin is released from CHT – NPs to concentrations high enough to inhibit bacterial growth. Moreover, both CHT – NPs systems D and E, loaded with 10 mg/ml and 20 mg/ml of ampicillin respectively, maintain growth inhibition with more dilutions with respect to ampicillin. It is worth noting that inhibition in samples A and C lower their effectiveness rapidly after the fifth dilution, but samples D and E have a much wider range of inhibition at lower concentrations for all remaining dilutions.

4. Conclusions

CHT-NPs synthesis was successfully achieved and the manipulation of CHT and TPP concentrations towards optimization for drug delivery resulted in smaller particles and increased surface charge. Besides, a detailed physicochemical characterization of CHT-NPs was obtained. CHT-NPs were able to encapsulate and release ampicillin. The encapsulation efficiency of ampicillin CHT-NPs exceeds 30% in most cases while antibiotic release maintains a relatively stable profile, ranging between 5 and 20% in a 24-hour period. A microbiological assay was used as proof of principle in order to verify the release of the antibiotic ampicillin from the CHT-NPs systems. Growth inhibition of the ampicillin-susceptible *E. coli* ATCC 25922 reference strain was achieved in a similar pattern compared with ampicillin alone.

In terms of the obtained release profile in a period of 24 h, several comparisons can be drawn in terms of other alternatives. The cumulative release profile, computed from data shown in **Figure** 7, indicates that less than 40% of the cargo was released after 24 h. Ampicillin-loaded electrospun poly(ε-caprolactone) nanofiber yarns have a much faster release profile, releasing more than 90% of its cargo in the same period [26]. By contrast, our formulation has a release profile faster than that reported for ampicillin-conjugated gum Arabic microspheres [27]. In the later cases, no additional antibiotic or bacteriostatic effect is present in the polymeric matrix. CHT-NPs loaded with ampicillin reported here are comparable to chitosan microgranules loaded with diclofenac sodium in contrast to chitosan beads loaded with diclofenac sodium in terms of the order of magnitude of released antibiotic [28]. Compared against ampicillin loaded methylpyrrolidinone chitosan and chitosan microspheres, our nanoparticles have a significantly lower encapsulation efficiency [29], yet the former are not suitable for sustained release. CHT-NPs are similar in size compared to existing literature on ampicillin-loaded nanoparticles and liposomes [30], with a release profile closer to that of liposomes. In the later study, liposomes had a larger inhibition halo at lower dilutions than nanoparticles of higher density and longer release time.

Finally, the synthesis protocol of CHT-NPs elaborated in this study constitutes a platform for the analysis of the encapsulation of other antibiotics with different structures as well as for assays on other bacteria.

Acknowledgements

The authors wish to acknowledge funding and support from the National Center for Advanced Technology Studies (CeNAT) and the National Council of Rectors for the period comprehended between 2012 and 2015.

Conflict of interest

The authors have no conflict of interest to declare.

Author details

Marilyn Porras-Gómez, Jose Vega-Baudrit* and Santiago Núñez-Corrales

*Address all correspondence to: jvegab@gmail.com

National Nanotechnology Laboratory, National Center for Advanced Technology Studies, San José, Costa Rica

References

[1] Liu H, Gao C. Preparation and properties of ionically cross-linked chitosan nanoparticles. Polymers for Advanced Technologies. 2009;**20**(7):613-619

[2] Cota-Arriola O, Cortez-Rocha MO, Burgos-Hernández A, Ezquerra-Brauer JM, Plascencia-Jatomea M. Controlled release matrices and micro/nanoparticles of chitosan with antimicrobial potential: Development of new strategies for microbial control in agriculture. Journal of the Science of Food and Agriculture. 2013;**93**(7):1525-1536

[3] Morris GA, Kök SM, Harding SE, Adams GG. Polysaccharide drug delivery systems based on pectin and chitosan. Biotechnology and Genetic Engineering Reviews. 2010; **27**(1):257-284

[4] Nasti A, Zaki NM, de Leonardis P, Ungphaiboon S, Sansongsak P, Rimoli MG, Tirelli N. Chitosan/tpp and chitosan/tpp–hyaluronic acid nanoparticles: systematic optimisation of the preparative process and preliminary biological evaluation. Pharmaceutical Research. 2009;**26**(8):1918-1930

[5] Gan Q, Wang T. Chitosan nanoparticle as protein delivery carrier—Systematic examination of fabrication conditions for efficient loading and release. Colloids and Surfaces B: Biointerfaces. 2007;**59**(1):24-34

[6] Ajun W, Yan S, Li G, Li H. Preparation of aspirin and probucol in combination loaded chitosan nanoparticles and in vitro release study. Carbohydrate Polymers. 2009; **75**(4):566-574

[7] Fàbregas A, Miñarro M, García-Montoya E, Pérez-Lozano P, Carrillo C, Sarrate R, Sánchez N, Ticó JR, Suñé-Negre JM. Impact of physical parameters on particle size and reaction yield when using the ionic gelation method to obtain cationic polymeric chitosan–tripolyphosphate nanoparticles. International Journal of Pharmaceutics. 2013; **446**(1):199-204

[8] Nguyen-Distèche M, Pollock JJ, Ghuysen J-M, Puig J, Reynolds P, Perkins HR, Coyete J, Salton MRJ. Sensitivity to ampicillin and cephalothin of enzymes involved in wall peptide crosslinking in *Escherichia coli* k12, strain 44. The FEBS Journal. 1974;**41**(3):457-463

[9] Blumberg PM, Strominger JL. Interaction of penicillin with the bacterial cell: penicillin-binding proteins and penicillin-sensitive enzymes. Bacteriological Reviews. 1974; **38**(3):291

[10] Neu HC. The crisis in antibiotic resistance. Science. 1992;**257**(5073):1064-1074

[11] Smet A, Martel A, Persoons D, Dewulf J, Heyndrickx M, Herman L, Haesebrouck F, Butaye P. Broad-spectrum β-lactamases among enterobacteriaceae of animal origin: Molecular aspects, mobility and impact on public health. FEMS Microbiology Reviews. 2010;**34**(3):295-316

[12] Pamela A Hunter, K Coleman, Jackie Fisher, and Dalveen Taylor. In vitro synergistic properties of clavu-lanic acid, with ampicillin, amoxycillin and ticarcillin. Journal of Antimicrobial Chemotherapy. 1980;**6**(4):455-470

[13] Ji J, Hao S, Danjun W, Huang R, Yi X. Preparation, characterization and in vitro release of chitosan nanoparticles loaded with gentamicin and salicylic acid. Carbohydrate Polymers. 2011;**85**(4):803-808

[14] Adams E, Coomans D, Smeyers-Verbeke J, Massart DL. Application of linear mixed effects models to the evaluation of dissolution profiles. International Journal of Pharmaceutics. 2001;**226**(1):107-125

[15] Ronghui X. Measuring explained variation in linear mixed effects models. Statistics in Medicine. 2003;**22**(22):3527-3541

[16] R Core Team. R: A Language and Environment for Statistical Computing. Vienna, Austria: R Foundation for Statistical Computing; 2013. Available from: http://www.R–project.org/

[17] Bates D, Maechler M, Bolker B, Walker S, Christensen RHB, Singmann H. Linear mixed-effects models using Eigen and S4. R package version 1.0-5. 2013

[18] Schaalje GB, McBride JB, Fellingham GW. Adequacy of approximations to distributions of test statistics in complex mixed linear models. Journal of Agricultural, Biological, and Environmental Statistics. 2002;**7**(4):512-524

[19] Sibaja B. Nanopartículas de Quitosano como Sistema de Liberación Controlada de Principios Activos [Master's thesis]. Heredia, Costa Rica: Facultad de Ciencias Extactas y Naturales, Universidad Nacional; 2011

[20] O'Brien RW et al. Electroacoustic studies of moderately concentrated colloidal suspensions. Faraday Discussions of the Chemical Society. 1990;**90**:301-312

[21] Hanaor D, Michelazzi M, Leonelli C, Sorrell CC. The effects of carboxylic acids on the aqueous dispersion and electrophoretic deposition of zro 2. Journal of the European Ceramic Society. 2012;**32**(1):235-244

[22] Gan Q, Wang T, Cochrane C, McCarron P. Modulation of surface charge, particle size and morphological properties of chitosan-tpp nanoparticles intended for gene delivery. Colloids and Surfaces B: Biointerfaces. 2005;**44**(2):65-73

[23] Hartigan JA, Wong MA. Algorithm as 136: A k-means clustering algorithm. Journal of the Royal Statistical Society. Series C (Applied Statistics). 1979;**28**(1):100-108

[24] Carbinato FM, de Castro AD, Evangelista RC, Cury BSF. Insights into the swelling process and drug release mechanisms from cross-linked pectin/high amylose starch matrices. Asian Journal of Pharmaceutical Sciences. 2014;**9**(1):27-34

[25] JB Patel, FR Cockerill, J Alder, PA Bradford, GM Eliopoulos, DJ Hardy, et al. Performance standards for antimicrobial susceptibility testing; twenty-fourth informational supplement. CLSI Standards for Antimicrobial Susceptibility Testing. 2014;**34**(1):1-226

[26] Liu H, Leonas KK, Zhao Y. Antimicrobial properties and release profile of ampicillin from electrospun poly (ε-caprolactone) nanofiber yarns. Journal of Engineered Fibers and Fabrics. 2010;**5**(4):10-19

[27] Nishi KK, Antony M, Jayakrishnan A. Synthesis and evaluation of ampicillin-conjugated gum arabic microspheres for sustained release. Journal of Pharmacy and Pharmacology. 2007;**59**(4):485-493

[28] Gupta KC, Ravi Kumar MNV. Drug release behavior of beads and microgranules of chitosan. Biomaterials. 2000;**21**(11):1115-1119

[29] Giunchedi P, Genta I, Conti B, Muzzarelli RAA, Conte U. Preparation and characterization of ampicillin loaded methylpyrrolidinone chitosan and chitosan microspheres. Biomaterials. 1998;**19**(1):157-161

[30] Fatal E, Rojas J, Roblot-Treupel L, Andremont A, Couvreur P. Ampicillin-loaded liposomes and nanoparticles: Comparison of drug loading, drug release and in vitro antimicrobial activity. Journal of Microencapsulation. 1991;**8**(1):29-36

8

An Overview of Chitosan-Xanthan Gum Matrices as Controlled Release Drug Carriers

Suha M. Dadou, Milan D. Antonijevic,
Babur Z. Chowdhry and Adnan A. Badwan

Abstract

Naturally occurring polysaccharides and/or their chemically modified derivatives have been widely investigated in relation to their use as components of controlled release systems for drug delivery. The aforementioned is due, in part, to their distinct properties such as abundant availability and biocompatibility as well as environmental and economic advantages. Chitosan (CS) and xanthan gum (XG) based matrices have received growing scientific/pharmaceutical interest as oral controlled release drug carriers. Herein, recent advances spanning the last two decades in CS-XG based drug delivery systems are reviewed with the emphasis being on oral tablet formulations, due to their versatility as pharmaceutical dosage forms. The mechanism of interaction between CS and XG, by means of computational and experimental approaches, is scrutinized. Results obtained from the literature establish the possibility of fabricating a controlled release drug delivery system based on CS and XG matrices. This can be achieved by monitoring and manipulating the physiochemical properties of the two polymers as well as the experimental variables affecting their drug retardation efficiency, without the need to employ special equipment or sophisticated experimental techniques/methodologies.

Keywords: drug delivery, controlled release, polymeric matrices, natural polysaccharides, xanthan gum, chitosan, polyelectrolyte complexes, molecular dynamics simulation

1. Introduction

The ultimate goal in drug design and development is to optimize a carrier that ensures the delivery of the active pharmaceutical ingredient(s) (APIs) to the systemic circulation in a safe and stable manner [1]. Patient compliance is a key aspect to consider when designing a new pharmaceutical dosage form [2]. Therefore, the way the drug will be introduced to the body should be optimized to ensure the availability of the drug at its site of action, at levels within the range of its therapeutic window (**Figure 1**).

Despite emerging advances in drug delivery, the oral route remains the predominant route of drug administration. It is the simplest route, non-invasive and provides ~200 m^2 of readily available surface area for drug absorption [3]. Conventional oral dosage forms usually release drugs immediately in the body, via first order release kinetics for both absorption and elimination processes [4]. Since the efficacy of the administered drug is limited to its residence time in plasma, frequent administration is required for APIs which exhibit a short biological half-life. As a result, low patient compliance and high fluctuation of drug levels in plasma is expected [5, 6]. In order to counter the foregoing drawbacks of conventional dosage forms, a new term in drug delivery was introduced; modified release dosage forms [7].

The United States Pharmacopeia defines modified release tablets as "coated or uncoated tablets that contain special excipients or are prepared by special procedures, or both, designed to modify the rate, the place or the time at which the active substance(s) are released". Modified release delivery systems can be divided into delayed release systems and prolonged/extended release systems. Extended release delivery systems can further be subdivided into sustained and controlled release delivery systems, which differ in the rate at which they deliver APIs to the human blood circulation. Sustained release formulations function by continuously releasing APIs for a prolonged period of time. On the other hand, controlled release (CR) delivery systems do not only retard the release of the drug, but they deliver the drug to the body at a predetermined release rate or location [8]. Consequently, constant drug levels can be achieved (**Figure 2**).

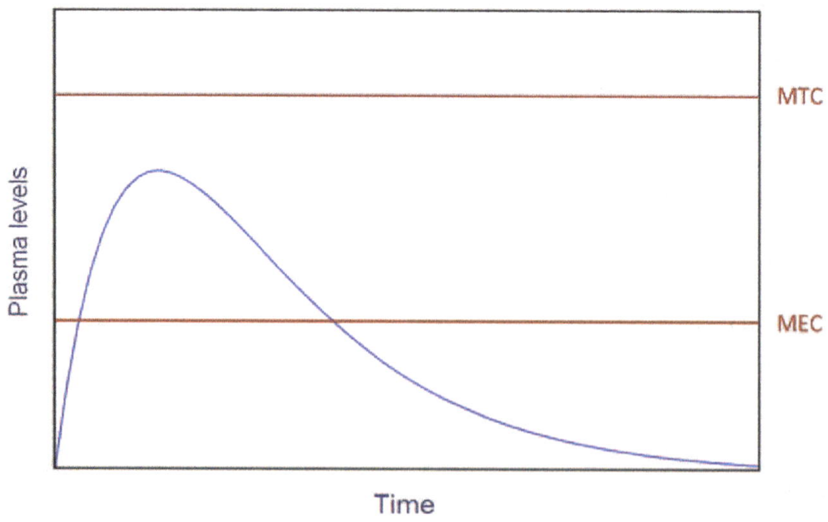

Figure 1. Plasma levels and therapeutic range of a drug following an oral administration of a single dose. MTC: minimum toxic concentration, and MEC: minimum effective concentration.

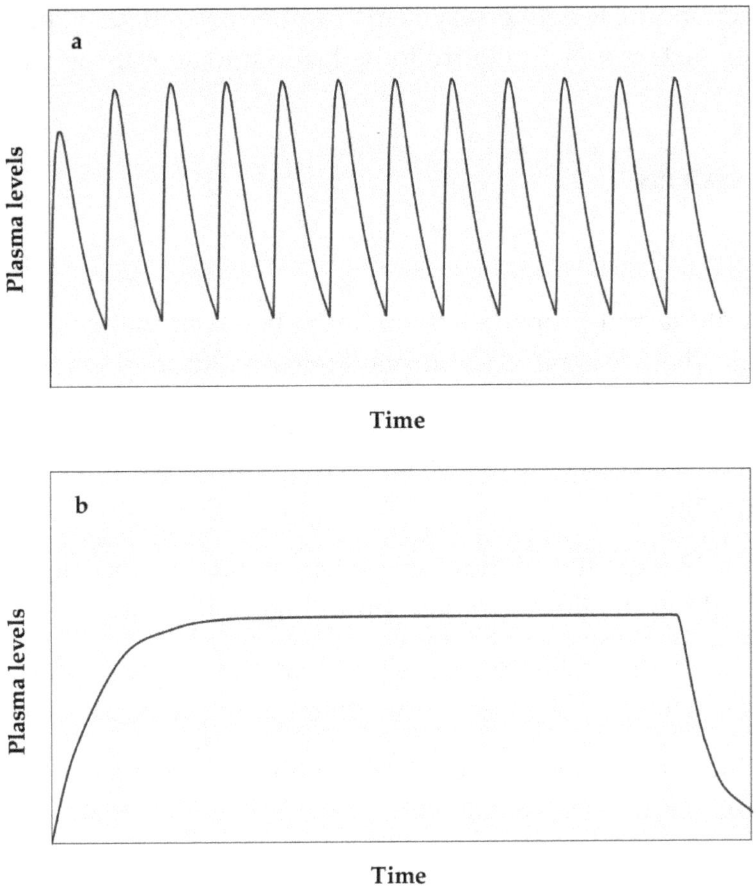

Figure 2. Comparison of plasma concentration-time profiles of drug release following multiple dosing from: (a) conventional, and (b) controlled-release dosage forms.

2. Controlled drug delivery

CR systems are composed of inactive pharmaceutical ingredient(s) that entrap the API(s) and release it/them at a time different from the immediate release form [4]. Researchers in the field of drug delivery have, and are currently still trying to acquire a better understanding of CR by attempting to integrate pharmaceutical technology with the relevant pharmacokinetic parameters associated with different drugs [9]. The rational underpinning controlled drug release includes, but are not limited to: masking the undesired side effects of drugs, attaining a constant drug release profile with minimal drug level fluctuations, and enhancing patient convenience by reducing administration frequency [10]. CR dosage forms are not only capable of extending the time over which drugs are released and providing constant drug levels but also with the potential of protecting therapeutic biomolecules such as peptides and proteins from enzymatic degradation in the gastrointestinal tract (GIT) [11]. CR systems can also be formulated to target the delivery of APIs to the desired site of action [12, 13].

Aside from the substantial need for CR formulations in drug delivery and the potential advantages they offer, the reproducibility and cost of equipment and techniques needed for the preparation of CR dosage forms on a large scale present a major obstacle towards the widespread production of CR delivery systems in pharmaceutical manufacturing. **Figure 3** summarizes the key factors which require to be taken into account when optimizing a new CR dosage form.

2.1. Design of CR systems

2.1.1. APIs

There are several criteria and properties that should be taken into consideration in the proposed use of an API when designing a controlled release formulation [14, 15].

- The elimination half-life of the drug should be short. Drugs with long half-lives, greater than 8 h, provide a sustained release profile without the need to be formulated in a controlled release system.
- Drugs with a wide therapeutic window are better candidates since higher doses need to be incorporated in CR formulations and dose dumping could occur.
- The absorption rate of a candidate drug should be high to make sure that the release of drug from the CR delivery system is the rate determining step, not the absorption rate.
- Drugs which exhibit high protein binding are retained in the plasma for a long time; thus, they do not require a CR delivery system.
- Drugs that undergo extensive first pass metabolism are poor candidates for CR, since releasing the drug at lower rates will decrease its bioavailability. APIs with a bioavailability index higher than 75% are preferable.

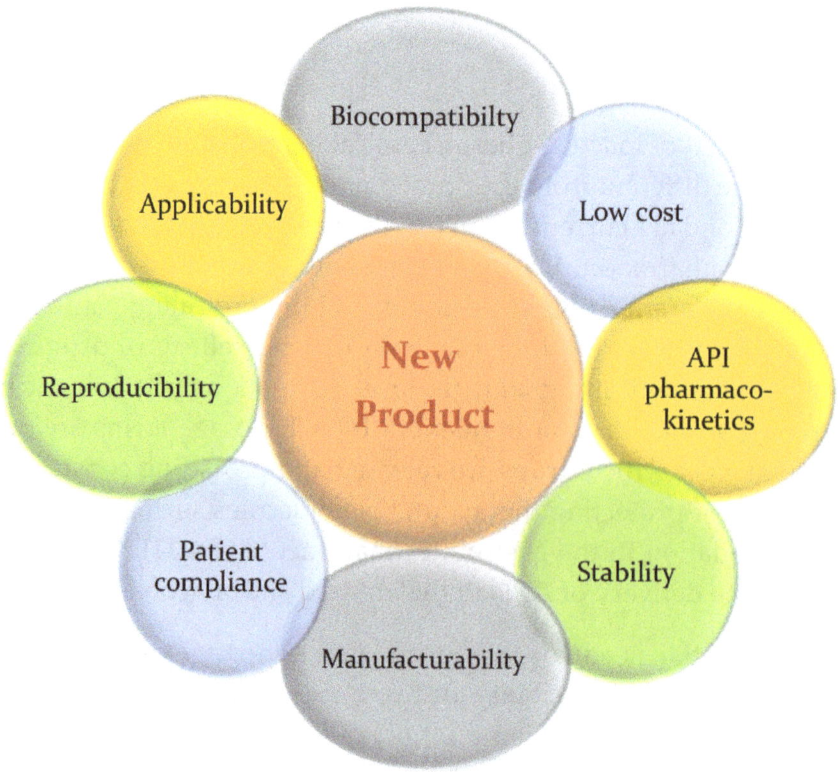

Figure 3. Key factors to be considered when developing a new dosage form in the pharmaceutical industry.

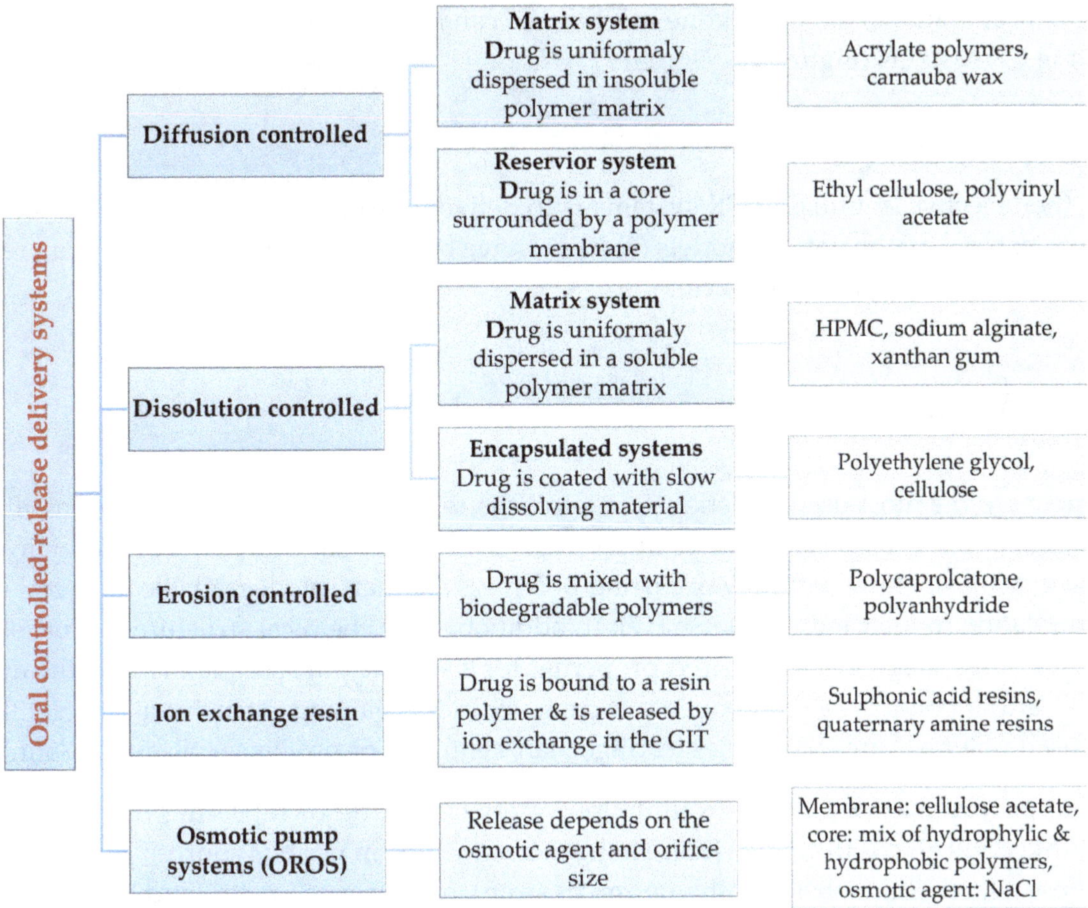

Figure 4. Classification of controlled release drug delivery systems combined with the main mechanisms involved in drug release and examples of inactive ingredients used to achieve CR.

Model	Equation	Mechanism of release
Zero order	$Q_t = Q_0 + K_0 t$	Release is independent of drug concentration within the matrix or device
First order	$\text{Ln}\, Q_t = \text{Ln}\, Q_0 + Kt$	Release is dependent on drug concentration within the matrix or device
Higuchi	$Q_t = K_H t^{1/2}$	Drug released via diffusion through an insoluble polymeric matrix
Hixon-Crowel	$Q_0^{1/3} - Q_t^{1/3} = K_{HC} t$	Drug release is dependent on drug dissolution rate in the media
Korsmeyer-Peppas	$Q_t/Q = K t^n$	This model is used when several mechanisms are involved in drug release from the system

*Where Q_0 is the initial amount of drug in the dissolution media, Q_t is the fraction released at time t, K is the rate release constant, and n is the release exponent.

Table 1. Mathematical models of drug release kinetics from CR formulations*.

2.1.2. Carriers and mechanism of drug release

Controlled drug release can be achieved by utilizing special techniques and devices. As the release of a drug from the delivery system is the rate limiting step in controlled release formulations, CR systems are classified according to the mechanism involved in drug release [3, 16].

In some preparations, more than one mechanism can be involved in the release of the API(s) from the CR systems (**Figure 4**).

2.1.3. In-vitro drug release kinetics

Since the objective of utilizing CR systems is to deliver a drug, or drugs, over a known time interval, several mathematical models (**Table 1**) have been suggested to describe drug release from the systems as a function of time.

3. Natural polysaccharides

Polymers are the most used materials to control the release of APIs. They can be classified as synthetic (silicons, polyesters and cellulose derivatives) and natural polymers (proteins and polysaccharides). Many naturally occurring polymers are inert, biodegradable, and cost-effective in relation to their industrial use [17]. In addition, their chemical structure can usually be easily modified to achieve the desired properties for a specific purpose [18]. Hence, the utilization of natural polymers as components of drug vehicles is gaining extensive attention [3]. The most used polymers are saccharides (carrageenan, cellulose) or proteins (collagen, gelatin) [19].

Natural polysaccharides are hydrophilic polymers consisting of repeating monosaccharide units linked via glycosidic bonds [20]. They are obtained from various sources, mainly vegetal (cellulose, starch), microbes (xanthan gum, dextran), crustaceans (chitin) and algae (alginate, carrageenan) [21, 22]. Depending on the identity of the constituent monomer(s), polysaccharides can be divided into homo-polysaccharides which are composed of the same repeating unit, such as cellulose, or hetero-polysaccharides which are built up from different saccharide units e.g., CS and XG [23, 24]. They can also be classified according to their ionic charge: non-polyelectrolyte (starch, cellulose), and polyelectrolyte polysaccharides. Polyelectrolytes are further sub-divided into negatively charged polymers; such as alginate and XG, or positively charged polymers, which are few in number, such as CS [25, 26].

The unique physicochemical characteristics of each polysaccharide are related to the type of monosaccharide building unit, position of the glycosidic bond, chain substitution and the overall molecular weight [25, 27]. Due to the presence of various functional groups attached to the polymer backbone (carboxyl —COOH, amine —NH2 and hydroxyl groups —OH), polysaccharides have the ability to form non-covalent bonds with a wide range of synthetic and biological molecules [28, 29]. Moreover, they can attach to body tissues and mucus layers and sustain the release of encapsulated active ingredients [13]. The aforementioned properties have attracted attention towards the usage of polysaccharides in major industries including food, agronomy, cosmetics, biochemical engineering and pharmaceutical manufacturing [30, 31].

3.1. Chitosan (CS)

CS is a linear polysaccharide produced by the *N*-deacetylation of chitin [32]. Chitin is found mainly in the exoskeleton of marine crustaceans as well as insects and fungi [33]. Glucosamine and *N*-acetyl glucosamine are the building units of CS. They are linked via β(1-4)glycosidic bonds (**Figure 5a**). The degree of acetylation and distribution of acetyl groups along the polymer

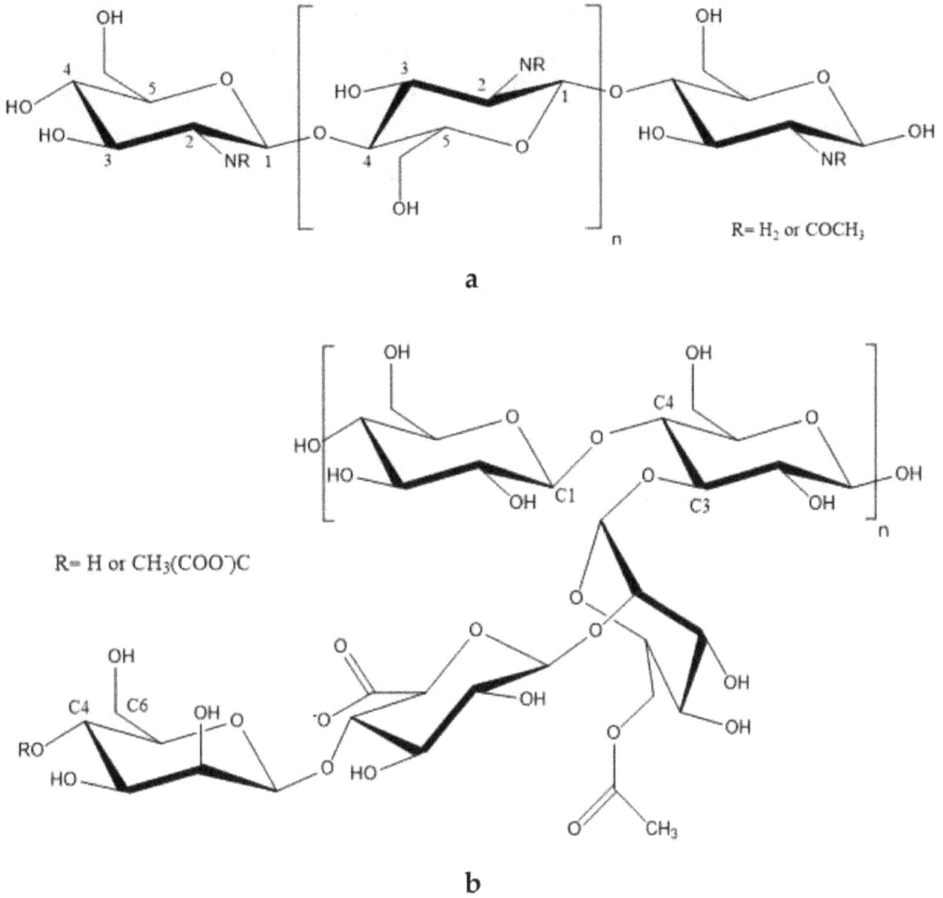

Figure 5. Schematic chemical structures of the building units of: (a) CS and (b) XG.

chain (either block or random distribution) are dependent on the duration of the deacetylation process and preparation method for CS [34, 35]. Following deacetylation of chitin, CS is (unlike chitin) soluble in acidic media. Moreover, the presence of primary amine groups leads to the unique properties of CS over all other natural polysaccharides [36]. It is the only saccharide possessing a high density positive net charge, which allows it to interact with a wide range of anionic polymers and biological molecules [32]. In addition, CS shows high mucoadhesion in the GIT which increases the residence time and enhances the permeation of active molecules [30]. Hence, CS is used commonly in the food industry, for pharmaceutical drug delivery and tissue engineering [37, 38].

3.2. Xanthan gum (XG)

XG is a branched, hetero-polysaccharide produced via microbial fermentation of the microorganism *Xanthomonas campestris* [39]. The primary unit of XG (**Figure 5b**) consists of a cellulosic backbone composed of two D-glucose units (1-4) β-linked to a side-chain of D-mannose and D-glucuronic acid units at a ratio of 2:1, respectively [40]. D-Mannose, which is connected to the main backbone, is attached to an acetyl group at O6, while approximately half of the terminal D-mannose forms a pyruvic acid group between carbons C4 and C6. This side-chain is found at the O3 atom of each alternate glucose unit on the backbone. Due to the presence of carboxylic groups in its structure, XG exhibits a net negative charge and can form complexes with cationic polymers [41]. In the last decade, the demand for XG, in industry, has been increasing at about 5–10% per annum [42]. It is used in a broad variety of industries, including cosmetics, agriculture, food, textiles and oil [43, 44]. This is due to its safety (non-toxic), desirable rheological properties, high stability over a wide range of pH and temperature, together with its high resistance against enzymatic degradation [45, 46].

4. CS and XG matrices as controlled release drug delivery systems

4.1. Advances and applications

Matrix systems, based on polyelectrolyte polysaccharides, used to retard the release of APIs have been reported in the literature and some of them have been commercialized [29, 47]. The long term instability of their corresponding preparations due to the existence of charged groups limits their application in pharmaceutical manufacturing [48]. Introducing a cross-linker, such as tripolyphosphate or glutaraldehyde, to neutralize the polymeric matrix is a necessary approach to confront such a shortcoming. Though, substituting the cross-linker with an oppositely charged copolymer aids and abbets the synergistic effect of drug release retardation. XG proved to be a potential CR drug carrier. In aqueous solutions, XG shows high viscosity and water uptake capacity encapsulating the drug inside a thick gel-like layer which hinders the release of the incorporated drug. XG has been used alone and with other polymers such as HPMC, karaya gum, guar gum, and polyvinylpyrrolidone (PVP), or ethyl cellulose [49–53]. Formulated matrices were able to sustain the release of caffeine, azithromycin, ibuprofen and propranolol HCl. XG demonstrates a high capability of generating a near zero drug release profile.

Being the only known positively charged natural polymer in aqueous solutions, CS has been extensively investigated as a potential drug vehicle. CS has the ability to preserve the stability of active biomolecules, namely insulin, and enhance their absorption from the GIT [54–56]. CS was mixed with various polymers with the aim of modifying the release of active ingredients, protect genes and therapeutic peptides in the GIT and improve their permeation across the intestinal epithelium and to immobilize antibodies [57–59]. Alginate, carrageenan, pectin, hyaluronic acid and XG are amongst the many natural polymers to be used with CS [47, 60, 61].

The combination of CS and XG was first used in the form of a polyelectrolyte complex (PEC) hydrogel [62]. The hydrogels formed displayed pH dependent swelling behavior and addressed the possibility of developing a gastrointestinal drug delivery system. PECs are formed due to the attractive ionic forces between the positively charged amino groups in CS and the negatively charged carboxyl groups in XG [63]. Therefore, features of the PECs produced can be controlled by manipulating the physicochemical properties of each polymer [64]. Molecular weight, degree of acetylation (DA) of CS, and pyruvic acid content in XG are amongst the most crucial factors to be addressed [47, 65, 66]. Complexation conditions (including concentration of each polymer, mixing ratios, and pH) have a significant influence on the behavior and stability of the resulting PEC [67]. The combination of CS and XG has been extensively studied as a platform for CR drug delivery, resulting in many patents and the publication of research articles.

4.1.1. Patents on CS-XG based controlled release drug delivery system

CS-XG hydrogels have been studied in order to immobilize biological materials. This is due to their insolubility and high stability in acidic medium allowing the system to preserve biological activity and release the materials at neutral pH. CS-XG hydrogels served as a promising candidate for sustained release dosage forms [68]. CS-XG hydrogels were capable of stabilizing and controlling the release of highly sensitive active ingredients such as vitamins, amino acids, nucleic acids and polypeptides when applied topically or orally as dietary supplements [69]. Moreover, the prepared hydrogels were shown to play a role in regulating the dissolution rate of poorly water-soluble drugs as disclosed by the patent WO 2002003962 where fenofibrate, ursodeoxycholic acid, nifedipine and indomethacin were used as models of poorly water-soluble APIs [70].

Tablets comprising CS and XG as a hydrophilic matrix for oral controlled release were first presented by Badwan et al. [71]. A wide range of basic drugs were tested (e.g. ambroxol, salbutamol, metoclopramide, anti-infective, non-steroidal and anti-inflammatory agents (NSAIDs)). Tablets formulated using CS and XG have been used to deliver basic APIs in a controlled release pattern. The drug to polymer ratio used was 1:3, respectively and the preferred XG to CS ratio was 1:1. When the system was studied in-vivo on human volunteers, it produced constant serum levels of ambroxol over a period of 24 h. This study paved the way for further research on approaches and mechanisms involving tablet formulations based on the foregoing combination. A tablet dosage form based on CS-XG for the treatment of hypercholesterolemia was prepared [72]. A combination of lycopene and *Monascus purpureus* were used as active ingredients. When testing the preparation on human volunteers, a significant decrease in the plasma levels of cholesterol, LDL and triglycerides was reported. Moreover, HDL values were reported to be increased.

4.1.2. Research articles on CS-XG based controlled release drug delivery system

The applicability of CS-XG combinations to a wide range of dosage forms with different routes of administration has been investigated. Examples of the preparations reported in the literature together with a brief description of preparation methods, application and examples of incorporated APIs are summarized in **Table 2**.

4.2. Ionic interaction between XG and CS

In spite of the significant amount of research work conducted on XG and CS based matrices, a lack of understanding of the nature of the interaction between the two polymers and their behavior at the molecular level still exists. It was suggested that physico-chemical conditions in the stomach are an ideal environment for the formation of insoluble gels between the two polymers, which retards the release of APIs resulting in a sustained drug release profile [87]. Moreover, in vitro residence time evaluation on porcine mucin and in vivo studies using sheep models have addressed the bioadhesive nature of CS-XG matrices [88].

In order to acquire an understanding of the interaction between XG and CS and factors governing it, a molecular dynamics simulation (MDs) study was conducted by Dadou et al. [89]. The contribution of the DA of CS and protonation was evaluated. The resulting trajectories

Dosage form	Preparation method	Application	Incorporated ingredient
Hydrogels	Solution mixing under heat	Drug delivery, tissue engineering, immobilization of biological active materials	Probiotics [73], enzymes [64]
Films	Solution casting	Drug delivery, tissue engineering, food industry	Wound healing [74], amoxicillin [75], scaffolds [76]
Capsules	Complex coacervation, encapsulation of physically mixed powder	Drug delivery	Theophylline [77], ciprofloxacin HCl [78]
Beads	Extrusion-dripping technique, complex coacervation mechanism	Drug delivery, immobilization of biological active materials	Probiotics [79], glipizide [80], antibodies [59]
Microspheres	Spray drying, ionotropic gelation method	Drug delivery	Meclizine HCl [81]
Micro-emulsions	Homogenization with oil phase	Drug delivery	Progesterone [82]
Liposomes (chitosomes)	Thin film hydration method, spray drying	Drug delivery	Rifampicin [83]
Cryogel	Freeze-drying	Immobilization of biological active materials	Enzymes [84]
Tablets	Direct compression, granulation, hot melt extrusion	Drug delivery	Metformin HCl [85], terbutaline sulfate [48], propranolol [86]

Table 2. Main applications of CS-XG based matrices.

and binding free energy calculations revealed that electrostatic forces (polar interactions, ΔE_{ele}) are the driving force for the interaction, and that the interaction occurs regardless of the DA and state of protonation of CS (free energy values are negative for all complexes). Protonation of CS molecules increases their penetration between the branched chains of XG and produces more stable complexes with lower free binding energy (**Table 3**). Intermolecular interactions (Van der Waals) showed a positive contribution to the formation of CS-XG PECs. This can be explained by the presence of a large number of hydroxyl groups along the chains of the polymers which can induce an instantaneous dipole attraction with the surrounding atoms. High positive solvation free energy (ΔG_{solv}) values justify the resultant insoluble PECs upon complexation between CS and XG in the laboratory. ΔG_{solv} increases with protonation, reaching a maximum value when CS is fully protonated, indicating a higher extent of interaction with XG.

4.2.1. Mixing ratio

Since the interaction between CS and XG is electrostatically driven, the properties of the resultant PECs can be modified by controlling the net charge density. This can be achieved either by altering the mixing ratios or the initial concentrations of the polymeric solutions. Films of CS and XG were prepared and examined by scanning electron microscope (SEM) for additional information relating to the behavior and the interaction between the two polymers in

CS	ΔE_{ele}	ΔE_{vdW}	ΔG_{sol}	ΔG
0% P*, 0% DA	−21.290	−14.03	21.890	−13.43
50% P, 0% DA	−227.53	−24.47	222.77	−29.22
100% P, 0% DA	−419.95	−23.27	412.57	−30.65
0% P, 50% DA	−25.080	−21.12	28.460	−17.74
50% P, 50% DA	−232.68	−23.96	227.36	−29.28
0% P, 100% DA	−25.070	−25.79	30.150	−20.71

Values presented are in kcal/mol.*P represents state of protonation.

Table 3. Binding free energy calculations for XG-CS complexes.

aqueous solutions at different mixing ratios [90]. SEM images (**Figure 6**) show the rough surface of CS, whilst XG films produce a smooth surface. Combining the two polymers resulted in a pronounced alteration in the surface morphology of the films. The resulting PECs form irregular and fibrous surfaces with a porous structure. PECs at a mixing ratio of 1:1 (w/v %) showed a dramatic change in the surface structure and it is suggested that they represent the maximum interaction between the two polymers.

4.2.2. *Initial concentration of XG*

Argin-Soysal et al., studied the effect of polymer solution concentration on the formation of stable capsules and their subsequent swelling behavior [67]. The initial concentration of the XG solution was found to be the determining factor in relation to complexation density, more than CS. This is due its high molecular weight and the highly viscous hydrogels it forms when in contact with water [91]. The physical cross-linking between XG and CS was complete when the concentration of XG was 1.5%, regardless of other experimental conditions. Consequently, the degree of swelling was shown to be dictated by the initial aqueous concentration of XG.

4.2.3. *pH and initial concentration of CS solutions*

Dumitri et al., found that the pH of CS solutions has a moderate effect on the extent of interaction between XG and CS [65]. PECs where readily obtained within a wide range of pH (3.6–8.0). At lower pH values, the carboxyl groups of XG become protonated (uncharged) while the amine groups in CS are fully charged; hence, the interaction between CS and XG is

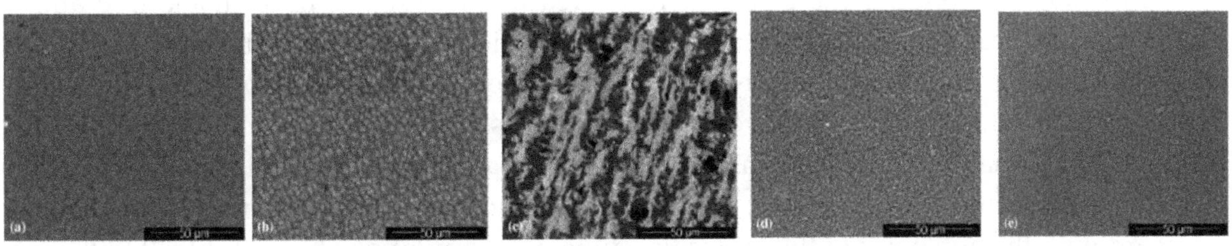

Figure 6. SEM images at magnification power of x2000 of: (a) CS, (b) CS-XG (2:1), (c) CS-XG (1:1), (d) CS-XG (1:2) and (e) XG films, from Eftaiha et al. [90].

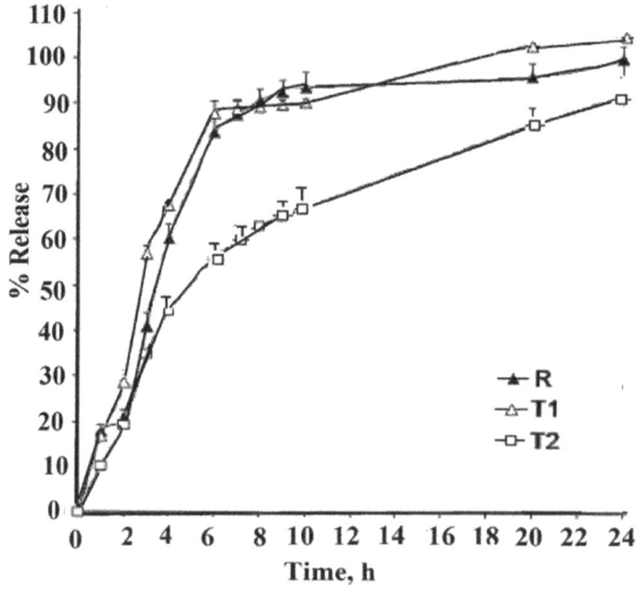

Figure 7. In-vitro release of ambroxol HCl from: (R) reference product, prepared tablets at a P:D of (T1) 1:1, and (T2) 3:1, as reported by Al Remawi et al. [94].

impeded and reduced drug retardation occurs. The effect of pH was more pronounced when preparing low concentration solutions of CS. A considerable increase in the degree of swelling with the pH of solutions at CS concentrations of 0.65–0.7% (w/v) occurs.

4.2.4. *Molecular weight (Mw) of CS*

The swelling capacity of CS-XG based PECs were found to be influenced by the Mw of CS [68]. Lower water retention capacity was achieved by using a higher Mw of CS. The absorption of water increased noticeably with around 1000% weight gain at lower Mw of CS. The increase in water absorption causes the formation of more PEC layers which results in potentially more drug retardation. The aforementioned claim was supported by the slow release of diclofenac sodium from low molecular weight CS tablets (13 and 30 kDa) [92]. AlAkayleh et al., found that the release rate of terbutaline sulfate from XG-low molecular weight CS tablets (viscosity 38 mPa s) was slower than XG-high molecular weight CS (70 mPa s) [48].

4.2.5. *DA of CS*

The PEC between XG and CS is formed due to the electrostatic attraction between oppositely charged groups. Increasing the DA content decreases the number of available free amine groups that are readily protonated. In addition, the rigidity of CS chains increased with DA owing to strong intramolecular hydrogen bonds dictated by amide groups [93]. As a consequence, the extent of interaction between the polymers is reduced. Release of propranolol HCl from an CS-XG matrix was studied as a function of the degree of deacetylation (DDA) of CS [86]. Release of drug from the matrix was faster from the acetylated form of CS. This result is in accord with the outcomes of the molecular dynamics simulation study (**Table 3**) [89].

4.2.6. Ionic strength of solution

Adding ionic species to the solution resulted in a large decrease in water uptake of CS-XG PECs. Competition takes place between free ions and water molecules for the hydroxyl groups of the polymers and reduces the hydration of CS and XG chains. Thus, the degree of swelling is lower which, in turn, will have an effect on the drug retardation capability of the system [63].

4.2.7. Concentration of incorporated API

Hydrophilic matrices need to be used at high polymer to drug ratios in order to exert their effect in sustaining the release of APIs [27]. Thus, their application is restricted to low strength drugs, as addressed by Badwan et al. [71]. Al Remawi et al. studied the effect of polymer to drug ratio (P:D) on the release of ambroxol HCl from CS-XG based tablets [94]. The release rate of ambroxol was highly dependent on the P:D ratio. Greater retardation of drug release was attained at higher polymer ratios (**Figure 7**).

5. Tablets comprising CS-XG

Oral solid dosage forms remain the most favorable choice to deliver APIs. The main reason is that they preserve the physicochemical stability of chemical entities more than liquid forms [95]. Additionally, tablets offer advantages for both manufacturers and patients which include ease of handling, low production cost, dose precision and self-administration capability [96].

Utilization of CS as an efficient excipient in tablet formulation is gradually increasing. CS powder exhibits a high surface area and porosity [93]. It produces tablets with high tensile strength that form a network-like structure when examined by microscopy [97]. The aim of using a combination of polymers, as tablet excipients, is to enhance compressibility and flowability properties. Furthermore, a polymeric mixture can increase the overall retardation performance of the system. CS-XG based tablets were formulated by compression of one layer and multi-layers; they were used solely or with other polymers such as galactomannan, seed gum or β-cyclodextrin [48, 85, 86, 98]. Moreover, they were used in immediate release, floating mucoadhesive and buccal tablets [99, 100]. According to Badwan et al., combining XG with CS has the advantage of improving the mechanical properties of both polymers [93].

5.1. Tablet preparation methods

5.1.1. Direct compression

Direct compression is a technique for formulating tablets which limits the use of solvents, temperature and equipment. It is the first choice whenever the API and inactive materials are suitable for direct compression and are stable at high pressure [101]. Powders of both active and inactive ingredients are mixed homogeneously, then sieved to the desired particle size. Finally, the prepared blend is compressed using a tablet press machine at a predetermined pressure [102]. CS-XG based tablets prepared via direct compression, showed a high potential towards sustaining the release of terbutaline sulfate and ambroxol [48, 103].

5.1.2. Dry granulation

Dry granulation is utilized to improve compaction properties of ingredients. It can influence flowability, stability, content uniformity of the powders and enhance the bioavailability of the API. This is attained by increasing the particle size of powder materials via aggregation of particles by either roller compaction or slugging and then milling to produce granules with the desired size [104].

5.1.3. Wet granulation

Wet granulation of tablet components is usually achieved using water, ethanol or a mixture of both. Following drying at an appropriate temperature, granules are mixed with other excipients if needed, passed through a sieve and finally compressed using a press machine at a predefined pressure [105]. Wet granulation is used to produce dust free granules, enhance flowability and cohesion. Eftaiha et al., investigated the ability of CS-XG tablets prepared by wet granulation using an aqueous solution of XG 1% (w/v) to modify the release of metronidazole [87]. The preparation was able to sustain the release of metronidazole, both in-vitro and in-vivo. A mucoadhesive behavior was observed when applying the tablets on sheep duodenum.

5.1.4. Hot melt extrusion (HME)

In HME the powdered API, functional polymers and any other excipients are blended in a mortar and pestle then fed into the hopper of a single or double screw extruder. Fukuda et al., prepared CS-XG tablets using HME to study the release of chlorpheniramine maleate [106]. The processing temperature was 90°C (zone 1), 95°C (zone 2), 105°C (zone 3) and 110°C (die) with a screw speed of 15 rpm. The processing time needed for powders inside the barrel of the extruder is usually ~3–4 min. The extruded materials were then manually cut into tablets of the desired weights. Chlorpheniramine release from the prepared CS-XG tablets occurred in a sustained manner and was independent of the pH and ionic strength of the dissolution media. HME offers the advantage of continuous processing and process analytical technology (PAT) which enables quality control testing throughout the process [107].

5.2. Mechanism of drug release

Drug release from CS-XG matrices is suggested to be governed by the dissolution rate of the drug and the polymers in the media as well as the diffusion of the drug from the matrices and erosion of the polymers. The data in **Figure 8** illustrates the processes of drug release from a tablet composed of CS and XG. When the tablet is first exposed to aqueous media, an insoluble gel layer forms on the top surface of the tablet, as a result of polyelectrolyte complexation between the two charged polymers [108]. Water molecules start to penetrate this layer towards the matrix owing to the high water uptake capability of XG. [91]. Accordingly, both polymers and drug are dissolved and a rubbery hydrated region is formed (white area) [27].

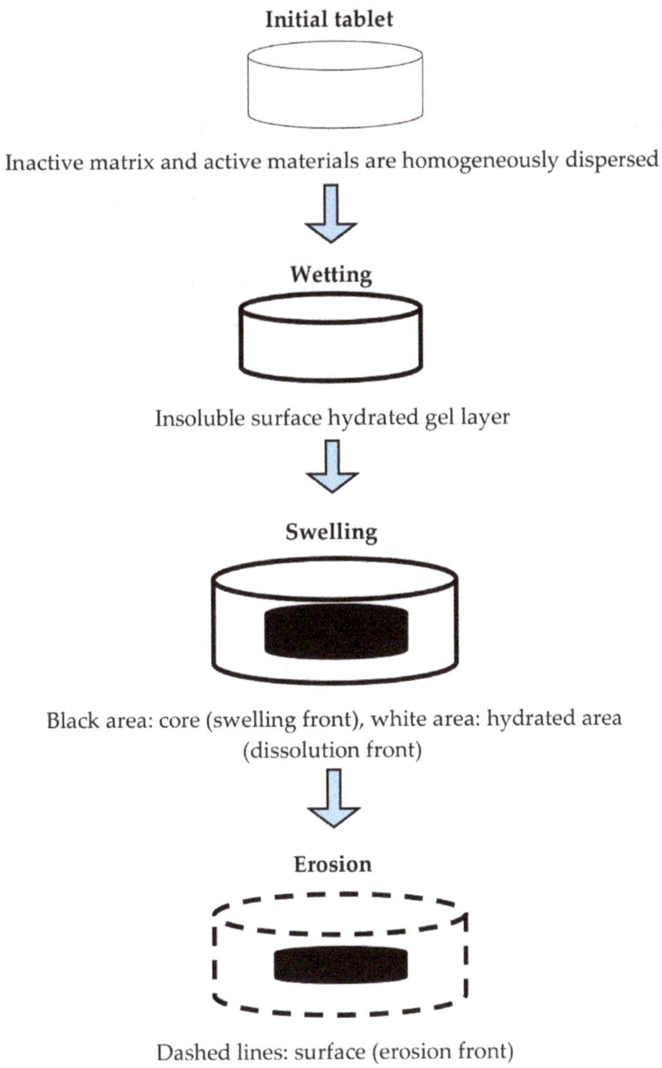

Figure 8. Schematic representation of the behavior of CS-XG tablets in an aqueous medium.

On the other hand, a non-hydrated glassy area is formed at the core of the tablet, where no water molecules reach the system (black area) [109]. As time lapses, further penetration of water molecules into the tablet occurs resulting in the polymers chains being solvated. Consequently, swelling of the matrix occurs [110]. At this stage, water molecules enter between the polymer chains, the radius of gyration of the polymers increases and the end-to-end distance of the polymer backbones also increases [111]. This phenomenon of polymer relaxation is referred to as "swelling of the matrix" [91]. As more water molecules pass into the matrix, the polymer concentration on the outer surface of the tablet decreases, losing its integrity, and starts to dissolve in the medium. This phenomenon is termed "polymer erosion" [112].

The rate of drug release from such a matrix could occur as a function of diffusion of the water molecules into the matrix, dissolution of both polymers and drug, polymer relaxation and erosion [91, 113]; this depends on the previously mentioned factors (Section 4.2) [111, 114].

6. Conclusions

Controlling the release of active ingredients is one of the fastest growing applications of CS-XG based matrices. Various drug delivery systems and newly emerging technologies have been developed in order to optimize the foregoing mixture. XG-CS matrices show a high potential towards controlling the release of a wide range of active biomolecules. The efficiency of CS-XG matrices to control the release of drugs can be reinforced by manipulating the physicochemical properties of CS and XG and the experimental conditions used. Thus, incorporation/use of expensive devices and the method of preparation can be kept to a minimum. With further optimization and the utilization of newly emerging computational and quality by design tools, relatively simple and straightforward CS-XG based matrices can be formulated as potentially universal carriers to control the release of APIs.

Conflict of interest

The authors declare no competing financial interests.

Author details

Suha M. Dadou[1], Milan D. Antonijevic[1], Babur Z. Chowdhry[1] and Adnan A. Badwan[2*]

*Address all correspondence to: dr.badwan@jpm.com.jo

1 Department of Pharmaceutical, Chemical and Environmental Science, Faculty of Engineering and Science, University of Greenwich, Kent, UK

2 Research and Innovation Centre, The Jordanian Pharmaceutical Manufacturing Co. (Plc), Naor, Jordan

References

[1] Tibbitt MW, Dahlman JE, Langer R. Emerging frontiers in drug delivery. Journal of the American Chemical Society. 2016;**138**(3):704-717. DOI: 10.1021/jacs.5b09974

[2] Bhowmik D, Gopinath H, Kumar BP, Duraivel S, Kumar KPS. Controlled release drug delivery systems. Pharma Innovation. 2012;**1**(10):24-32

[3] Rosen H, Abribat T. The rise and rise of drug delivery. Nature Reviews. Drug Discovery. 2005;**4**(5):381-385. DOI: 10.1038/nrd1721

[4] Perrie Y, Rades T. Chapter 1— Controlling drug delivery. In: Pharmaceutics—Drug Delivery and Targeting. 2nd ed. PhP; 2012. pp. 1-24

[5] Kushal M, Monali M, Durgavati M, Mittal P, Umesh S, Pragna S. Oral controlled release drug delivery system: An overview. International Research Journal of Pharmacy. 2013;**4**(3):70-76. DOI: 10.7897/2230-8407.04312

[6] Mrsny RJ. Oral drug delivery research in Europe. Journal of Controlled Release. 2012;**161**(2):247-253. DOI: 10.1016/j.jconrel.2012.01.017

[7] Park K. Controlled drug delivery systems: Past forward and future back. Journal of Controlled Release. 2014;**190**:3-8. DOI: 10.1016/j.jconrel.2014.03.054

[8] Crommelin DJA, Florence AT. Towards more effective advanced drug delivery systems. International Journal of Pharmaceutics. 2013;**454**(1):496-511. DOI: 10.1016/j.ijpharm.2013.02.020

[9] Hoffman AS. The origins and evolution of "controlled" drug delivery systems. Journal of Controlled Release. 2008;**132**(3):153-163. DOI: 10.1016/j.jconrel.2008.08.012

[10] Lee PI, Li JX. Evolution of oral controlled release dosage forms. In: Wen H, Park K, editors. Oral Controlled Release Formulation Design and Drug Delivery: Theory to Practice. Hoboken, NJ: John Wiley & Sons, Inc.; 2010. pp. 21-31. DOI: 10.1002/9780470640487.ch2

[11] Patel G, Misra A. Oral delivery of proteins and peptides: Concepts and applications. In: Ambikanandan M, editors. Challenges in Delivery of Therapeutic Genomics and Proteomics. London: Elsevier; 2011. pp. 481-529. DOI: 10.1016/B978-0-12-384964-9.00010-4

[12] Siegel RA, Rathbone MJ. Overview of controlled release mechanisms. In: Siepmann J, Siegel RA, Rathbone MJ, editors. Fundamentals and Applications of Controlled Release Drug Delivery. New York: Springer; 2012. pp. 19-43. DOI: 10.1007/978-1-4614-0881-9_2

[13] Sinha VR, Kumria R. Polysaccharides in colon-specific drug delivery. International Journal of Pharmaceutics. 2001;**224**(1-2):19-38. DOI: 10.1016/S0378-5173(01)00720-7

[14] Chrzanowski F. Preformulation considerations for controlled release dosage forms: Part I selecting candidates. AAPS PharmSciTech. 2008;**9**(2):635-638. DOI: 10.1208/s12249-008-9068-2

[15] Mahler S, Roy I. Advances in drug delivery. Journal of Chemical Technology and Biotechnology. 2015;**90**(7):1167-1168. DOI: 10.1002/jctb.4689

[16] Holowka EP, Bhatia SK. Controlled-release systems. In: Drug Delivery: Materials Design and Clinical Perspective. New York: Springer; 2014. pp. 7-62. DOI: 10.1007/978-1-4939-1998-7

[17] Hovgaard L, Brondsted H. Current applications of polysaccharides in colon targeting. Critical Reviews in Therapeutic Drug Carrier Systems. 1996;**13**(3-4):185-223

[18] Posocco B, Dreussi E, De Santa J, et al. Polysaccharides for the delivery of antitumor drugs. Materials (Basel). 2015;**8**(5):2569-2615. DOI: 10.3390/ma8052569

[19] Malafaya PB, Silva GA, Reis RL. Natural-origin polymers as carriers and scaffolds for biomolecules and cell delivery in tissue engineering applications. Advanced Drug Delivery Reviews. 2007;**59**(4-5):207-233. DOI: 10.1016/j.addr.2007.03.012

[20] Zhang N, Wardwell PR, Bader RA. Polysaccharide-based micelles for drug delivery. Pharmaceutics. 2013;**5**(2):329-352. DOI: 10.3390/pharmaceutics5020329

[21] Raemdonck K, Martens TF, Braeckmans K, Demeester J, De Smedt SC. Polysaccharide-based nucleic acid nanoformulations. Advanced Drug Delivery Reviews. 2013;**65**(9):1123-1147. DOI: 10.1016/j.addr.2013.05.002

[22] Yang J, Han S, Zheng H, Dong H, Liu J. Preparation and application of micro/nanoparticles based on natural polysaccharides. Carbohydrate Polymers. 2015;**123**:53-66. DOI: 10.1016/j.carbpol.2015.01.029

[23] Debele TA, Mekuria SL, Tsai HC. Polysaccharide based nanogels in the drug delivery system: Application as the carrier of pharmaceutical agents. Materials Science and Engineering: C. 2016;**68**:964-981. DOI: 10.1016/j.msec.2016.05.121

[24] Gulrez SKH, Al-Assaf S, Phillips GO. Hydrogels: Methods of preparation, characterisation and applications. In: Carpi A, editors. Progress in Molecular and Environmental Bioengineering—From Analysis and Modeling to Technology Applications. Rijeka, Croatia: InTech; 2011. pp. 117-150. DOI: 10.5772/24553

[25] Liu Z, Jiao Y, Wang Y, Zhou C, Zhang Z. Polysaccharides-based nanoparticles as drug delivery systems. Advanced Drug Delivery Reviews. 2008;**60**(15):1650-1662. DOI: 10.1016/j.addr.2008.09.001

[26] Alvarez-Lorenzo C, Blanco-Fernandez B, Puga AM, Concheiro A. Crosslinked ionic polysaccharides for stimuli-sensitive drug delivery. Advanced Drug Delivery Reviews. 2013;**65**(9):1148-1171. DOI: 10.1016/j.addr.2013.04.016

[27] Maderuelo C, Zarzuelo A, Lanao JM. Critical factors in the release of drugs from sustained release hydrophilic matrices. Journal of Controlled Release. 2011;**154**(1):2-19. DOI: 10.1016/j.jconrel.2011.04.002

[28] Rinaudo M. Non-covalent interactions in polysaccharide systems. Macromolecular Bioscience. 2006;**6**(8):590-610. DOI: 10.1002/mabi.200600053

[29] Shelke NB, James R, Laurencin CT, Kumbar SG. Polysaccharide biomaterials for drug delivery and regenerative engineering. Polymers for Advanced Technologies. 2014;**25**(5):448-460. DOI: 10.1002/pat.3266

[30] Klein S. Polysaccharides in oral drug delivery—Recent applications and future perspectives. In: Polysaccharide Materials: Performance by Design. Vol. 1017. ACS Symposium Series. American Chemical Society; 2009. pp. 1-13. DOI: 10.1021/bk-2009-1017.ch001

[31] Goodarzi N, Varshochian R, Kamalinia G, Atyabi F, Dinarvand R. A review of polysaccharide cytotoxic drug conjugates for cancer therapy. Carbohydrate Polymers. 2013;**92**(2):1280-1293. DOI: 10.1016/j.carbpol.2012.10.036

[32] Kushwaha Swatantra KS, Rai Awani K, Satyawan S. Chitosan: A platform for targeted drug delivery. International Journal PharmTech, Research. 2010;**2**(4):2271-2282

[33] Boonsongrit Y, Mitrevej A, Mueller BW. Chitosan drug binding by ionic interaction. European Journal of Pharmaceutics and Biopharmaceutics. 2006;**62**(3):267-274. DOI: 10.1016/j.ejpb.2005.09.002

[34] Lu B, Wang C-F, Wu D-Q, Li C, Zhang X-Z, Zhuo R-X. Chitosan based oligoamine polymers: Synthesis, characterization, and gene delivery. Journal of Controlled Release. 2009;**137**(1):54-62. DOI: 10.1016/j.jconrel.2009.03.004

[35] Chang KLB, Tsai G, Lee J, Fu WR. Heterogeneous N-deacetylation of chitin in alkaline solution. Carbohydrate Research. 1997;**303**(3):327-332. DOI: 10.1016/S0008-6215(97)00179-1

[36] Brugnerotto J, Desbrieres J, Heux L, Mazeau K, Rinaudo M. Overview on structural characterization of chitosan molecules in relation with their behavior in solution. In: Macromolecular Symposia. WILEY-VCH Verlag GmbH; 2001;**168**(1):1-20. DOI: 10.1002/1521-3900(200103)168:1<1::AID-MASY1>3.0.CO;2-W

[37] Amidi M, Mastrobattista E, Jiskoot W, Hennink WE. Chitosan-based delivery systems for protein therapeutics and antigens. Advanced Drug Delivery Reviews. 2010;**62**(1):59-82. DOI: 10.1016/j.addr.2009.11.009

[38] Qinna NA, Akayleh FT, Al Remawi MM, Kamona BS, Taha H, Badwan AA. Evaluation of a functional food preparation based on chitosan as a meal replacement diet. Journal of Functional Foods. 2013;**5**(3):1125-1134. DOI: 10.1016/j.jff.2013.03.009

[39] Christensen BE, Smidsrød O. Hydrolysis of xanthan in dilute acid: Effects on chemical composition, conformation, and intrinsic viscosity. Carbohydrate Research. 1991;**214**(1):55-69. DOI: 10.1016/S0008-6215(00)90530-5

[40] Arendt O, Kulicke W-M. Determination of the viscoelastic properties of a homologous series of the fermentation polymer xanthan gum. Die Angewandte Makromolekulare Chemie. 1998;**259**(1):61-67. DOI: 10.1002/(SICI)1522-9505(19981001)259:1<61::AID-APMC61>3.0.CO;2-V

[41] García-Ochoa F, Santos VE, Casas JA, Gómez E. Xanthan gum: Production, recovery, and properties. Biotechnology Advances. 2000;**18**(7):549-579. DOI: 10.1016/S0734-9750(00)00050-1

[42] Lo YM, Yang ST, Min DB. Effects of yeast extract and glucose on xanthan production and cell growth in batch culture of Xanthomonas campestris. Applied Microbiology and Biotechnology. 1997;**47**(6):689-694. DOI: 10.1007/s002530050996

[43] Benny IS, Gunasekar V, Ponnusami V. Review on application of xanthan gum in drug delivery. International Journal PharmTech, Research. 2014;**6**(4):1322-1326

[44] Faria S, De Oliveira Petkowicz CL, De Morais SAL, et al. Characterization of xanthan gum produced from sugar cane broth. Carbohydrate Polymers. 2011;**86**(2):469-476. DOI: 10.1016/j.carbpol.2011.04.063

[45] Bueno VB, Petri DFS. Xanthan hydrogel films: Molecular conformation, charge density and protein carriers. Carbohydrate Polymers. 2014;**101**(1):897-904. DOI: 10.1016/j.carbpol.2013.10.039

[46] Leela JK, Sharma G. Studies on xanthan production from Xanthomonas campestris. Bioprocess Engineering. 2000;**23**(6):687-689. DOI: 10.1007/s004499900054

[47] Luo Y, Wang Q. Recent development of chitosan-based polyelectrolyte complexes with natural polysaccharides for drug delivery. International Journal of Biological Macromolecules. 2014;**64**:353-367. DOI: 10.1016/j.ijbiomac.2013.12.017

[48] Al-Akayleh F, Al Remawi M, Rashid I, Badwan A. Formulation and In vitro assessment of sustained release terbutaline sulfate tablet made from binary hydrophilic polymer mixtures. Pharmaceutical Development and Technology. 2013;**18**(5):1204-1212. DOI: 10.3109/10837450.2011.620968

[49] Munday DL, Cox PJ. Compressed xanthan and karaya gum matrices: Hydration, erosion and drug release mechanisms. International Journal of Pharmaceutics. 2000;**203**(1-2): 179-192. DOI: 10.1016/S0378-5173(00)00444-0

[50] Santos H, Veiga F, Pina ME, Sousa JJ. Compaction, compression and drug release properties of diclofenac sodium and ibuprofen pellets comprising xanthan gum as a sustained release agent. International Journal of Pharmaceutics. 2005;**295**(1-2):15-27. DOI: 10.1016/j.ijpharm.2005.01.014

[51] Mughal MA, Iqbal Z, Neau SH. Guar gum, xanthan gum, and HPMC can define release mechanisms and sustain release of propranolol hydrochloride. AAPS PharmSciTech. 2011;**12**(1):77-87. DOI: 10.1208/s12249-010-9570-1

[52] Talukdar MM, Plaizier-Vercammen J. Evaluation of xanthan cum as a hydrophilic matrix for controlled-release dosage form preparations. Drug Development and Industrial Pharmacy. 1993;**19**(9):1037-1046. DOI: 10.3109/03639049309062999

[53] Verhoeven E, Vervaet C, Remon JP. Xanthan gum to tailor drug release of sustained-release ethylcellulose mini-matrices prepared via hot-melt extrusion: In vitro and in vivo evaluation. European Journal of Pharmaceutics and Biopharmaceutics. 2006;**63**(3):320-330. DOI: 10.1016/j.ejpb.2005.12.004

[54] Elsayed A, Al-Remawi M, Qinna N, Farouk A, Al-Sou'od KA, Badwan AA. Chitosan–sodium lauryl sulfate nanoparticles as a carrier system for the in vivo delivery of oral insulin. AAPS PharmSciTech. 2011;**12**(3):958-964. DOI: 10.1208/s12249-011-9647-5

[55] Rekha MR, Sharma CP. Synthesis and evaluation of lauryl succinyl chitosan particles towards oral insulin delivery and absorption. Journal of Controlled Release. 2009;**135**(2):144-151. DOI: 10.1016/j.jconrel.2009.01.011

[56] Elsayed A, Remawi MA, Qinna N, Farouk A, Badwan A. Formulation and characterization of an oily-based system for oral delivery of insulin. European Journal of Pharmaceutics and Biopharmaceutics. 2009;**73**(2):269-279. DOI: 10.1016/j.ejpb.2009.06.004

[57] Bhattarai N, Gunn J, Zhang M. Chitosan-based hydrogels for controlled, localized drug delivery. Advanced Drug Delivery Reviews. 2010;**62**(1):83-99. DOI: 10.1016/j.addr.2009.07.019

[58] Chen MC, Mi FL, Liao ZX, et al. Recent advances in chitosan-based nanoparticles for oral delivery of macromolecules. Advanced Drug Delivery Reviews. 2013;**65**(6):865-879. DOI: 10.1016/j.addr.2012.10.010

[59] Albarghouthi M, Fara DA, Saleem M, El-Thaher T, Matalka K, Badwan A. Immobilization of antibodies on alginate-chitosan beads. International Journal of Pharmaceutics. 2000;**206**(1-2):23-34. DOI: 10.1016/S0378-5173(00)00470-1

[60] Li L, Wang L, Shao Y, Ni R, Zhang T, Mao S. Drug release characteristics from chitosan-alginate matrix tablets based on the theory of self-assembled film. International Journal of Pharmaceutics. 2013;**450**(1-2):197-207. DOI: 10.1016/j.ijpharm.2013.04.052

[61] Volod'Ko AV, Davydova VN, Chusovitin E, et al. Soluble chitosan-carrageenan polyelectrolyte complexes and their gastroprotective activity. Carbohydrate Polymers. 2014;**101**(1):1087-1093. DOI: 10.1016/j.carbpol.2013.10.049

[62] Chu CH, Sakiyama T, Yano T. Ph-sensitive swelling of a polyelectrolyte complex gel prepared from xanthan and chitosan. Bioscience, Biotechnology, and Biochemistry. 1995;**59**(4):717-719. DOI: 10.1080/bbb.59.717

[63] Martínez-Ruvalcaba A, Chornet E, Rodrigue D. Viscoelastic properties of dispersed chitosan/xanthan hydrogels. Carbohydrate Polymers. 2007;**67**(4):586-595. DOI: 10.1016/j.carbpol.2006.06.033

[64] Dumitriu S, Chornet E. Immobilization of xylanase in chitosan-xanthan hydrogels. Biotechnology Progress. 1997;**13**:539-545. DOI: 10.1021/bp970059i

[65] Magnin D, Dumitriu S, Chornet E. Immobilization of enzymes into a polyionic hydrogel: ChitoXan. Journal of Bioactive and Compatible Polymers. 2003;**18**(5):355-373. DOI: 10.1177/088391103038375

[66] Qinna AN, Karwi GQ, Al-Jbour N, et al. Influence of molecular weight and degree of deacetylation of low molecular weight chitosan on the bioactivity of oral insulin preparations. Marine Drugs. 2015;**13**(4):1710-1725. DOI: 10.3390/md13041710

[67] Argin-Soysal S, Kofinas P, Lo YM. Effect of complexation conditions on xanthan–chitosan polyelectrolyte complex gels. Food Hydrocolloids. 2009;**23**(1):202-209. DOI: https://doi.org/10.1016/j.foodhyd.2007.12.011

[68] Dumitriu S, Chornet E, Vidal P. Polyionic Insoluble Hydrogels Comprising Xanthan and Chitosan. U.S. Patent No. 5,620,706. Washington, DC. April 15, 1997. http://www.google.com/patents/US5620706

[69] Chornet E, Dumitriu S. Polyionic hydrogels based on xanthan and chitosan for stabilising and controlled release of vitamins. January 2000. http://www.google.com/patents/WO2000004086A1?cl=en

[70] Chornet E, Ishizawa C, Dumitriu S. Drug delivery system for poorly water soluble drugs. January 2002. https://www.google.ch/patents/WO2002003962A2?cl=en

[71] Badwan AA, Al-Remawi M, Salem M. Universal controlled-release composition comprising chitosan. Eur Pat EP1512394. 2008

[72] Mangiapane A. Combination of substances for the treatment of patients with hypercholesterolemia and related disorders. May 2012. https://www.google.ch/patents/EP2455073A1?cl=en

[73] Chu C-H, Kumagai H, Nakamura K. Application of polyelectrolyte complex gel composed of xanthan and chitosan to the immobilization of corynebacterium glutamicum. Journal of Applied Polymer Science. 1996;**60**(7):1041-1047. DOI: 10.1002/(SICI)1097-4628(19960516)60:7<1041::AID-APP15>3.0.CO;2-3

[74] Bellini MZ, Caliari-Oliveira C, Mizukami A, et al. Combining xanthan and chitosan membranes to multipotent mesenchymal stromal cells as bioactive dressings for dermo-epidermal wounds. Journal of Biomaterials Applications. 2014;**29**(8):1155-1166. DOI: 10.1177/0885328214553959

[75] Thakur A, Monga S, Wanchoo RK. Sorption and drug delease studies from semi-interpenetrating polymer networks of chitosan and xanthan gum. Chemical and Biochemical Engineering Quarterly. 2014;**28**(1):105-115

[76] Veiga IG, Moraes ÂM. Study of the swelling and stability properties of chitosan-xanthan membranes. Journal of Applied Polymer Science. 2012;**124**(Suppl. 1):E154-E160. DOI: 10.1002/app.35526

[77] Popa N, Novac O, Profire L, Lupusoru CE, Popa MI. Hydrogels based on chitosan-xanthan for controlled release of theophylline. Journal of Materials Science. Materials in Medicine. 2010;**21**(4):1241-1248. DOI: 10.1007/s10856-009-3937-4

[78] Verma A, Bansal A, Ghosh A, Pandit J. Low molecular mass chitosan as carrier for a hydrodynamically balanced system for sustained delivery of ciprofloxacin hydrochloride. Acta Pharmaceutica. 2012;**62**(2):237-250. DOI: 10.2478/v10007-012-0013-2

[79] Fareez IM, Lim SM, Mishra RK, Ramasamy K. Chitosan coated alginate-xanthan gum bead enhanced pH and thermotolerance of Lactobacillus plantarum LAB12. International Journal of Biological Macromolecules. 2015;**72**:1419-1428. DOI: 10.1016/j.ijbiomac.2014.10.054

[80] Kulkarni N, Wakte P, Naik J. Development of floating chitosan-xanthan beads for oral controlled release of glipizide. International Journal of Pharmaceutical Investigation. 2015;**5**(2):73-80. DOI: 10.4103/2230-973X.153381

[81] Kharshoum RM, Aboutaleb HA. Formulation, development and evaluation of meclozine hydrochloride microspheres. Journal of Bioequivalence and Bioavailability. 2016;**8**(1):27-32. DOI: 10.4172/jbb.1000262

[82] Cornas F, Dolz M, Herraez M, Diez-Sales O. Rheological properties of progesterone microemulsions: Influence of xanthan and chitosan biopolymer concentration. Journal of Applied Polymer Science. 2008;**110**(2):1225-1235. DOI: 10.1002/app.28657

[83] Manca ML, Manconi M, Valenti D, et al. Liposomes coated with chitosan-xanthan gum (chitosomes) as potential carriers for pulmonary delivery of rifampicin. Journal of Pharmaceutical Sciences. 2012;**101**(2):566-575. DOI: 10.1002/jps.22775

[84] Liu H, Nakagawa K, Kato DI, Chaudhary D, Tadé MO. Enzyme encapsulation in freeze-dried bionanocomposites prepared from chitosan and xanthan gum blend. Materials Chemistry and Physics. 2011;**129**(1-2):488-494. DOI: 10.1016/j.matchemphys.2011.04.043

[85] Corti G, Cirri M, Maestrelli F, Mennini N, Mura P. Sustained-release matrix tablets of metformin hydrochloride in combination with triacetyl-β-cyclodextrin. European Journal of Pharmaceutics and Biopharmaceutics. 2008;**68**(2):303-309. DOI: 10.1016/j.ejpb.2007.06.004

[86] Phaechamud T, Ritthidej GC. Sustained-release from layered matrix system comprising chitosan and xanthan gum. Drug Development and Industrial Pharmacy. 2007;**33**(6):595-605. DOI: 10.1080/03639040601015521

[87] Eftaiha AF, Qinna N, Rashid IS, et al. Bioadhesive controlled metronidazole release matrix based on chitosan and xanthan gum. Marine Drugs. 2010;**8**(5):1716-1730. DOI: 10.3390/md8051716

[88] Mura P, Cirri M, Mennini N, Casella G, Maestrelli F. Polymeric mucoadhesive tablets for topical or systemic buccal delivery of clonazepam: Effect of cyclodextrin complexation. Carbohydrate Polymers. 2016;**152**:755-763. DOI: 10.1016/j.carbpol.2016.07.075

[89] Dadou SM, El-Barghouthi MI, Alabdallah SK, Badwan AA, Antonijevic MD, Chowdhry BZ. Effect of protonation state and N-acetylation of chitosan on its interaction with xanthan gum: A molecular dynamics simulation study. Marine Drugs. 2017;**15**(10):298. DOI: 10.3390/md15100298

[90] Eftaiha AF, El-Barghouthi MI, Rashid IS, Al-Remawi MM, Saleh AI, Badwan AA. Compressibility and compactibility studies of chitosan, xanthan gum, and their mixtures. Journal of Materials Science. 2009;**44**(4):1054-1062. DOI: 10.1007/s10853-008-3186-9

[91] Ghori MU, Conway BR. Hydrophilic matrices for oral control drug delivery. American Journal of Pharmacology and Toxicology. 2015;**3**(5):103-109. DOI: 10.12691/AJPS-3-5-1

[92] Alakayleh F, Rashid I, Al-Omari MMH, Al-Sou'od K, Chowdhry BZ, Badwan AA. Compression profiles of different molecular weight chitosans. Powder Technology. 2016;**299**:107-118. DOI: 10.1016/j.powtec.2016.05.019

[93] Badwan AA, Rashid I, Al Omari MMH, Darras FH. Chitin and chitosan as direct compression excipients in pharmaceutical applications. Marine Drugs. 2015;**13**(3):1519-1547. DOI: 10.3390/md13031519

[94] Al Remawi M, Al-Akayleh F, Salem MS, Al Shami M, Badwan A. Application of an excipient made from chitosan and xanthan gum as a single component for the controlled release of Ambroxol. Journal of Excipients and Food Chemicals. 2013;**4**(2):48-57

[95] Gad SC, editors. Pharmaceutical Manufacturing Handbook: Production and Processes. New Jersey: John Wiley & Sons, Inc.; 2007. DOI: 10.1002/9780470259818

[96] Patel JS. A review on bilayer tablets. Journal of Drug Discovery and Therapeutics. 2013;**1**(3):40-48

[97] Picker-Freyer KM, Brink D. Evaluation of powder and tableting properties of chitosan. AAPS PharmSciTech. 2006;**7**(3):75. DOI: 10.1208/pt070375

[98] Nogueira CC dos S, Cabral LM, Santos TC dos, Marucci A, Alhaique F. Evaluation of new polysaccharides networks for extended-release purposes: Mesquite seed gum (MSG), xanthan gum and chitosan. Revista Brasileira de Ciências Farmacêuticas. 2003;**39**:273-288. DOI: 10.1590/S1516-93322003000300007

[99] Azhar SA, Rajesh Kumar P, Vivek S, Somashekar S. Studies on directly compressed ondansetron hydrochloride mucoadhesive buccal tablets using gelatin, chitosan and xanthan gum along with HPMC K4M. Journal of Applied Pharmaceutical Science. 2012;**2**(5):100-105. DOI: 10.7324/JAPS.2012.2517

[100] Panigrahy RN, Mahale AM, Dhaked PS. Formulation and in vitro evaluation of combined floating-mucoadesive tablet of Metoprolol succinate. International Journal of Pharmacy and Pharmaceutical Sciences. 2011;**3**(2):221-226

[101] Bushra R, Shoaib MH, Aslam N, Hashmat D. Formulation development and optimization of ibuprofen tablets by direct compression method. Pakistan Journal of Pharmaceutical Sciences. 2008;**21**(2):113-120. DOI: 10.1002/adv.21536

[102] Al-Remawi MMA, Badwan AA, Salem MBAWS. Universal controlled-release composition comprising chitosan. March 2005. http://www.google.com/patents/EP1512394A1?cl=en

[103] Al-Akayleh F, Al Remawi M, Salem MS, Badwan A. Using chitosan and xanthan gum mixtures as excipients in controlled release formulations of ambroxol HCl—In vitro drug release and swelling behavior. Journal of Excipients and Food Chemicals. 2014;**5**(2):140-148

[104] Kleinebudde P. Roll compaction/dry granulation: Pharmaceutical applications. European Journal of Pharmaceutics and Biopharmaceutics. 2004;**58**(2):317-326. DOI: 10.1016/j.ejpb.2004.04.014

[105] Shao Y, Li L, Gu X, Wang L, Mao S. Evaluation of chitosan-anionic polymers based tablets for extended-release of highly water-soluble drugs. Asian Journal of Pharmaceutical Sciences. 2015;**10**(1):24-30. DOI: 10.1016/j.ajps.2014.08.002

[106] Fukuda M, Peppas NA, McGinity JW. Properties of sustained release hot-melt extruded tablets containing chitosan and xanthan gum. International Journal of Pharmaceutics. 2006;**310**(1-2):90-100. DOI: 10.1016/j.ijpharm.2005.11.042

[107] Patil H, Tiwari RV, Repka MA. Hot-melt extrusion: From theory to application in pharmaceutical formulation. AAPS PharmSciTech. 2016;**17**(1):20-42. DOI: 10.1208/s12249-015-0360-7

[108] Tiwari SB, Rajabi-Siahboomi AR. Modulation of drug release from hydrophilic matrices. PharmTech Europe. 2008;**20**(9):24-32

[109] Nokhodchi A, Raja S, Patel P, Asare-Addo K. The role of oral controlled release matrix tablets in drug delivery systems. BioImpacts: BI. 2012;**2**(4):175-187. DOI: 10.5681/bi.2012.027

[110] Rohindra D, Nand A, Khurma J. Swelling properties of chitosan hydrogels. South Pacific Journal of Natural and Applied Sciences. 2004;**22**(1):32-35. DOI: 10.1071/SP04005

[111] Huanbutta K, Sriamornsak P, Limmatvapirat S, et al. Swelling kinetics of spray-dried chitosan acetate assessed by magnetic resonance imaging and their relation to drug release kinetics of chitosan matrix tablets. European Journal of Pharmaceutics and Biopharmaceutics. 2011;**77**(2):320-326. DOI: 10.1016/j.ejpb.2010.11.019

[112] Ferrero C, Muñoz-Ruiz A, Jiménez-Castellanos MR. Fronts movement as a useful tool for hydrophilic matrix release mechanism elucidation. International Journal of Pharmaceutics. 2000;**202**(1-2):21-28. DOI: 10.1016/S0378-5173(00)00407-5

[113] Kim H, Fassihi R. Application of binary polymer system in drug release rate modulation. 2. Influence of formulation variables and hydrodynamic conditions on release kinetics. Journal of Pharmaceutical Sciences. 1997;**86**(3):323-328. DOI: 10.1021/js960307p

[114] Mitchell K, Ford JL, Armstrong DJ, Elliott PNC, Rostron C, Hogan JE. The influence of concentration on the release of drugs from gels and matrices containing Methocel®. International Journal of Pharmaceutics. 1993;**100**(1-3):155-163. DOI: 10.1016/0378-5173(93)90086-U

9

Chitin/Chitosan's Bio-Fertilizer: Usage in Vegetative Growth of Wheat and Potato Crops

Boukhlifi Fatima, Mamouni Fatima Zahrae and R. Razouk

Abstract

This chapter consists of valuing the chitosan to create bio-fertilizers as fertilizers without going through the composting process because of their richness in the nutrient base elements of plants: nitrogen and phosphorus. Physicochemical analyses of the chitosan focused on pH, dry matter, organic matter, nitrogen, phosphorus and potassium as well as IR and XRD. The samples thus prepared were monitored for 15 days. PH, temperature and conductivity were monitored daily. According to the physicochemical analyses of waste (nitrogen, phosphorus and potassium) and the nutritional needs of our selected crop (soft wheat, Arrehane variety which are 90-90-50 U/ha), several doses are then determined for the purpose of the optimal formula after their application on the crop. An application of bio-fertilizer on the potato was also undertaken. Follow-ups were carried out during this study, such as the monitoring of the vegetative growth of wheat and the mineralization of the soil via its physicochemical analyses. The results show that our bio-fertilizer is rich in nitrogen with 4.98% and phosphorus with 1.42% and mineralizes quickly on the ground while leaving the soft wheat to absorb its nutrients effectively and improving its growth properties, then giving good yields.

Keywords: chitosan, vegetative growth, wheat, potato, crops, bio-fertilizer

1. Introduction

The consumption is always higher and more diverse, which leads to a growing production of wastes in quality and quantity. This growth causes huge danger on the environment and hence on the human health [1]. So many organic wastes are generated then constantly to the world by the domestic and the halieutic industry [2]. The sector of fishing is a part of the strategic sectors in the world. It plays an important role in the global economy [3, 4]. The development of this sector is related to environmental issues, in particular to waste management. Indeed, the quantities of the halieutic waste are considered at several thousand, tons of waste a year as researched by Chabbar [5] and Afilal et al. [6].

These are randomly put in the existing systems of evacuation causing big problems [7]. One of these problems is the negative effect on the environment and human health El Moutawakila [8] and Benzakour et al. [9]. However, several studies have been interested in the evaluation and in the treatment of this waste. Some of them had studied the evaluation of their potential polluting [10].

Thus, the process of biotransformation for this type of waste seems to be the most suitable to resorb these problems. It corresponds to the elaboration of beneficial products, of natural origin, usable as bio-fertilizing for grounds [4] in substitution of artificial fertilizers. Moreover, the excessive application of artificial fertilizers for one of the most important agricultural processes in the world, the volatilization of ammonia in the air, the pollution of water resources causing their eutrophication, the degradation of the ground by their pollution attack of the cultures by phytopathogenic diseases [11–13].

This biotransformation consists then of the spreading of bio-fertilizers, which are in the form of dried, crushed and spread waste in agricultural plots.

Indeed, direct spreading is the simplest method of valuation requiring the least investment; it provides nutrients, improves soil quality with water retention and stimulates microbial activity. However, composting does not seem to be favorized in the logistics (time and local manufacturing, odors, etc.), due to the limited supply of carbon materials and environmental constraints [14] such as the attraction of insects and plants, pests and the risk of weeds in crops [15].

In this context, the bio-fertilizing potential of seafood, that is, chitin/chitosan, is explored for wheat and potato crop. They represent in fact a rich organic source for organic farming. The objective of this chapter is to valorize seafood waste, considering it as a source of bio-fertilizers and not only as a source of pollution, thanks to a simple and inexpensive process by spreading them directly in the agricultural environment. Their valorization always allows the protection of the environment and the acquisition of a new economic source.

2. Study of the use of raw chitin-chitosan as a bio-fertilizer

2.1. Inconveniences of nitrogenous chemical fertilization

In agriculture, chemical fertilizers are administered to increase crop yield. They provide the nutrients that plants need to grow. There are several chemical forms of nitrogen fertilizers in the market and the distinction between them is made possible through the various conventional chemical tests. The choice of one form over another is often difficult because of the contradictions in the published results on the composition of measures of agronomic efficiency [16].

The trio nitrogen, phosphate and potassium (NPK) is the basis of all these products, and they are also responsible for massive soil pollution but are especially the major cause of pollution of groundwater, the main reservoirs of drinking water. If they change their environment and make the water unsafe, there are certain dangers listed in **Table 1** [17].

The dangers of nitrogen fertilizers (nitrates)
- When they are not consumed by plants, they easily seep into the soil and progressively reach the groundwater.
- They are soluble in water.
- Before the 1950s, the nitrate content per liter of water did not exceed 1 mg. Nowadays, it easily exceeds 50 mg/l.
- Contribute to the phenomena of eutrophication [18].
- They are degraded by a bacterium and turn into nitrites that can poison the blood by oxidizing hemoglobin. The fluid then poorly fixes oxygen and causes respiratory problems.

Table 1. Some dangers of chemical fertilizers.

2.2. Contribution of waste to soils

The return to the ground of waste has been practiced by man since always. There are two reasons to explain this ancestral practice: first, the fertilizing value of this waste and then the capacity of the soil to purify the effluents, in particular, liquids, which makes it possible to protect the deep and surface waters against all risk of pollution. Strengthening regulations on the protection of the environment should make it possible to sustain this agricultural recycling while preserving the quality of the receiving soils, crops and water [13]. Waste spread in agricultural fields comes primarily from agriculture itself or from industries directly linked to it. Depending on their chemical composition, waste can be brought to the soil to provide fertilizer amounts equivalent to mineral fertilization (**Table 2**).

Some residual products are brought to the soil as an amendment. An amendment is a contribution to soils for the main purpose of improving their physico-chemical and biological properties. The organic matter content of a soil is one of the key elements leading to a stable structure and helping to limit the risk of soil erosion, especially in soils with a silty texture [19, 20]. However, the use of waste as fertilizer or amendment can only be accepted if their negative environmental impacts are minimal compared to their positive effects. With increasing public awareness of the need to preserve soil quality as well, the risks associated with land application now include not only the plant and water aspects but also those of soil and even air.

2.3. Physico-chemical preparation and characterization of the raw chitin

The waste of shrimp (DC) was naturally collected and dried. These samples are crushed and sieved in 2 mm (**Figure 1**).

Type of waste	% Material dries	N (g/kg)	P (g/kg)	K (g/kg)
Solid sewage sludge	55.2	24.8	8.3	1.7
Foams of sugar refinery	29.0	7.4	4.0	0.9
Liquid manure of poultry	13.3	10.2	1.8	5.4
Fertilizer of cattle	28.4	6.2	1.4	5.9
Fertilizer of sheep (mutton)	29.3	8.6	1.8	11.0
Compost of fertilizer of bovine	35.2	7.6	1.3	6.1

Table 2. Average contents of fertilizing elements in waste spread in agriculture [18].

Figure 1. Shrimp waste after drying and grinding "DC".

The analyses focused on pH, nitrogen (N), phosphorus (P), potassium (K) and organic matter [17]. The characterization of raw chitin is performed using IR spectroscopy and X-ray diffraction (XRD).

2.4. Results and discussions

The results of the physico-chemical analyses made on the chitin [17] are summarized in **Table 3**.

These measures are in accordance with the International Standards for AFNOR, the NF U44-051 standard approved in 2006 for fertilizers of plant and/or animal origin, the amendment of which allows the soil to be maintained or stockpiled of its existing organic material as well as the improvement of the physical, chemical and biological properties of the soil [21, 22]. This standard stipulates the following contents:

N $<$ 3% on MB, $P_2O_5 < 3\%$ on MB, $K_2O < 3\%$ on MB and $N + P_2O_5 + K_2O < 7\%$ on MB.

MS $\geq$ 30% MB and MO $\geq$ 20% MB depending on the type designation and C/N $>$ 8.

MB: raw material; MS: dry matter; MO: organic matter.

According to the results obtained, our bio-fertilizer ratifies almost all the values of the AFNOR standard quoted above.

Indeed, the sum of the percentages on N, P and K is less than 7%, it is 6.45%. The percentages of phosphorus and potassium are 3% lower except that in nitrogen, the MO content is

Waste/parameter	pH	DM (%)	OM (%)	C (%)	N (%)	P (%)	K (%)	C/N
SW	8.55	26.13	56	28	4.98	1.42	0.05	5.62

Table 3. Results of physico-chemical analyses of the raw chitin.

greater than 20%. The MS is close to the norm. On the other hand, the C/N ratio is lower than the norm.

According to the analyses, shrimp waste is rich in nitrogen. The composition of its waste (carapaces, viscera, small portions of flesh attached and associated water) is characterized by its high nitrogen content, which is granted with several previous works [15, 23, 24].

This richness is due to the high crude protein content; a factor of 6.25 was used to convert total nitrogen into protein. The percentage of protein in our sample is therefore 31.12%; this value is approximate to those of other research: protein contents of 52 [25], 44.12 ± 0.79 [26] and 47, 43 and 47.75% [23].

Ravichandran et al. also reported [27] that the percentage of crude protein in the dry matter of raw chitin is 24.03%. Similarly, a percentage close to these values of 29.3% was found by Prameela et al. in 2010 [28]. Also, Khan and Nowsad [29], in 2012, found high percentages ranging from 40 to 50% of proteins in shrimp shells.

These results showed that shrimp waste is high in phosphorus, which is in agreement with many authors who have found that this content in the head and shell is, respectively, 0.017 and 0.029% [23]. This phosphorus richness is attributed to its contribution to the formation of crustaces structures and their strengthening when phosphorus is combined with calcium.

The shrimp waste had an alkaline pH (8.55); this value was found by Mohan et al. which is 8.10 ± 0.10 [28, 30, 31].

2.4.1. Infrared analysis

Infrared spectrum of raw chitin.

Figure 2 shows the Fourier transform infrared (FTIR) absorption spectra of crude chitin. The positions of the various bands observed and their attributions are summarized in **Table 4**. The spectrum shows broad vibration bands located at 3100–3500 cm^{-1} corresponding to the —NH and —OH elongation vibrations including the hydrogen bonds. Two absorption bands appear in the two 1557 and 1652 cm^{-1} spectra due to the —CO—NH_2 elongation vibrations. These bands of amide I and amide II of chitin are more easily identifiable in the case of chitin. The spectrum of chitin shows bands in the region 500–900 cm^{-1} called the region sensitive to the structure.

The infrared spectrum is used to illustrate the presence of nitrogen in the fertilizer matrix. It is almost similar to the spectrum of chitin found by the author Boukhlifi et al. [30, 31].

2.4.2. DRX analysis

Figure 3 represents the diffractogram DRX of the waste of shrimps.

The observation of the X-ray diffraction spectra of shrimp waste shows the presence of two intense peaks at 2θ = 9.9° and at 2θ = 19.9°; these results are identical to those obtained by Ahlafi et al. [31], with pure chitin; these researchers showed that α-chitin has two peaks of diffraction, 2θ = 9.3° and 2θ = 19.4°. Other authors, Liu S at al. 2012, [32] showed that chitin

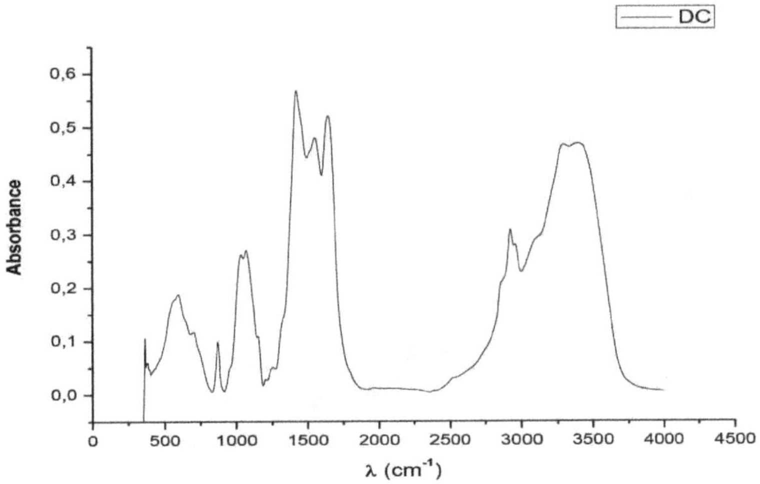

Figure 2. Infrared specter of the waste of shrimp.

Band cm^{-1}	Nature of vibration or rotation
500–900	Region sensitive to the structure
1028	Vibration of strain of the C—O—C of the cycle glucosidique
1089	Vibration of strain of —OH
895 and 1153	Liaison glucosidique $\beta(1\rightarrow4)$
1254	Vibration of distortion of O—H
1423	Symmetric distortion of CH_3 and CH_2
1557	Amide Il
1652	Amide I
523, 741 and 1652	Identifies the chitin
2880 and 2923	Strain of —CH and CH_2
3100–3500	Strain NH and —OH, including the binding of hydrogens

Table 4. Characteristic vibration bands of chitin and chitosan.

has a strong reflection at 2θ = [9–10°] and at 2θ = [20–21°] and a minor reflection at 2θ = 26.4°; we can conclude therefore that the characteristic peaks of chitin exist in the analyzed waste and appear in our spectrum; we can still see in the spectrum that there is an intense peak at 2θ = 30°, which is due to the calcite present in shrimp shells and in the chemical fertilizers used. The diffractogram also shows that the mineral part of our sample contains a mixture of two varieties: calcium carbonate and $CaCO_3$/calcite, syn [17, 19].

2.5. Choice of the culture of execice

This chapter consists of adding raw chitin to a bio-fertilizer while applying it to soft wheat (*Triticum aestivum*), Arrehane variety and potatoes. A comparison with two witnesses was made, the first is the commercial chemical fertilizer (ammonitrate, 21%) the most consumed

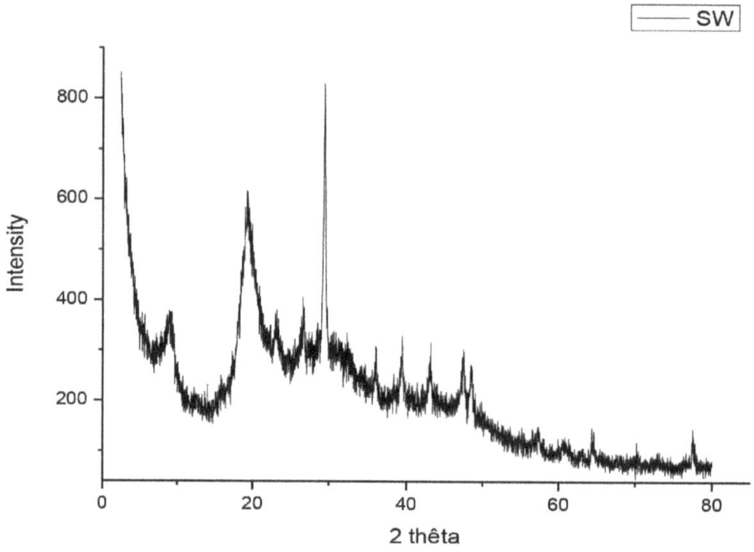

Figure 3. Diffractogram DRX of the waste of shrimp.

in Morocco with 27% [16] and the second is no fertilizer. This is to improve the biofertility of the soil, which is a fundamental value for organic pioneers.

The analyses of the waste used and the soil samples of the test were carried out at the National Institute of Agronomic Research to determine the fertilizing power of the waste and to follow the evolution of the mineralization of the soil.

2.6. Application on soft wheat

2.6.1. *Cultivation of soft wheat*

The determination of organic inputs for crude chitin in kg/t in terms of N, P and K was made on the basis of the results of physicochemical analyses of waste and the requirements of common wheat in these same elements are 49.8 kg/t [17].

Soft wheat, variety Arrehane, was sown at a rate of 15 seeds per pot with a surface area of 0.07 m^2, simulating a seeding rate of 214 seeds per square meter. The sowing was in November 2014, the growth period started and ended at the end of June and the harvest took place in July 2015. The temperature and lighting are natural ambient. Watering was done as needed with well water.

The test is done in pots with one-third of sand and two-third of soil (**Figure 4**). The organic input doses (in g/pot) (**Table 5**) were calculated based on the shrimp waste content in N, P and K elements as well as the requirements for soft wheat in these elements, which are 90–90–50 kg/ha [20]. Four treatments were predetermined on the basis of nitrogen fertilizer content (N); the tested treatments are 100, 150, 200 and 300%. This method is similar to that of Yadav et al. [21]. A chemical fertilizer (EC) treatment is also applied to wheat with the same treatment and an absolute control where the soil has received no fertilizer.

The monitoring of the crop is carried out there after measuring the growth parameters of the wheat until it reaches maturity. Each sample then consists of the ears of wheat harvested in

F.Z.MAMOUNI, 2015

Figure 4. Some steps of substrate preparation and sowing.

Treatment waste (%)	SW (g/pot)	EC (g/pot)
100	14	3
150	21	4.5
200	28	6
300	42	9

Table 5. Organic and chemical input rates for wheat cultivation.

all pots of the test. These ears were then shredded with the electric thresher and the recovered kernels were weighed for determination of estimated crop yield.

2.6.2. Soil tests

Soil samples of 0–20 cm were taken three times during the wheat growth cycle at the time of spawning, tillering and at maturity in all pots of the experimental set using a stainless steel tool. The samples were put in the laboratory for measuring the fresh weight, then are dried in an oven for 48 h at a temperature of 60°C for carrying out analyses of the elements P, K, organic matter and pH and conductivity measurement according to the internal protocols of INRA.

a Content in potassium

The extraction of the potassium in the ground was made by the addition of extract of the ground (acetate of ammonia). In every sample to extract all the elements of the ground by means of Wheaton-Omnispense more and make shake flasks in an agitator goes and comes hanging (AGITELEC) 30 min. After these stages, samples are filtered. The reading of the content of filtrates in potassium is made on the photometer for flame:

$$K_2O\ (\text{ppm}) = \text{ppm} \times 10 \times 1.2 \quad (1)$$

b Phosphorus content

Phosphorus analysis was performed using the 0.5 M sodium bicarbonate ($NaHCO_3$) extraction solution at pH = 8.5. The same filtration process is thus carried out, extract was taken and put in Erlenmeyer flasks and then sulfuric acid (5 N) was added to acidify the solution. The staining solution was then added. This supplemented with distilled water; the solutions of the calibration range were also prepared in order to plot the calibration curve to deduce the phosphorus concentrations. The intensity of the blue color of the solutions is read at 820 nm after 15–30 min.

The calculation formula is as follows:

$$\text{P (ppm; soil)} = \text{reading (ppm)} \times 20 \tag{2}$$

c Soil content of organic matter

To know the amount of organic matter present in the soil, 1 N of potassium dichromate, sulfuric acid (H_2SO_4), distilled water and concentrated phosphoric acid (H_3PO_4) were used.

In the last step, a few drops of the indicator (diphenyl amine) were added. After homogenization, the excess of $K_2Cr_2O_7$ with 1 N $FeSO_4$ was titrated to bright green and the volume of $FeSO_4$ was recorded under the same conditions as a blank solution without a soil sample.

The calculation is done by the following formula:

$$\text{OOC \%} = \frac{((V(\text{white}) - V(\acute{e}ch.titr\,\acute{e})) \times 2 \times 0.3 \times 0.5}{\text{weight of the ground}} \tag{3}$$

$$\text{TOC \%} = 1334 \times \text{OOC\%} \tag{4}$$

$$\text{\%M.O} = 1724 \times \text{TOC} = 2.3 \times \text{OOC} \tag{5}$$

d pH analysis

A quantity of soil is suspended in a double volume of distilled water (5 g of soil/10 ml of water). The mixture is stirred with a glass rod. The mixture was allowed to stand for 30 min, stirred 5 or 6 times during this period, and then the pH was measured.

2.6.3 Results of growth monitoring of common wheat and comparison of root volume and weight

At the end of the test, roots and residues were extracted by a stream of water that caused the soil to settle downward and the exclusion of floating roots upward. The roots of the wheat were then removed from the soil, washed and cleared of soil debris and then oven-dried at a temperature of 60°C for 48 h.

The growth of common wheat was followed throughout the season by length measurement. At harvest, several parameters were measured, namely:

- final length of plants (cm)
- weight of grains/pot (g)
- grain yield estimate (ton)/ha
- root volume and weight

a Wheat growth for each dose

Figure 5 shows the growth of wheat, based on the doses of treatments that are 100, 150, 200 and 300%. In general, the average growth data showed that wheat grew well in all the increasing doses provided by the biological treatment (raw chitin) ending up to almost 99 cm in length (150% treatment) and exceeds the growth of wheat receiving the chemical fertilizer at all these doses and the witness which did not undergo any contribution. For the 100% dose, the wheat amended by the raw chitin exceeds in length that is amended by the chemical fertilizer with 6.33 cm and the control with 13.53 cm, which is the same for the other doses of 150, 200 and 300%, successively, with 10.27, 5.37 and 1.1 cm for the chemical fertilizer and with 20.73, 19.5 and 12.16 cm for the witness.

These results do not corroborate those found in the study of the effect of raw chitin on radish cultivation [15] where the length of radishes fertilized by mineral fertilizers exceeds that of radishes amended by shrimp residues with 1.1 cm. On the other hand, they are corroborated with the work of Taiek et al. where they demonstrated that the fish waste allied to malting releases allowed for an optimal growth of barley and tomato, better even than the commercial fertilizers [5].

b Comparison between doses

The growth of wheat fertilized by doses 150 and 200% of the gross chitin exceeds that of the other two doses; the wheat fertilized by the 300% dose is good growth but the latter has declined toward the end of the growth cycle (from April 29 to May 06, 2015). This is due to the excess of nitrogen in the soil and therefore is absorbed by the wheat plants, which causes a delay in the maturity phase [33] unlike, the others; it is concluded that the 300% dose is limited to growth. From there, we can recommend that the following doses corresponding to treatments 150 and 200% of our bio-fertilizer are equivalent successively to 3 and 4 t of N/ha which is equivalent to 135 and 180 U N/ha (**Figure 6**).

The wheat grows very closely with regard to all the doses of the chemical fertilizers brought. At the end of the growth cycle, the wheat amended with the 200% dose outgrows the other wheat plants. That is the dose that corresponds to 180 U of nitrogen per hectare, and this is the recommended dose per hectare by the INRA in Morocco which varies between 160 and 200 U/ha [34].

c Yield results of soft wheat

At maturity, soft wheat was harvested manually from ground level, and the harvested biomass was weighed later. The grains were separated from the straw with a drummer, and the grain yield was recorded after weighing.

Applying our only bio-fertilizers significantly increased wheat yields compared to chemically fertilized wheat; (**Figure 6**) it is a maximum yield of 30 q/ha for bio-fertilizers against 16.18 q/ha for commercial fertilizer treatment 200% corresponding to the contribution of 180 U of

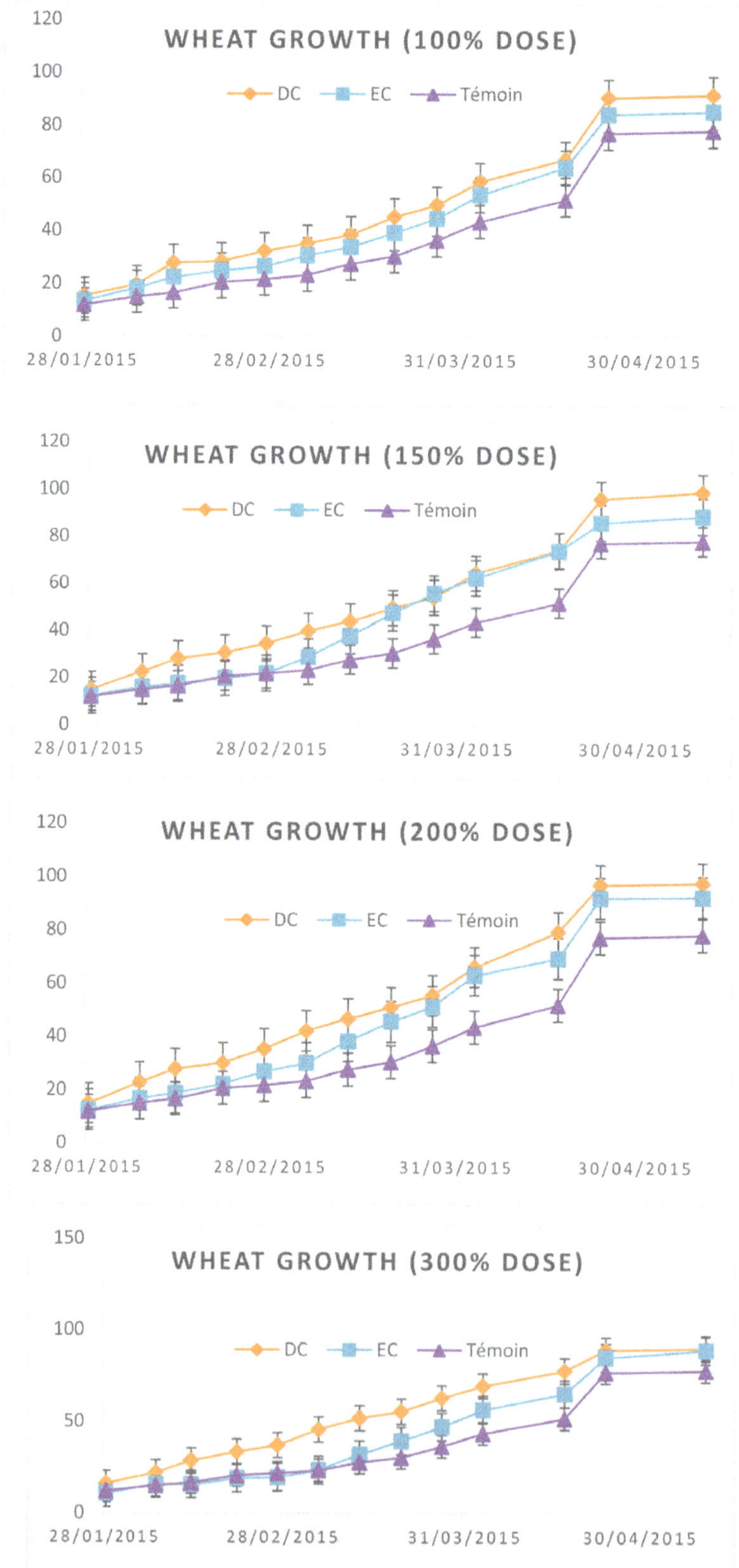

Figure 5. Comparison between bio-fertilizer and chemical fertilizer in terms of wheat growth for each dose: (a) 100, (b) 150, (c) 200, and (d) 300%.

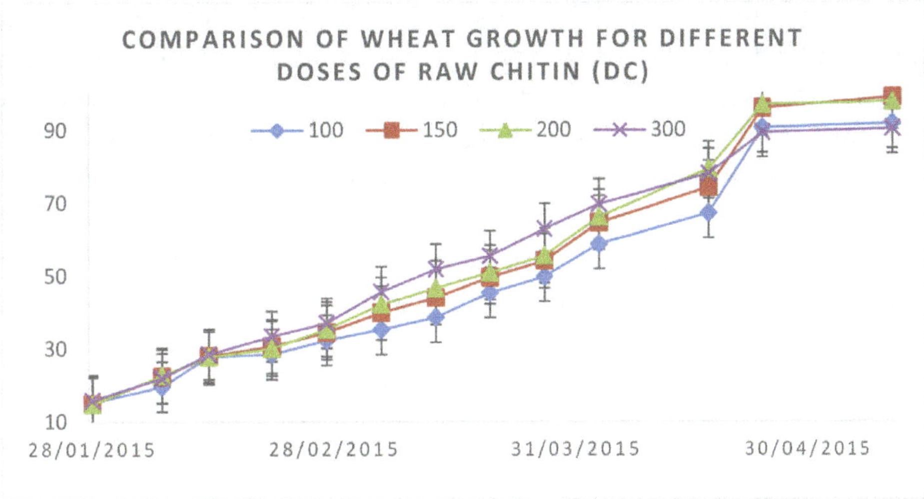

(a)

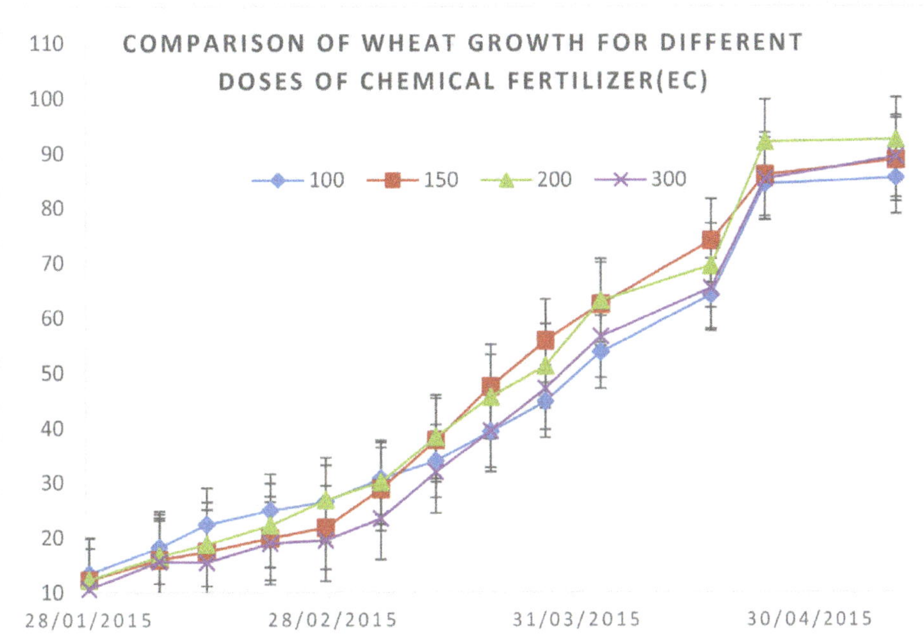

(b)

Figure 6. Comparison of growth of the wheat for all the doses of the used fertilizers (a) DC and (b) EC.

nitrogen per hectare, which approves a correlation with growth well developed in wheat fertilized by the gross chitin with the same treatment above. According to the Ministry of Agriculture at the national level, the average yield of common wheat has increased, between the periods 2000–2007 and 2008 and 2015, from 14.3 to 19.2 q/ha, a value lower than that found for our organic fertilizers [35].

In addition, application of marine residues has been shown to have positive effects on crop yields. Abdel-Mawgoud [36] has shown that a foliar application of chitosan on strawberry plants helps to increase the height of the plants, the number of leaves and even the yield of strawberries. Abdel-Mawgoud et al. [37] also noted an increase in fall triticale yield following the early spring application of mussel residues. Also, Karine has shown that treatment with pure marine residues, dried shrimp waste, has produced the best radish yields

[14]. In addition, the authors state that composting reduces the availability of nitrogen from marine residues. Indeed, some of the nitrogen is lost through volatilization during the composting process. After fermentation, the decomposition of the fresh residues generates toxic compounds (volatile fatty acids, lactic and acetic acids, etc.). Some studies have highlighted another benefit that corresponds to the safener effect of marine residues or marine residue extracts on various diseases. In 2006, ADAS [38] stated that the addition of crustacean residues stimulates soil microbial activity, which can promote competition between soil microorganisms at the expense of pathogenic microorganisms.

d Results of comparison between volume and weight racinaires

The role of roots is very important for the absorption and the transport of the water and the mineral elements toward the air parts of plants [39]. Their development intervenes in the evolution of the properties of the ground and more particularly its structure and its content in organic matter [40]. The roots of the wheat fertilized by the bio-fertilizer exceed in weight and volume, though some wheat is fertilized chemically; (**Figures 7, 8 and 9**) our results confirm the results

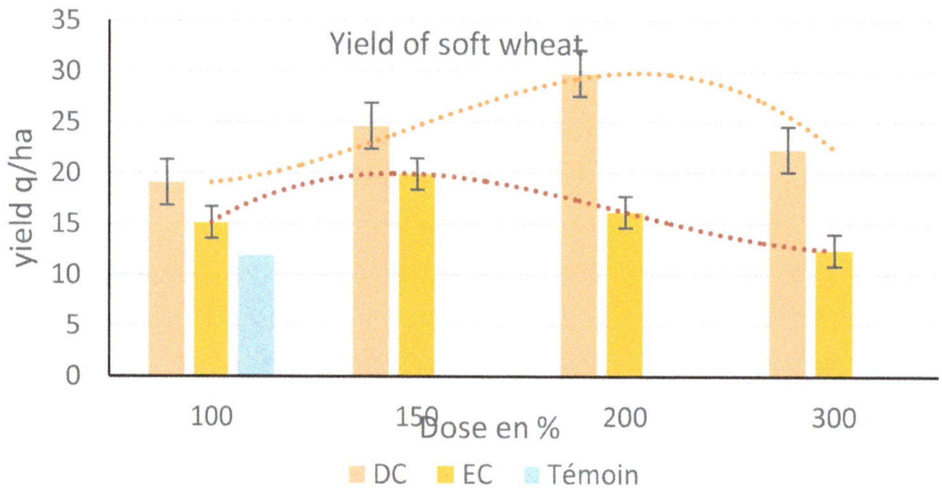

Figure 7. Yield on the common wheat.

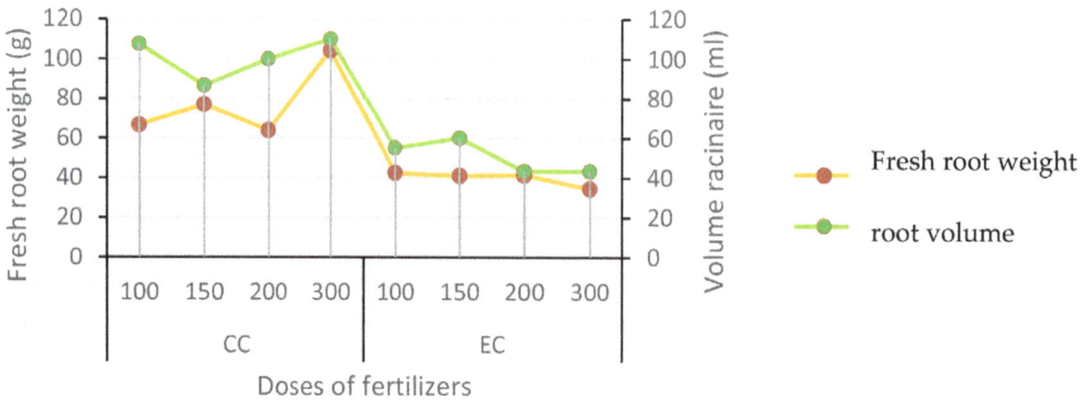

Figure 8. Comparison between the bio-fertilizer and the fertilizing in terms of volume and weight racinaires some common wheat.

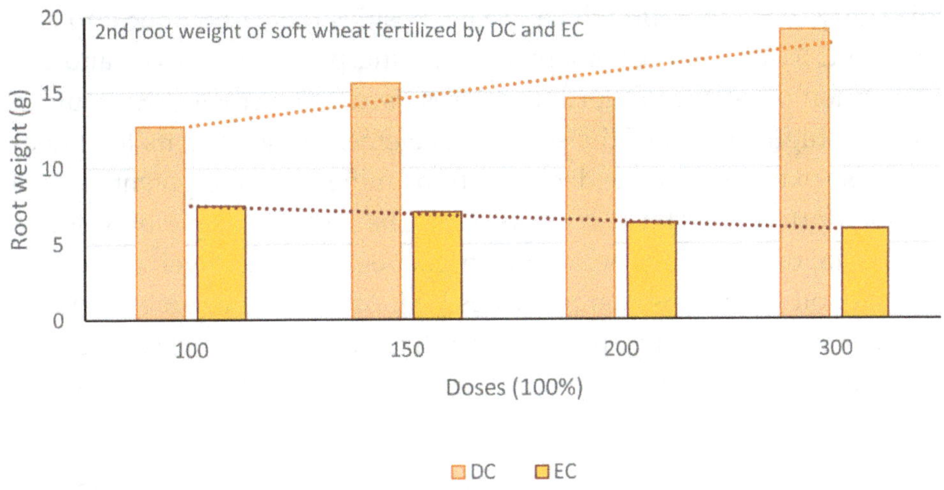

Figure 9. Secondary root weight of soft wheat.

of Karine [14] where the diameter and the biomass of roots for radish fertilized by the waste of shrimps are superior with regard to those of radish which is fertilized by the artificial fertilizer.

e Evolution of soil mineralization

Table 6 represents the mineralization of total organic carbon (percentage) during the growth phases of soft wheat.

The mineralization of organic matter is a process of degradation. Its main consequences are the decrease of the organic matter content in a soil and the selective disappearance of certain compounds [41, 42]. The decomposition of organic matter is expressed in terms of mineralized carbon and types of molecules present in the soil over time [43].

However, the effects of bio-fertilizers on biological activity and mineralization of nitrogen in the soil depend in particular on their nitrogen concentration and their C/N ratio [44], which further confirms the results found for our bio-fertilizer from shrimp waste where the nitrogen content is 4.98%. The bio-fertilizers with lower C/N ratios are mineralized rapidly in the soil, releasing significant amounts of nitrogen absorbed by subsequent crops [14, 45–47] that also

Dose	DC			EC		
	After lifting-tallage	Tillering-montaison	Bolting-heading	After lifting-tallage	Tillering-montaison	Bolting-heading
100	23.84	96.28	−48.14	17.91	98.95	−66.86
150	28.20	113.66	−53.49	24.88	98.95	−48.14
200	31.22	108.31	−45.47	27.62	100.29	−50.81
300	35.17	100.29	−56.16	18.49	82.91	−34.77
Témoin	40.87	73.55	−34.77	40.78	73.55	−34.77

Table 6. Total organic carbon mineralization (%).

states, for this study, the C/N ratio which is 5.62. Besides the carbon decomposition rate has peaked at 31.22 and 35.17% at post emergence-tillering phase and 113.66 and 108.31% at tillering-stem extension phase for bio-fertilizer doses used, respectively, at about 200 and 300% which explains the high content of the soil in this element following its total decomposition. This increase is also explained by the decrease in the mineralization potential, which corroborates the work of Martel et al. [48] and Karine [14]. The regular supply of residues or manure can increase the total organic C of the soil to a higher equilibrium level, related to the balance between C inputs and decomposition processes [48]. And it is the same—the decomposition of the carbon element in the soil for the wheat fertilized by the chemical fertilizer is maximum for the doses of 150 and 200%, successively at 24.88 and 27.62% (after the emergence-tillering phase) and 98.95 and 100.29% (tillering-stem extension) for doses of successively 100 and 200%.

Subsequent to the cycle, we find that 53.49 and 56.16% of the carbon element was mineralized during the stem extension-heading period; those are maximum values for the bio-fertilizer doses of 150 and 300%, respectively. For the chemical fertilizer, 66.86 and 50.81% values of the mineralized carbon (maximum values) are determined for doses of 150 and 300% too. This is where the microbial communities convert the carbon element provided by the bio-fertilizer and the chemical fertilizer into stable C in the soil, as studied by Kallenbach et al. [49].

2.7. Application of the bio-fertilizer on potatoes

2.7.1. Preparation of the ground

The slightly acidic pH (pH = 5.5 or 6) of the soil can give good yield. Excess alkalinity of the soil can cause development of the common tuberous gall.

The high salinity of the soil can block the absorption of water by the root system, so to complete in combination with these conditions, the soil preparation is an essential step for the best fertilizer application. The preparation techniques consist of ensuring good contact between the tubers and the soil because the development of the root system will usually be delayed if the soil is poorly prepared. The soil should be prepared to a depth of at least 25–30 cm; such a loose layer promotes aeration of the soil, ensures good root development and facilitates ridging.

A good seedbed can be done as follows:

- medium plowing (25–30 cm) with plow;
- spreading of dry waste and phospho-potassium fertilizers with doses well respected according to the needs of the potato/ha;
- a good preparation of the first 10 cm of the soil allows a good cover of the plant.

Fertilization remains one of the most important factors for good potato production. For a production of 25 tons/ha (tubers + haul), the requirements of the amount of essential elements for the plant include 160 kg (N)/ha, 45 kg (P_2O_5)/ha, 275 kg (K_2O)/ha, 50 kg (MgO)/ha and 70 kg (CaO)/ha.

The potato is very demanding in organic manure, the needs are of the order of 30 T/ha and this dose can be doubled in soil low in organic matter. The maximum nitrogen absorption takes place at the time of 50–80 days after planting. Nitrogen can be applied in the form of sulfate of ammonia; the phosphorus is hardly absorbed by the plant for it must be applied before planting and in the most assimilable form. The doses should be divided into three periods; emergence, first hump and second hump.

The recommended doses are only average and must be adapted according to the richness of the soil. A preliminary soil analysis is necessary to evaluate the level of soil fertility. Nitrogen should be located at the ridges while avoiding direct contact between the plant and the fertilizer.

2.7.2. *Plantation of the potato*

The tubers of the potato are classified according to the following sizes: from 28 to 35 mm, from 35 to 45 mm, from 45 to 55 mm and calibers greater than 55 mm. Planting density of 15 to 20 stems/m^2 gives good land use; pre-germinated 35–55 mm of plant produces approximately 5–6 main stems, usually 4 plants/m^2 with a distance of 70 cm between lines and 30 cm between plants; the seed requirements/ha equals about 2000 up to 2500 kg. Planting tubers at a uniform depth that depends on the type of soil, climatic conditions, and the physiological age of the plants gives a homogeneous culture.

The choice of planting at 5–6 cm of depth is applied in moist, heavy soil as the mother tubers may run out before the germs reach the soil surface, but for a textured soil, where a slight or a risk of drying out is to be feared, the planting is often done at a depth of about 10 cm.

2.7.3. *Irrigation of the potato*

The organization of activated irrigation promotes and ensures the mechanisms of transport of mineral elements, synthetic products, transpiration and thermal regulation at the leaf level because the potato is very sensitive to both deficit of water and excess water; it requires water which is evaluated between 400 and 600 mm according to the climatic conditions, the type of soil and the length of the cycle. According to the following frequency:

- During germination, the amount of water needed is low.
- For 60 days up to 90 days after planting; irrigation should be carried out at very short intervals (6 or 7 days) in light soil and for 12 or 15 days in heavy soil up to 10–20 days before harvest.

The quality of irrigation is measured by the rate of salts which must be less than 4 g.

2.7.4. *Earthing-up*

The operation of earthing-up consists of returning the earth to drive the mound; the purpose of this operation is of cover the superficial roots of the plant, the nitrate and potassium fertilizers applied during the culture, and prevent the culture of the moth (decreased the attack of insects).

The first planting is done 2 to 3 weeks after the late emergence of the plant, the length of the plant must be at least 10cm above the ground. It is necessary to stay up during harrowing not to affect the system racinaire and the recently formed tuber, this operation is important because it consists in taking all the weed. The yellowing of the lower leaves, the drying of stalks and the firmness of the skin of tubers are factors of the harvest after 3–4.5 months. The lifting must be made in dry weather and tubers should not be left exposed too much to the sun to avoid the development of the black spots and the attack by the moth.

2.7.5. Hilling

The humping operation is to bring the previously loosened earth to the ridge to form the hill; the purpose of this operation is to

cover the superficial roots of the plant, the nitrogenous and potassic fertilizers applied during the cultivation and prevent the cultivation of ringworm.

The first hilling is done 2–3 weeks after the lifting, after a hoeing operation, so as to cover at least 10 cm of soil, and then the operation is repeated every 2–3 weeks.

Care should be taken during hoeing not to touch the root system and newly formed tubers, this is important because it involves taking all the weeds.

Yellowing of the lower leaves, drying of the stems and firmness of the tuber skin are factors in the harvest after 3–4.5 months.

Tearing should be done in dry weather and care should be taken not to leave the tubers too exposed to the sun to avoid the development of black spots and ringworm attacks.

2.7.6. Application of bio-fertilizer: raw chitin

The application of raw chitin on the potato in this study is performed qualitatively in bags.

- initialization of the bottom of the bags by 2 cm sand sole to seeds of large pores to facilitate the permeability of excess water during irrigation;
- preparation of a mixture of sand-clay-soil (one-third of the sand + two-thirds of the clay soil);
- addition of a 5 cm layer of soil that we have prepared so as to not lose fertilizer doses by the permeability of water;
- tuber planting in another layer of soil mixed with 50% of the initially determined doses to a depth of 7 cm;
- monitoring of buddies by irrigation according to the needs of the water needed for the growth of the potato during planting and after so as to keep the soil moist;
- the soil after the first hilling must be mixed with one of the 25% of the doses;
- the same operation for what concerns the second hilling with the last 25% of the doses, after having the germs on the surface and the formation of the stems.

Time (J), after survey	Number of main stems (petioles)	Length of petiole max (cm)	Length of compound leaf max (cm)	Number of compound leaves	Length of secondary leaf (leaflet) max (cm)	Width of secondary leaf (leaflet) max (cm)	The length of the plant (cm)	Number of secondary petioles
9	3	5.5	9266	7666	4166	3233	9.5	
12	3	8	14.333	13.666	6566	5733	13.333	
16	3333	13.666	25.666	23	8	6266	23.5	
19	3333	19.666	28	24	9.5	6,4	31.666	
23	3333	28	29	27	10.433	9666	38	
26	3333	31.333	30	35.333	10.93	9,5	40.666	14
30	3333	33.833	30.5	32.333	10.933	9633	41.833	15
33	3333	35.666	30.666	33.666	11	9833	45.5	17
37	3333	37	31.5	35.333	11.133	9,9	46.666	19
40	3333	39.666	33	37.666	11.333	10	48	19

Table 7. Potato monitoring in soil contains 300% of N requirements in raw chitin.

Figure 10. Growth of the potato according to the fertilizer used.

2.7.7. Study and follow-up of the factors indicating the good growth of the potato

The length of the plant, the length of the petioles, the length of the compound leaves and the length and width of the leaflets are all more important factors that promote photosynthesis that allows the transformation of mineral matter into plant tissue. They increase the volume of the flowering and improve its precocity, help to increase the weight of the seeds, lengthen the plant and increase the protein content of the seeds, so all these factors are important to obtain at the end of the harvest a good performance. Thus, in **Table 7**, the analyses and measurements performed on these growth factors are grouped together. We also give pictures (**Figure 10**) of some pots on which we made our measurements.

Figure 10 illustrates well that the potato in the case of the fertilization by the raw chitin is well pushed compared to the case of fertilization by the chemical fertilizer. This proves that there is a significant degradation of chitin which brings a significant amount of nitrogen to the plant.

3. Conclusion

As far as this study is concerned, the dehydrated marine residues, such as the shrimp waste, mainly raw chitin, represent an interesting source of nitrogen and carbon for the agricultural valorization. Consequently, their treatment produces wheat yields significantly higher than those fertilized with chemical fertilizers. The raw chitin is more effective than chemical fertilizer and more effective than the test.

So, we recommend the dose to bring from 200% corresponding to 180 U of nitrogen of our bio-fertilizer which is the raw chitin for the cultivation of soft wheat, variety Arrehane. The richness of shrimp was also proved by the quality and quantity of potatoes obtained after fertilization with raw chitin. The biological degradation of chitin and the important contribution of nitrogen have been proved by vegetative growth.

The use of bio-fertilizers from organic waste as chitin/chitosan remains a useful and sustainable practice. It is necessary for the reduction of these environmentally harmful wastes' rate to increase the agricultural yields and to have an effective maintenance of soil fertility.

Author details

Boukhlifi Fatima[1]*, Mamouni Fatima Zahrae[1,3] and R. Razouk[2]

*Address all correspondence to: boukhlifi1@yahoo.fr

1 Equip "Materials and Applied Catalysis", Faculty of Science, University Moulay Ismail, Meknes, Morocco

2 Equip Research Agronomy and Plant Physiology, Regional Center for Agricultural Research, Meknes, Morocco

3 Equip "Science of Water and Environmental Engineering", Faculty of Science, University Moulay Ismail, Meknes, Morocco

References

[1] Afilal ME, Auriol M, Filali-Meknassi Y. Chapitre du livre "Les énergies renouvelables au Maroc—Le débat est lancé". UNESCO; 2009. 196 pp. ISBN: 9954-8068-2-2

[2] Le réseau régional d'échange d'informations et d'expertises dans le secteur des déchets solides dans les pays du Maghreb et du Machreq. 2014

[3] Rapport sur Les Marchés Internationaux des Produits de la Mer Marocains (3ème version) Par Roland Wiefels. 17 Octobre 2014

[4] Taiek T, Boutaleb N, Bahlaouan B, El Jaafari A, Khrouz H, Safi A, El Antri S. Valorisation de déchets de poisson alliés à des rejets brassicolesen vue d'obtenir un biofertilisant. Déchets Sciences et Techniques. N°68. Novembre 2014

[5] Chabbar AZ. Technologie de l'ensilage de poisson et disponibilité de la matière première au Maroc. Mémoire de 3ème cycle, IAV Hassan II, Rabat, Maroc. 1996

[6] Afilal ME, Elasri O, Merza Z. Caractérisations des déchets organiques et évaluation du potentiel Biogaz (Organicwastecharacterization and evaluation of itspotentialbiogas). Journal of Materials and Environmental Science. 2014;**5**(4):1160-1169

[7] Angilella M, Böhmer J, Kdadri O, Köhler R, Ali AM, Reis A, Runge N, Sutterer N. Etude sur les potentiels de biomasse pour la région Souss-Massa-Drâa et la province d'Essaouira. 2010

[8] El Moutawakil A. Problèmes de la décharge de Kénitra. In: Actes du Séminaire sur le Compostage et la Gestion des Décharges; Kénitra, Maroc. 1995. pp. 103-112

[9] Benzakour M, Saissi R, Didier G. Situation actuelle et évolution de la gestion des déchets solides au Maroc. Expérience pilote de la mise en place d'une décharge contrôlée à Essaouira. 2003. http://doc.abhatoo.net.ma/doc/IMG/pdf/gestion_decharge.pdf

[10] Afilal ME, Belkhadir N, Merzak Z. Global Journal of Science Frontier Research Biological Sciences (USA). 2013;**13**(1):96-102. Online ISSN: 2249-4626 and Print ISSN: 0975-5896

[11] Chen J-H. The combined use of chemical and organic fertilizers and/or biofertilizer for crop growth and soil fertility. In: International Workshop on Sustained Management of the Soil-Rhizosphere System for Efficient Crop Production and Fertilizer Use. 16-20 October 2006

[12] Xu M-G, Li D-C, Li J-M, Qin D-Z, Kazuyuki Y, Yasukazu H. Effects of Organic Manure Application with Chemical Fertilizers on Nutrient Absorption and Yield of Rice in Hunan of Southern China. 2008

[13] Lhoussaine M. Les engrais minéraux, Caractéristiques et Utilisations. Transfert de technologie en agriculture. 2000

[14] Karine B. Valorisation des matières résiduelles issues des usines de transformation de la crevette Volet 1: résidus solides séchés. 2013

[15] Inckel M, de Smet P, Tersmette T, Veldkamp T. La fabrication et l'utilisation du compost. 2005

[16] Organisation Des Nations Unies Pour L'alimentation Et L'agriculture (FAO), Rome. Utilisation des engrais par culture au Maroc, Service de la gestion des terres et de la nutrition des plantes, Division de la mise en valeur des terres et des eaux. 2006

[17] Mamouni FZ, Razouk R, Kajji A, Daoui K, El Ouali A, Boukhlifi F. Characterization of Organic Waste Used as Bio-Fertilizers: Case of Potato Peelings. Almond Hulls and Shrimp Shells; 2016

[18] Graham Solomons TW, Craig B. Fryhle Organic Chemistry. 9th ed. Chapter 2. 2000

[19] Martini S, Herrera MX. X-ray diffraction and crystal size. JAOCS. 2002;**79**(3):315-316

[20] Fertial Spa, Laboratoire Agronomique. Analyses: Terre-Eau-Foliaires, in partnership with International Finance Corporation. Manuel: Utilisation des engrais, Grandes cultures, Arboriculture, Cultures maraîchères et industrielles. 2010

[21] Yadav RL, Dwivedi BS, Prasad K, Tomar OK, Shurpali NJ, Pandey PS. Yield trends, and changes in soil organic-C and available NPK in a long-term rice-wheat system under integrated use of manures and fertilisers. 2000

[22] ADEME. Collecte sélective et traitement biologique des biodéchets des ménages, méthodes et outils de mesure des indicateurs de qualité et de performance, 1ère édition. Vol. 2. 2000. p. 12

[23] Ibrahim HM, Salama MF, El Banna HA. Shrimp's waste: Chemical composition, nutritional value and utilization. 1999;**43**(6):418-423

[24] Gildberg A, Stenberg E. A New Process for Advanced Utilization of Shrimp Waste. 2000

[25] Revanker GD. Sea Food Export Journal 10. Avi Publishing Company, Inc.; 1978

[26] Shahidi F. In: Shahidi F, Botta JR. Sea Foods: Chemistry, Processing Technology and Quality. London, Glasgow, Weinheim, New York, Tokyo, Melbourne, Madras: Blackie Academic and Professional; 1994. pp. 320-333

[27] Ravichandran S, Rameshkumar G, Rosario Prince A. Biochemical composition of shell and flesh of the Indian white shrimp Penaeusindicus. American-Eurasian Journal of Scientific Research. 2009;**4**:191-194. ISSN: 1818-6785

[28] Prameela K, Mohan ChM, Smitha PV, Hemalatha KPJ. Bioremediation of shrimp biowaste by using natural probiotic for chitin and carotenoid production an alternative method to hazardous chemical method. International Journal of Applied Biology and Pharmaceutical Technology. 2010;**45**:220-226

[29] Khan M, Nowsad AKMA. Development of protein enriched shrimp crackers from shrimp shell wastes. Journal of the Bangladesh Agricultural University. 2012;**10**(2):367-374. ISSN: 1810-3030

[30] Boukhlifi F, Bencheikh A, Ahlafi H. Characterisation and adsorption propriety of chitin toward copper Cu2+. Physical and Chemical News. 2011;**58**:67-72

[31] Ahlafi H, Moussout H, Boukhlifi F, Echetna M. Kinetics of N-deacetylation of chitin extracted from shrimp shells collected from Coastal Area of Morocco. Mediterranean Journal of Chemistry. 2013;**2**:503-513

[32] Liu S, Sun J, Yu L, Chushu Z, Jie B, Zhu F, Mingjing Q, Jiang C, Yang Q. Extraction et caractérisation de la chitine de la coléoptère HolotrichiaparallelaMotschulsky. 2012. pp. 4604-4611

[33] Sajomsang W, Gonil P. Preparation and characterization of α-chitin from cicada sloughs. Materials Science and Engineering: C. 2010;**30**:357-363

[34] Soltner. Les basses des productions végétales. Ed 23ème T1: le sol et son amélioration. 2003. 464 p

[35] Aït Houssa A, Oubaki L, Reda-Fathmi K, Drissi S, Lamghari M, Benbella M, Chraibi H. Eléments agro-économiques pour réussir la culture du blé tendre en Bour, 2016

[36] Ministère d'agriculture du Maroc, La dernière note stratégique n°106 du ministère de l'Agriculture et de la pêche maritime: Détails de l'évolution de la filière céréalière au Maroc au fil des années. 2017

[37] Abdel-Mawgoud AMR, Tantawy AS, El-Nemret MA, Sassine YN. Growth and yield responses of strawberry plants to chitosan application. European Journal of Scientific Research. 2010;**39**(1):170-177

[38] MacLeod JA, Kuo S, Gallant TL, Grimmett M. Seafood processing wastes as nutrient sources for crop production. Canadian Journal of Soil Science. 2006;**86**:631-640

[39] ADAS UK Ltd. Review of the application of shellfish by-productsto land. 2006

[40] Russell EW. Soil Conditions and Plant Growth. 9th ed. London: Longmans; 1961. 688 p

[41] Deffontaine JP. Une méthode d'appréciation du système racinaire sous prairie. Fourrages. 1964;**19**:91-96

[42] Abiven S. Thèse de doctorat. Relations entre caractéristiques des matières organiques apportées, dynamique de leur décomposition et évolution de la stabilité structurale du sol. Rennes. Agrocampus

[43] Grosbellet C. Thèse de doctorat, Evolution et effets sur la structuration du sol de la matière organique apportée en grande quantité. 2008. 241 p

[44] Frankenberger WT Jr, Abdelmagid HM. Kinetic parameters of nitrogen mineralization rates of leguminous crops incorporated into soil. Plant and Soil. 1985;**87**:257-271

[45] Janzen HH, Bole JB, Biederberk VO, Slinkard AE. Fate of N applied as green manure or ammonium fertilizer to soil subsequently cropped with spring wheat at three sites in Western Canada. Canadian Journal of Soil Science. 1990;**70**:313-323

[46] Mahler RL, Hemanda H. Evaluation of nitrogen fertilizer value of plant materials to spring wheat production. Agronomy Journal. 1993;**85**:305-309

[47] Kuo S, Sainju UM, Jellum E. Winter cover cropping influence on nitrogen mineralization, presidedress soil nitrate test, and corn yields. Biology and Fertility of Soils. 1996;**22**:310-317

[48] Martel S, Desmeules X, Bilodeau M, Côté J. Valorisation agricole des résidus d'usine de transformation de poissons. Rapport final, Agrinova, Alma (Québec). Septembre 2012;**2012**:29

[49] Kallenbach CM, Frey SD, Stuart Grandy A. Direct evidence for microbial-derived soil organic matter formation and its ecophysiological controls. Nature communications. 2016;**7**:13630

10

Blended Composites of Chitosan: Adsorption Profile for Mitigation of Toxic Pb (II) Ions from Water

Asha H. Gedam, Prashil K. Narnaware and Vrushali Kinhikar

Abstract

An environmental pollution is the unfavorable alteration of surrounding toxicity due to heavy metals, organic pollutants, radioactive materials, pesticides, dyes, pigments, fatty/oil impurities and minerals that are responsible for crucial ecological and health concerns. The indiscriminate industrial and anthropological activities render water resources unsuitable for consumptions. Percolations of synthetic pollutants in water are responsible for detrimental effects on aquatic flora and fauna. Environmental contamination of water poses the major challenge to develop efficient water treatment techniques based on usage of biopolymers. Hence, chitosan (de-acetylated chitin: *β-(1 → 4) D-glucosamine*) biosorbent is preferred which is cheap, biodegradable, and biocompatible for the mitigation of few heavy metals from water. Chitosan's flexible skeleton was modified by doping few organic/inorganic moieties to yield biocomposites for adsorption of varied pollutants. In this chapter, the batch adsorption of toxic Pb (II) ions from water using graphite doped chitosan composite (GDCC) as an adsorbent is discussed. Maximum Pb (II) ions adsorption capacity was 6.711 mg/g (from Langmuir) at optimum pH 6 with dosage of 1 g/L in 120 min. Biosorption mechanism is emphasized in context with wastewater cleanup procedures.

Keywords: chitosan, graphite, GDCC, Pb (II) ions, water, adsorption

1. Introduction

Chitin (pronounced as Kite-in) is the precursor of chitosan (pronounced as Kite-o-san) was first discovered in 1811 by a Frenchman named Henri Braconnot as a result of extraction

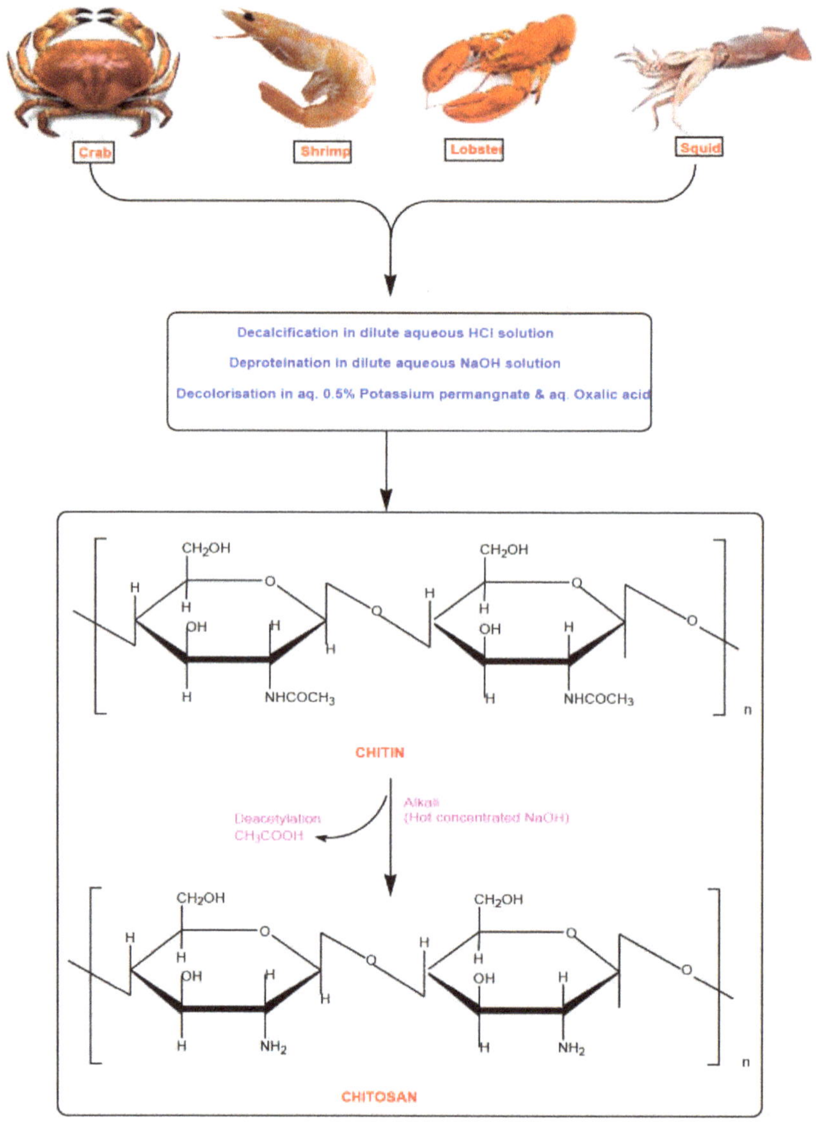

Figure 1. Chemical structures of chitin and chitosan.

from the fungus. Later on in the year 1820, it was derived from the skin of insects. In 1859, C. Roughet obtained "modified chitin" from chitin treated with alkaline sodium hydroxide solution. However, in 1894, F. Hoppe-Seyler treated chitin with different alkaline solution of potassium hydroxide at 180°C and the product called as Chitosan [1]. In 1902, Frankel and Kelly recognized the chemical composition of chitin and chitosan, and in the 1950s, several researchers such as H. Sponsler and W.H. Dore have determined their chemical structures through X-ray experiments.

Chitosan is a hydrophilic natural polymer produced by alkaline deacetylation of chitin, which is the most abundant biopolymer occurring in nature, after cellulose and an exoskeleton part of crustaceans such as shrimp, lobsters, and crab shells [2]. Chitin and chitosan are commercially produced in countries like India, Japan, Poland, Norway and Australia. Chitin and chitosan are nitrogenous polysaccharides that consist of acetyl glucosamine and glucosamine units. Chemically these two polymers are β-(1–4)-2-acetamido-2-deoxy-*D*-glucan and β-(1-4)-2-amino-2-deoxy-*D*-glucan, respectively (**Figure 1**). The difference between the chitin and chitosan lies in the degree of deacetylation (DD) and their solubility in 0.1 M dilute acidic medium. If the degree of deacetylation of chitin falls to 50%, then it becomes soluble in aqueous acidic media such as acetic acid, propionic acid, and so on. [3]. The insolubility of chitosan in water, alkaline medium and organic solvents is due to the presence of hydrogen bonds between its molecules; however, the protonation of amine group renders its solubility in acidic solutions [4]. The DD affects the adsorption capacity of the chitosan. High DD generally results from the presence of high amounts of amino groups, and it can increase the adsorption capacity of chitosan by protonation [5].

Chitin and chitosan possess molecular weight up to several million g/mol. Commercially available chitosan has an average molecular mass ranging from 3800 to 500,000 g/mol and its degree of *N*-acetylation is 2–40% [6]. Deacetylation of chitosan ensures the presence of free amino groups that can be easily protonated in an acidic environment, making chitosan as cationic polyelectrolyte (pKa ≈ 6.5) and water soluble below the pH of 6.5 [7]. It shows high affinity for water pollutants adsorption due to the presence of amine ($-NH_2$) and hydroxyl (–OH) functional groups that act as chelating sites. This integral amine functionality (primary, secondary and tertiary) acquires positive charge in acidic condition and thus become sorption site for anions. Chitin and chitosan are of commercial interest due to their high nitrogen content (7.21%) compared to synthetically substituted cellulose (1.25%) and their excellent properties such as biocompatibility, biodegradability, non-toxicity and adsorptive abilities. It was found that chitosan is highly selective with respect to the uptake of metal ions. It shows an uptake of transition and post-transition metal ions and does not allow the sorption of alkali and alkali earth metal ions from the aqueous solution [8]. These selective adsorption properties have been used for environmental cleanup viz. uptake of heavy metals ions, pesticides, dyes/pigments, radionuclide, and so on from the polluted water resources.

2. Chitosan modification

Although of its several broad spectrum advantages, it has severe limitations viz. lower chemical and mechanical stabilities (due to hydrophilic nature), high pH sensitivity, and solubility in most organic acids, non-porosity and low specific surface area. These inadequacies limit its usage in wastewater treatment applications. Thereby to overcome all of these pitfalls and to tailor it for the specific wastewater treatment application, several attempts for its physical and chemical modification to achieve their biocomposites/nanocomposites have been conducted. The physical and chemical modification of chitosan was adopted to derive the desired adsorbent's characteristics and to improve its adsorption kinetic parameters feasible for pollutants removal.

Physical modification of chitosan has been carried out by various techniques to obtain accomplished polymer as powders, beads, flakes, nanoparticles, hydrogels, films, fibers membranes, sponge, honeycomb, and so on (**Figure 2**). Chitosan beads and fibers of various porosities can be prepared by neutralization methods (**Figure 3**) where chitosan is treated with acetic acid and mixture added drop wise to 1 M NaOH by using microsyringe [9].

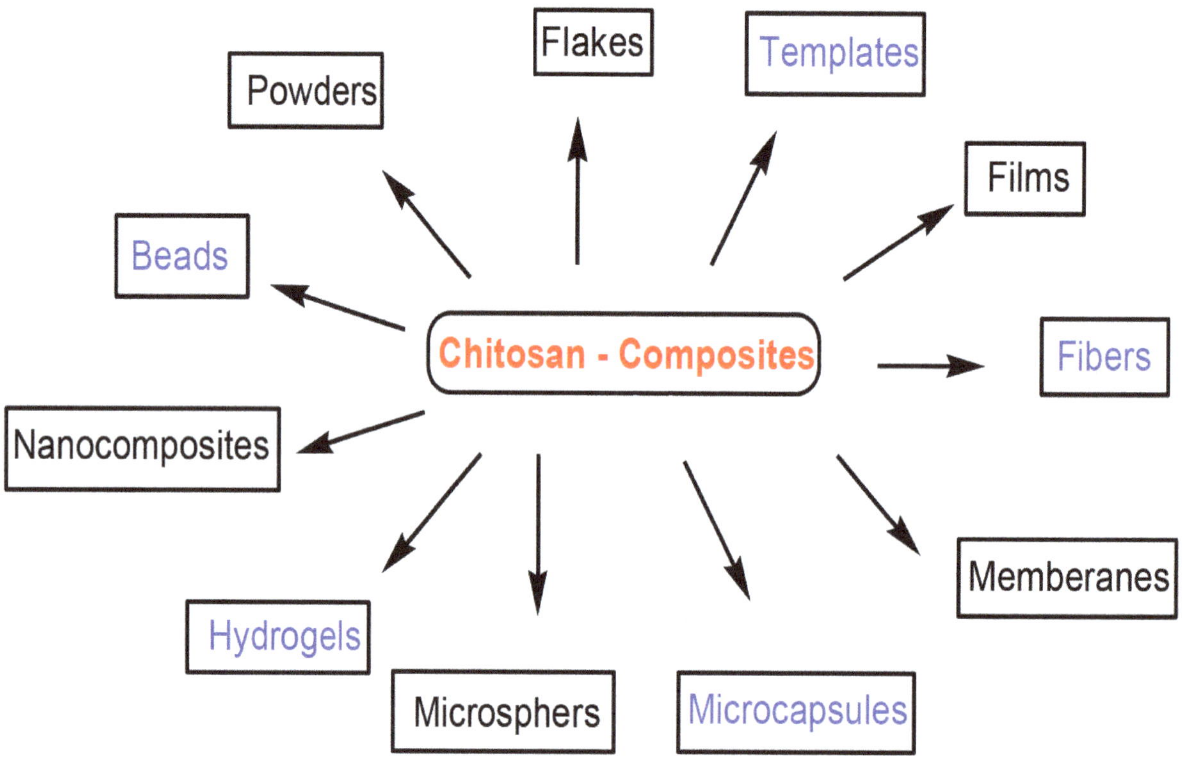

Figure 2. Various physical forms of chitosan bio-composites.

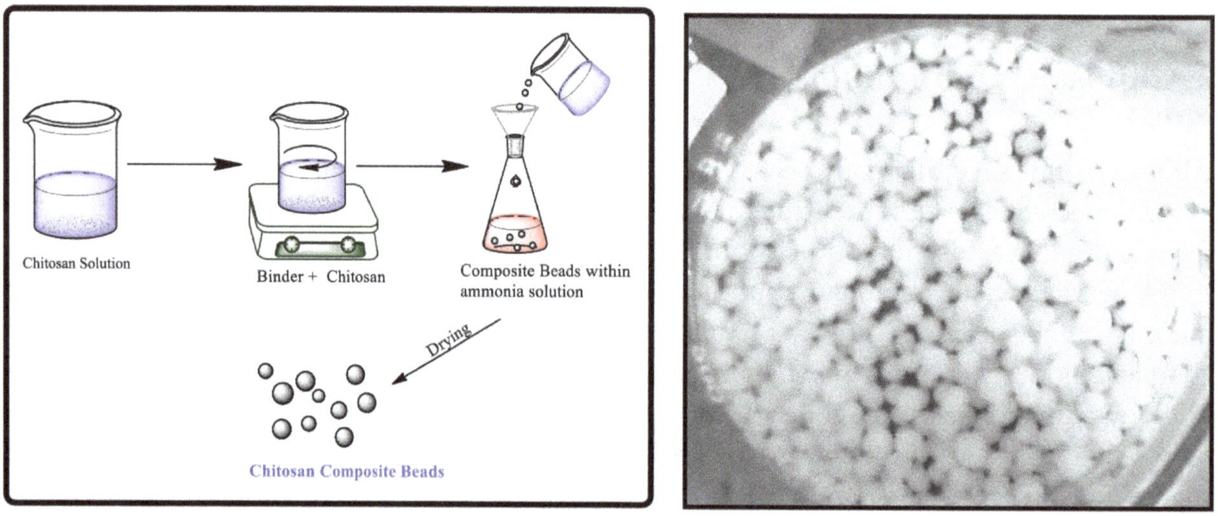

Figure 3. Synthesis of chitosan bio-composites beads.

Chitosan membranes can also be synthesized by treating chitosan with acetic acid where the solution is poured into a Petri dish and once the solvent evaporated, the membrane is neutralized with sodium hydroxide [10]. Moreover, chitosan sponges with different porosities can be prepared by freeze-drying techniques where chitosan solutions or gels are frozen followed by lyophilization [9]. One of the attractive ways of physical modification is to provide new desirable characters to chitosan and to synthesize chitosan-based biocomposites by mixing or blending of chitosan with the support or reinforcement matrix [11]. In blending, at least two polymers are mixed to obtain a new material with different physical properties [12]. At thermodynamic equilibrium, the two polymers of amorphous nature appear to be as a single phase or homogeneous on blending with a new set of improved properties from the individual components. The miscibility and compatibility between the blended polymers are decided by their mechanical and thermal properties [13]. The method of blending is effective in practical application due to its simplicity in operation and availability of various organic compounds and natural polymers. Blending permits the wide range of properties by union of both the components viz. chitosan and the reinforcement matrix to achieve physically and chemically stable biopolymers required for the specific applications.

Chemical modification is the application of various chemical treatments such as cross-linking, sulfonation, carboxymethylation, depolymerization, nitration, alkylation, sulfonation, phosphorylation, xanthations, Schiff's base formation, alkylation, acylation, hydroxylation and graft copolymerization. Various extensive novel chitosan derivatives can be obtained by chemical modifications. The chemical modification of chitosan has two main objectives: (1) to enhance the metal adsorption properties and (2) to improve the stability of chitosan in water or acidic medium. The chemical modification incorporates the various functional groups in the chitosan may involve the $-NH_2$ group at the C-2 position (specific reactions) or –OH groups at the C-3 and C-6 positions (nonspecific reactions) [3]. The various cross-linking reagents such as glyoxal, formaldehyde, glutaraldehyde, epichlorohydrin, ethylene glycol, diglycidyl ether and isocyanates are commonly used to modify the chitosan but are not preferred due to their toxicity. Cross-linking decreases the adsorption efficiency particularly in case of chemical reactions involving amino groups, but it provides mechanical strength and enhances the stability of chitosan against acidic and basic solutions [14].

Graft copolymerization is also the promising method that allows the covalent bonding between the grafted functional groups onto the chitosan backbone. The objectives of grafting new functional groups on chitosan are to alter the pH range and adsorption sites so that to enhance adsorption selectivity for the target pollutant.

Enzyme-modified chitosan is also one of the attractive methods of chitosan modification due to the reaction specificity. The alteration of the surface and rheological properties of chitosan using hexyloxyphenol which was grafted onto chitosan mediated by tyrosinase [15] has been studied for the modification of chitosan to achieve desired characteristics for the particular applications (**Figure 4**) .

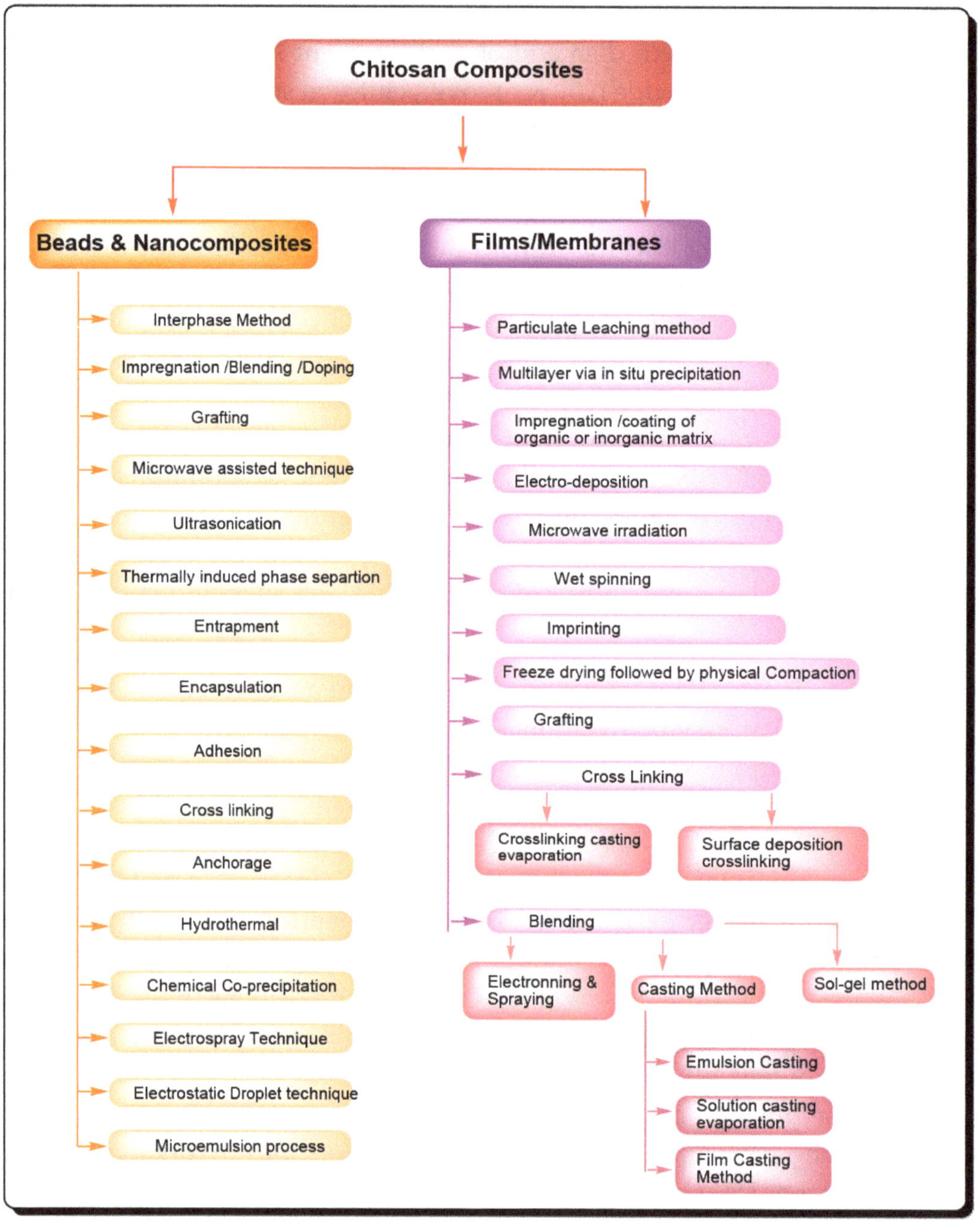

Figure 4. Various synthetic methods of chitosan biocomposites.

3. Applications of chitosan

Chitosan has wide and vast variety of applications ranging from biomedical and cosmetics products to agriculture and wastewater treatment [16]. The applications of chitosan in various fields and its specific properties responsible for specific applications are tabulated as follows (**Table 1**).

Applications of chitosan	Properties of chitosan
Cosmetics	Fungicidal and Fungistatic in nature. Facilitates the interaction with common integuments (skin covers) and hair. Chitin, chitosan and its derivatives offers uses in three areas of cosmetics: Hair, skin and oral care.
Paper industry	Chitosan molecules greatly resemble those of cellulose the main constituent of plant walls. It is used in the production of toilet paper, packaging/wrapping paper and cardboard.
Textile industry	Antistatic and soil repellant properties of chitin derivatives are used in textile industries.
Food processing	Chitosan used in food industry as it is nontoxic to warm blooded animals. Microcrystalline chitin shows good emulsifying properties, superior thickening and gelling agent for stabilizing foods.
Agriculture	It acts as plant growth enhancer.
Photography	Scratch resistance, optical and film forming property.
Chromatographic separation	The presence of free $-NH_2$ group and primary and secondary –OH groups enables its use as chromatographic support by using HPLC.
Solid-state batteries	Solubility in acetic acid facilitates ionic conductivity.
LED application	Dyes containing chitosan gel have been used as potential component in laser and other light emitting devices (LEDs).
Biomedical applications	Such as artificial skin, artificial kidney, wound dressing, drug delivery system and space filling implants. Chitosan has been found to have an accelerator effect on the tissue engineering process owing to its polycationic nature. Chitosan used for burn treatment since it can form tough water absorbent, biocompatible films and shows excellent oxygen permeability.
Opthalmology	It possesses all the properties required for an ideal contact lens, optical clarity, mechanical stability, sufficient optical correction, gas permeability, wettability and immunological compatibility.
Pharmaceutical application	Since chitin and chitosan do not cause any biological hazard and are inexpensive, these polymers might be suitable for use in the preparation of commercial drugs.

Table 1. Applications of chitosan and its composites in various fields.

4. Wastewater treatment application

The toxic heavy metal ions must be detained before its percolation into the water resources to protect the aquatic flora, fauna, human beings and consequently the environment.

To accomplish the increased stringent environmental regulations and maximum permissible limit of contaminants in water, a wide range of treatment technologies such as chemical precipitation, coagulation flocculation, flotation, ion exchange, membrane filtration, electrochemical treatment technologies, adsorption/bioadsorption [17], and so on are most frequently examined. Among the aforementioned technologies, adsorption has been preferred due to its flexible operation, generation of high-quality treated effluent and regeneration of

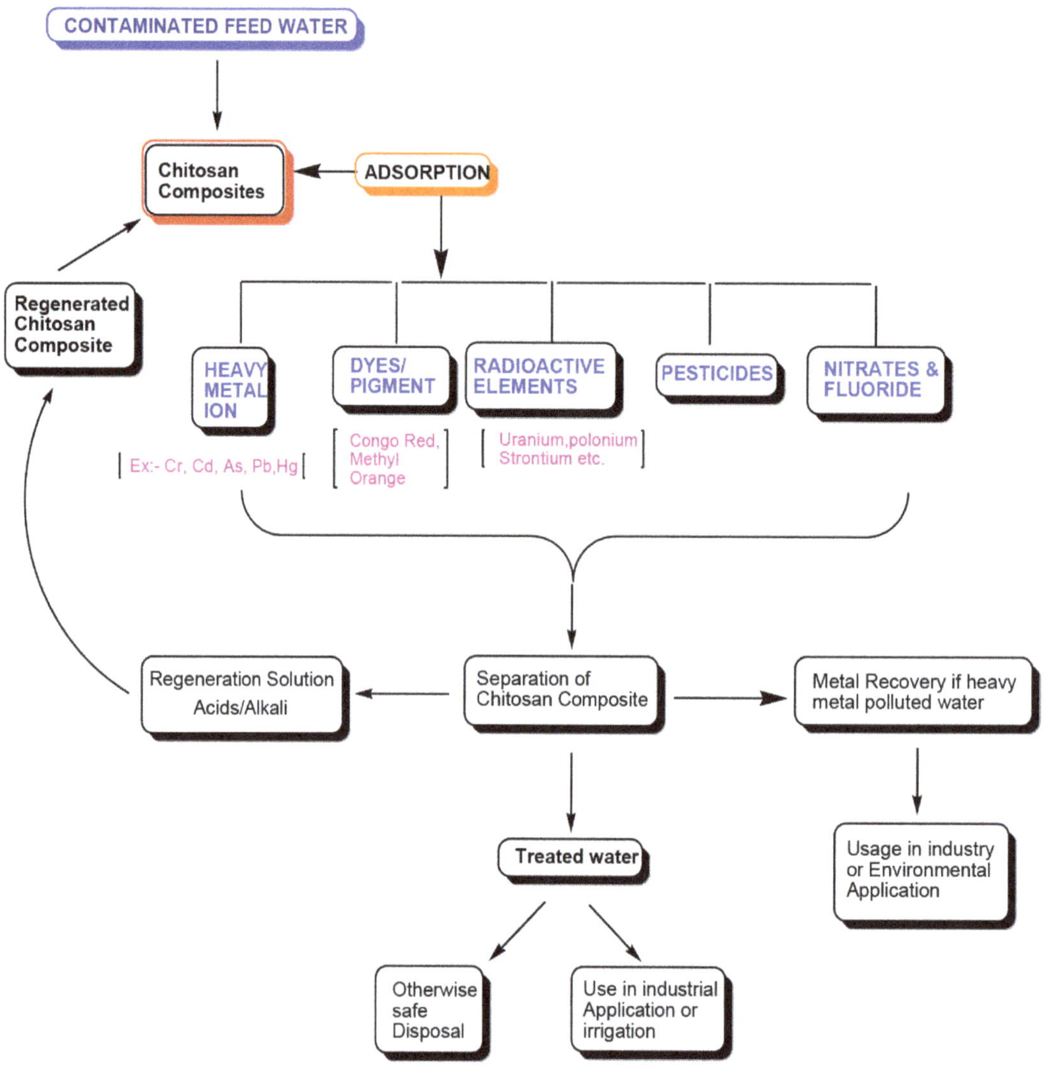

Figure 5. Schematic representation of chitosan composite/biocomposite in wastewater treatment application.

adsorbent by desorption. In chitosan, the integral $-NH_2$ and –OH functional groups acts as a chelating sites for the adsorption of various water pollutants viz. heavy metal ions, dyes/pigments, pesticides, and so on (**Figure 5**).

5. Heavy metal contamination status in India

The Central Pollution Control Board (**CPCB**) [18] carried out a major groundwater quality survey and the report recognized about 20 critical sites of ground water pollution in various states of India. CPCB found that industrial effluents are the primary and major cause for ground water pollution. The major heavy metals contamination sites including lead metal in Indian scenario are given in **Table 2**.

The heavy metal lead (Pb) was considered as father of all metals during Roman era. Much of its gratitude was due to its huge availability, consequently used in daily life by people across

Sr. No.	Area	Industrial activities	Groundwater quality problems
1	Digboi (Assam)	Oil refinery	Fe and Mn ions were more than permissible limit. Ni, Zn, Cd, Cr, Pb were also reported.
2	Howrah (West Bengal)	Foundries and Electroplating	Heavy metals viz. Pb, Cd, Cr were very high and Zn, Cu, were within limit. Hg, Fe, Mn and pesticides were also very high, CN and phenolic compounds in traces.
3	Botharam Patncheru (AP)	Pesticides, Pharmaceuticals	Phosphates, Hg, As, Cd, Fe, Mn and Pb were beyond limit, Pesticides, coliform, TDS, were also exceeded the desirable limit.
4	North Arcot (TN)	Tanneries and dyeing units	Hg, Cd, Pb and As were in traces. Zn, Cu, Cr, Fe and Mn beyond limit at several locations. Total Coli form and fecal Coli form were also on higher side.
5	Ratlam, Nagda (MP)	Distillery Dye, Pharmaceuticals	TDS, Hg, Pb were on higher side. Pesticides and fecal Coli forms were also present.
6	Mandi Gobindgarh (Punjab)	Wooden, chemicals, electro-plating and other steel metals units.	Pb, Cu, Cd exceeded the desirable limit of drinking water. Phenol compounds and cyanide were also present on higher side.
7	Parwanoo (HP)	Ancillary, fruit proceeding plant, air pesticides.	The presence of Cd, Pb, Fe, and Mn was observed on higher side. Pesticides and phenol were above the toxic limit.
8	Kala Amb (HP)	Paper mills	Heavy metals like Cd, Pb and Mn and Phenol compound were higher than the toxic limit. Pesticides, Coli forms were also present.
9	Pali (Rajasthan)	Textile, dyes	Pb and Zn, F, TDS and Cl found higher.
10	Jodhpur (Rajasthan)	Textile, steel, engineering foundry, chemicals, oil, pulses and rubber.	Heavy metals such as Fe, Cr, Mn, Pb were also on higher side. Na, TDS exceeded the limit.
11	Angul Talcher	Thermal power station, fertilizers, chemicals, mining activities.	Cr, Fe, Cd, Pb & F, NO_3 all were found in concentration level exceeding standards limits.

Table 2. Pb (II) ions contamination status in India.

all classes. The special attention needs to be given on its mitigation from contaminated water bodies. Lead was ranked second on the Comprehensive Environmental Response.

Compensation and Liability Act (1980) (CERCLA) Priority List of Hazardous Substances in 1999 and 2001 (after arsenic #1, and before mercury - #3). The priority list is prepared by the Agency for Toxic Substances and Disease Registry (ATSDR) and Environmental Protection Agency (EPA) and is based on the frequency of occurrence of particular contaminants at National Priorities List (NPL) sites and their potential threat to human health. Pb (II) ions can be found in effluents from battery recycling plants, lead mining and electronic assembly plants. Lead metal elucidates destructive effects almost on every organ systems viz. nervous, blood circulation, reproductive, digestive, kidneys as become highly toxic and carcinogenic even at low concentration. World Health Organization (WHO) prescribed the maximum permissible limit (MPL) of lead metal in drinking water as 50 ppb initially during 1995 that was further decreased to 10 ppb in 2010. However, more recently, an EPA document recommended a zero lead value in a national primary drinking water standard [19].

In recent years, the chitosan blended biocomposites have been synthesized by impregnation with graphite [20], iodate [21] activated carbon of *Luffa cylindrica* [22], and so on and were utilized for Pb (II) ions mitigation from water. The choice of these materials was concerned with its high adsorption efficiency, safe and simple to use, easy to maintain, minimal production of residual mass, low capital cost and nontoxicity. The resultant adsorbents viz. graphite doped chitosan composite (GDCC), Iodate doped chitosan composite, and activated carbon of *Luffa cylindrica* doped chitosan biocomposite satisfy all these requirements during their usage as a bioadsorbents. In this chapter, the synthesis, characterization and batch adsorption of Pb (II) ions by using GDCC are explained.

6. Synthesis of graphite doped chitosan composite (GDCC)

Chitosan dissolved in acetic acid and heated at 50°C to obtain gel followed by the addition of powdered graphite in (1,1 w/w) ratio. Mixture was then agitated magnetically (800 rpm) at 27°C) for 5–6 h and dropped in aqueous ammonia to obtain beads. Finally, it was filtered, washed with double distilled water and dried in oven at 70–80°C. The GDCC was grounded, and the particle size recorded in range of 176–246 μm. (**Figure 6B**).

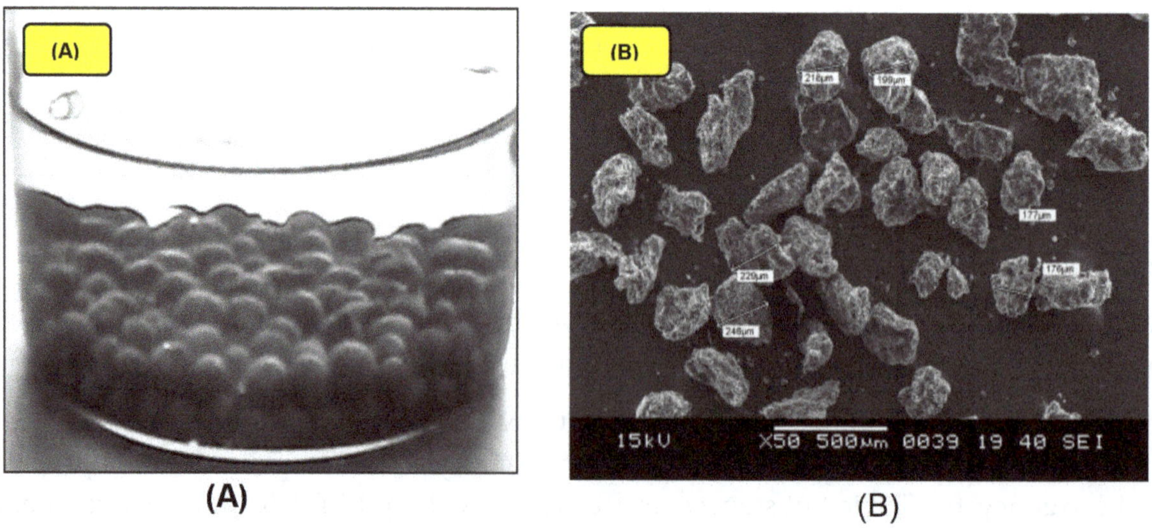

Figure 6. (A) GDCC beads and (B) SEM image of powdered GDCC, with particle size in range of 176–246 μm.

7. Results and discussion

7.1. Physicochemical characterization of GDCC

The results of the proximate and elemental analysis of GDCC is shown in **Table 3**

7.1.1. TGA and DSC analysis of GDCC

Thermogravimetric/Differential Scanning Calorimetry (TGA/DSC) was used to evaluate the thermal stability and to determine the decomposition temperature of the adsorbents. TGA and DSC analysis of chitosan (CS), Graphite (Gr), and Graphite doped chitosan composite (GDCC) are shown in **Figure 7(A)** and **(B)** respectively.

From TGA curve, it was observed that CS showed two steps of degradation. The initial degradation occurred at around 30–100°C and displayed 5% weight loss. This degradation may correspond to the loss of adsorbed and bound water or moisture vaporization. Initial decomposition around 100°C can be attributed to the strong water adsorptive nature of CS. The second decomposition stage was at 270°C and continued up to 312°C with 46.28%

Adsorbent	Proximate analysis (%)				Elemental analysis (%)				
	Moisture	Volatile matter	Ash	Fixed Carbon	C	H	N	S	O
CS	5.3		6.2		38.5	7.8	7.1	0.19	46.45
GDCC	7.2	55	4	33.8	57.69	3.78	4.03	0.23	34.27

Table 3. Proximate and elemental analysis of GDCC.

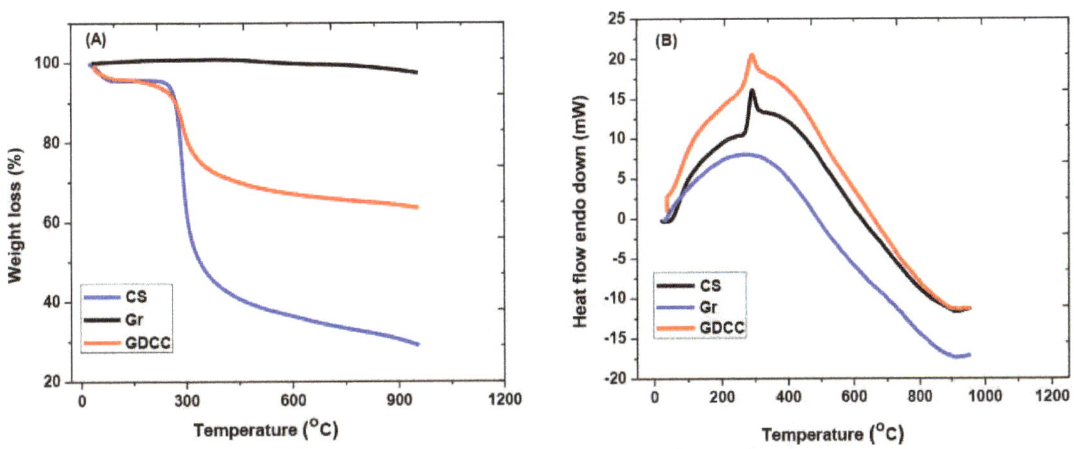

Figure 7. (A) Thermogravimetric analysis (B) differential scanning calorimetric analysis of chitosan (CS), graphite (Gr) and graphite doped chitosan composite (GDCC).

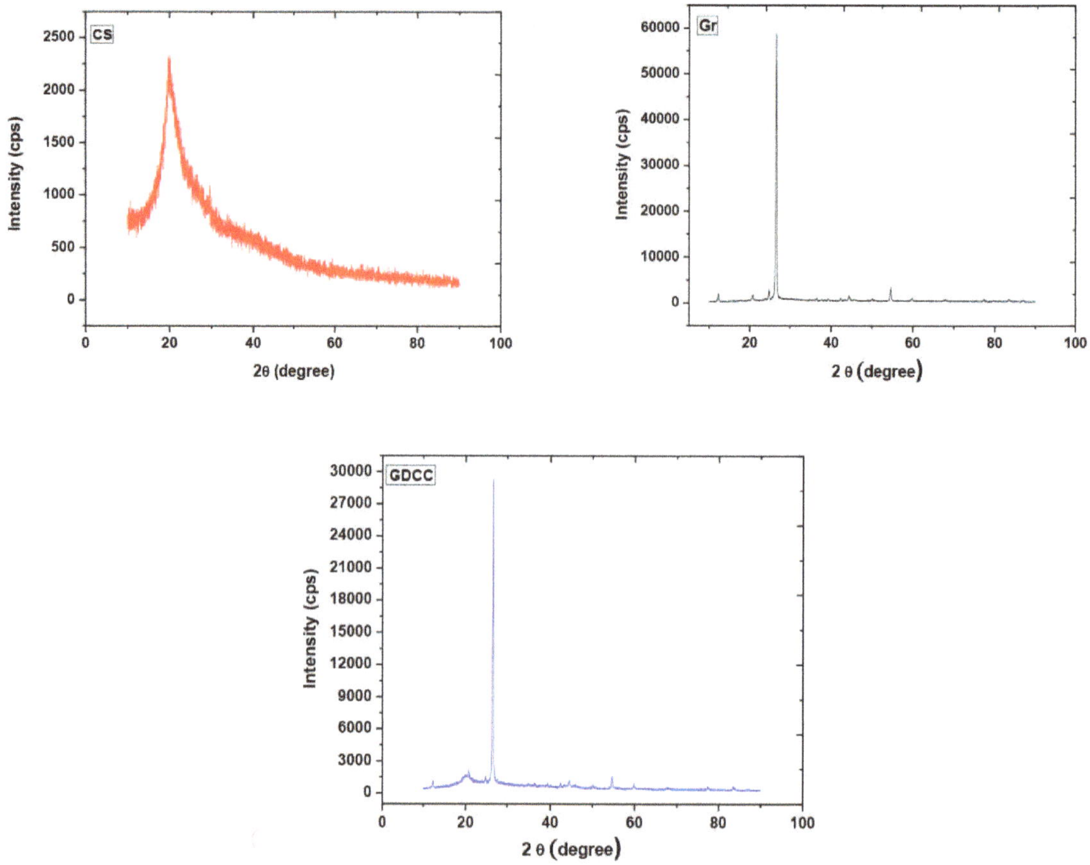

Figure 8. XRD analysis of chitosan (CS), powdered graphite (Gr) and GDCC.

weight loss. The temperature at which maximum degradation occurred was 288.35°C. At the end of 955.1°C, the total weight loss of CS was 70%. TGA analysis of powdered graphite shown high thermal stability up to 700°C and displayed only 2.5% weight loss at the end of 955°C.

TGA analysis of GDCC also exhibited two steps of degradation. First stage decomposition occurred between 38.01 and 200°C which showed about 5% weight loss due to evaporation of water. The second stage of decomposition showed a weight loss of 18.37% in the temperature range of 265.15–321.6°C. The temperature at which the maximum degradation occurred was 288.55°C. At the end of 955°C, the total weight loss was 35%. The TGA analysis revealed that with respect to powdered graphite, the GDCC became less thermally stable, whereas with respect to the CS, the composite became more thermally stable. These observations showed a good miscibility between CS and graphite to achieve GDCC or biocomposite.

The DSC curve of CS and GDCC both shows one exothermic peak at 292 and 291.37°C, respectively. For CS, the onset of exothermic peak was at 276.89°C and continued up to 311.49°C with Δ H = −149 J/g and for GDCC the onset of exothermic peak was at 270.64°C and continued up to 311.86°C with Δ H =.–60.1211 J/g.

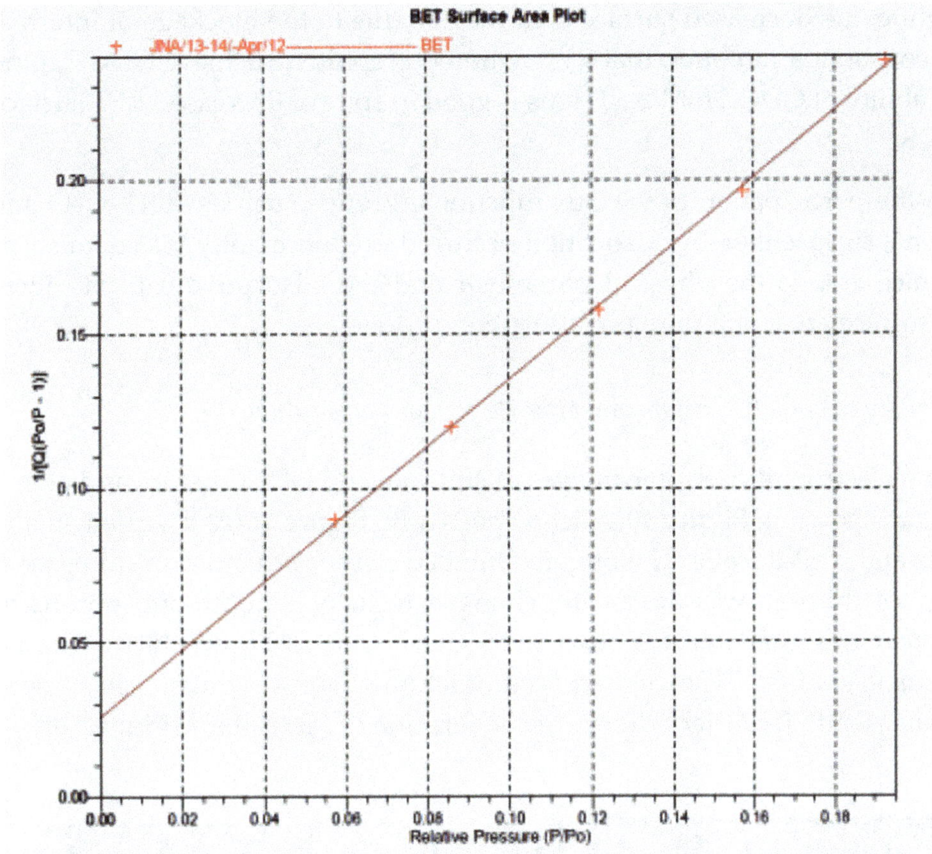

Figure 9. BET surface area plot of GDCC.

7.1.2. XRD analysis of GDCC

The XRD pattern of powdered graphite, chitosan and GDCC is shown in **Figure 8**. X-ray diffraction pattern of CS exhibited broad diffraction peak at 2θ = 20° with d-spacing of 4.2 Å is characteristic of semi crystalline chitosan [23]. The peaks are broadened due to amorphous nature of chitosan polymer. The diffraction peak appeared at 2θ = 26.5° which indicated d-spacing of about 3.35 Å is a characteristic of graphite peak [24]. The XRD pattern of the GDCC indicated the formation of homogeneous/single phase composite, and the peaks were obtained at 2θ value 26.5°. The broad peak at around 2θ = 20° which was due to CS decreased in intensity after doping with graphite which confirms that graphite is doped on the surface of chitosan. A predominant peak of graphite along with small peak of chitosan appeared in GDCC showed that the incorporation of graphite in matrix was successful and effectively provided a support to the chitosan.

7.1.3. BET surface area analysis of GDCC

The BET surface area plot of GDCC is shown in **Figure 9**. The BET surface area of GDCC adsorbent was 3.89 m^2/g, whereas for CS, it was 9.923 m^2/g. Thus, it was observed that the BET surface area of GDCC composite was decreased with respect to CS. During modification of

CS by graphite, the decreased surface area may be due to the blockage of internal porosities of CS by incorporated modifier, that is, powdered graphite to achieve GDCC composite. The adsorptive ability of GDCC for Pb (II) ions is good in spite of decreased BET surface area with respect to CS.

It is due to the participation of various functional groups such as –OH, C=O and $-NH_2$ on the adsorbents surface thereby adsorption occurred predominantly via chemisorption mechanism. Surface area is the physical parameter and the adsorptive capacity increases with increasing surface area for a pure physisorption process.

7.1.4. SEM analysis of GDCC before and after Pb (II) ions adsorption

The scanning electron microscopic images of chitosan and GDCC before and after Pb (II) ions adsorption are shown in **Figure 10(A-C)** respectively. **Figure 10(A)** revealed small amount of round voids and well-developed elongated bilobed porous structure in chitosan. The surface morphology of chitosan was drastically changed to flaky, smooth and porous nature with some voids/cavities after impregnation with graphite as in **Figure 10(B)**. The morphology of Pb (II) ions loaded GDCC exhibited accumulation of shiny, whitish, sharp needle shaped crystalline mass onto its surface due to the adsorption of Pb (II) ions (**Figure 10**C).

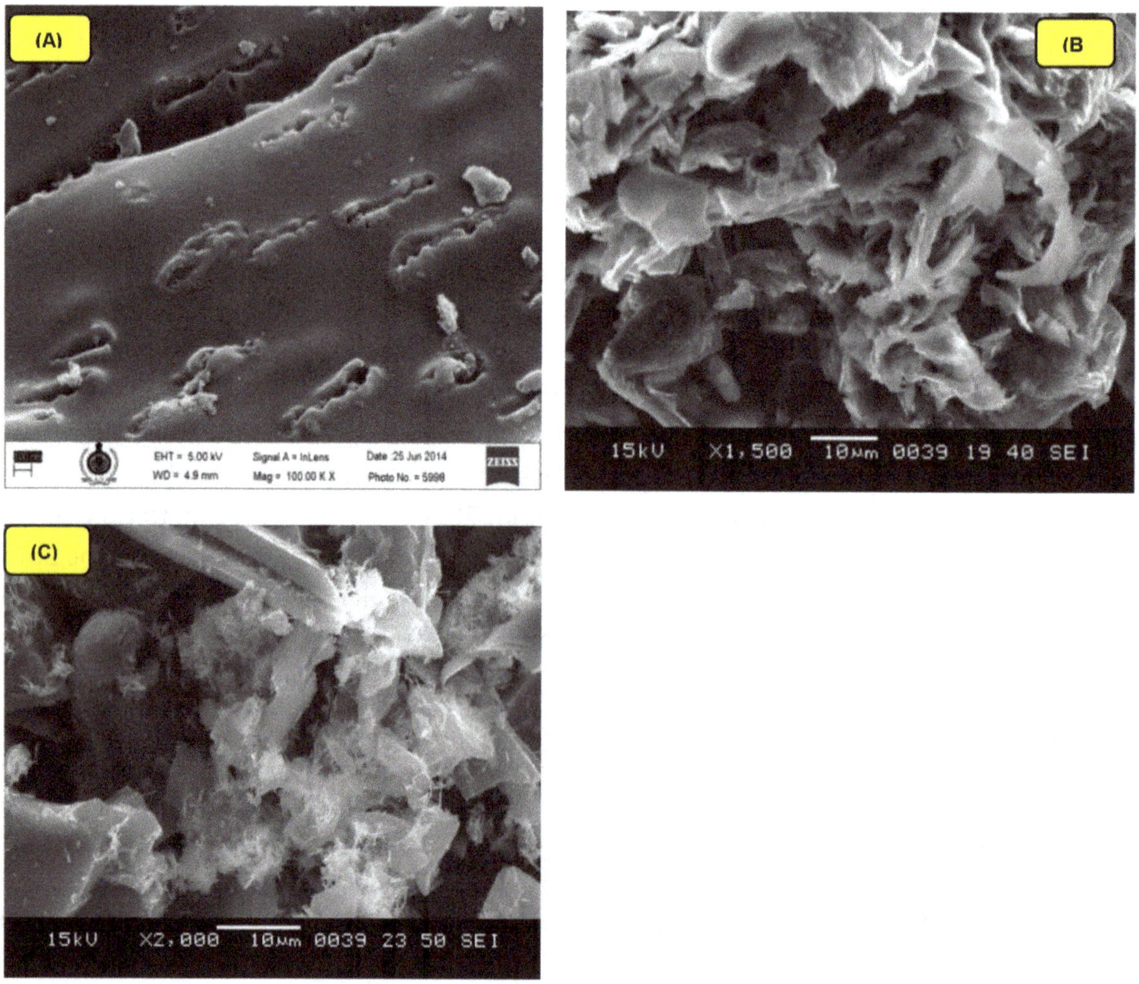

Figure 10. SEM image of (A) CS, (B) GDCC before adsorption, and (C) GDCC after adsorption.

8. Adsorption mechanism

The bioadsorbents possess various functional groups like carboxyl, hydroxyl, amino, phosphate, and so on that can provide an active binding site for the adsorption of heavy metal ions. The mechanism of bioadsorption is quiet complicated due to assorted structure of the bioadsorbents. The factors that affect the efficient bioadsorption onto the surface of biosorbents are the availability of number of active binding sites, the affinity of pollutant for the bioadsorbent surface and the presence of variety of functional groups that can exhibit an acceptor-donor interaction with the heavy metal ions. The adsorption process is a combination of ion exchange, complexation, precipitation, and so on and greatly influenced by the solution pH. Similarly, in order to understand the adsorption mechanism, it is also necessary to determine the pH of point zero charge (pHpzc) of the adsorbent. pHpzc is of prime importance in the field of environmental science. It determines how easily and adsorbent adsorbs toxic ions. The difference between the initial pH (pHi) and final pH (pHf) values is plotted against initial pH (pHi). The point of intersection of the resulting curve at which difference between pH = 0 is noted as pH of point zero charge. The cationic adsorption is favored at pH > pHpzc while anionic adsorption is favored at pH < pHpzc [25] This is due to the fact that at low pH values, hydronium ions concentration increases that competes with cationic pollutants for the adsorption sites on the adsorbent. While at higher pH, hydroxide ions concentration increases and the adsorbent surface becomes negatively charged thereby increases the attraction between the cationic pollutants with the adsorbent surface. Thus, pH > pHpzc is favorable for the cationic adsorption. In anionic adsorption, solution pH should be less than the pH pzc so that the adsorbent surface becomes positively charged to enhance the anionic pollutants adsorption onto the adsorbent surface. The general adsorption mechanistic representation is shown in **Figure 11**.

On the basis of results obtained from the analytical and spectroscopic data, the schematic representation of Pb (II) ions adsorption mechanism onto GDCC is represented in scheme as below.

The amino group of chitosan plays a major role in removal of Pb (II) ions via adsorption as it acts as coordination site for metal ions. The amino group gets protonated in acidic medium due to reaction with H^+ ions and is chemically represented as below.

$$RNH_2 + H^+ \longrightarrow \underset{\text{Protonated Chitosan}}{RNH_3^+}$$

The FTIR analysis revealed the corresponding prominent changes of –NH bending vibrations after doping of graphite with chitosan indicated that –NH vibration is affected as a result of modification. Thus, the doping of graphite with chitosan results in formation of nonprotonated chitosan-graphite composite and is represented as follows.

$$\underset{\text{(Protonated Chitosan)}}{RNH_3^+} + \text{Graphite} \longrightarrow \underset{\text{(GDCC composite)}}{(RNH_2\text{ - Graphite})} + H^+$$

The abovementioned relationship suggests that the acidic pH can enhance the complexation between chitosan and graphite.

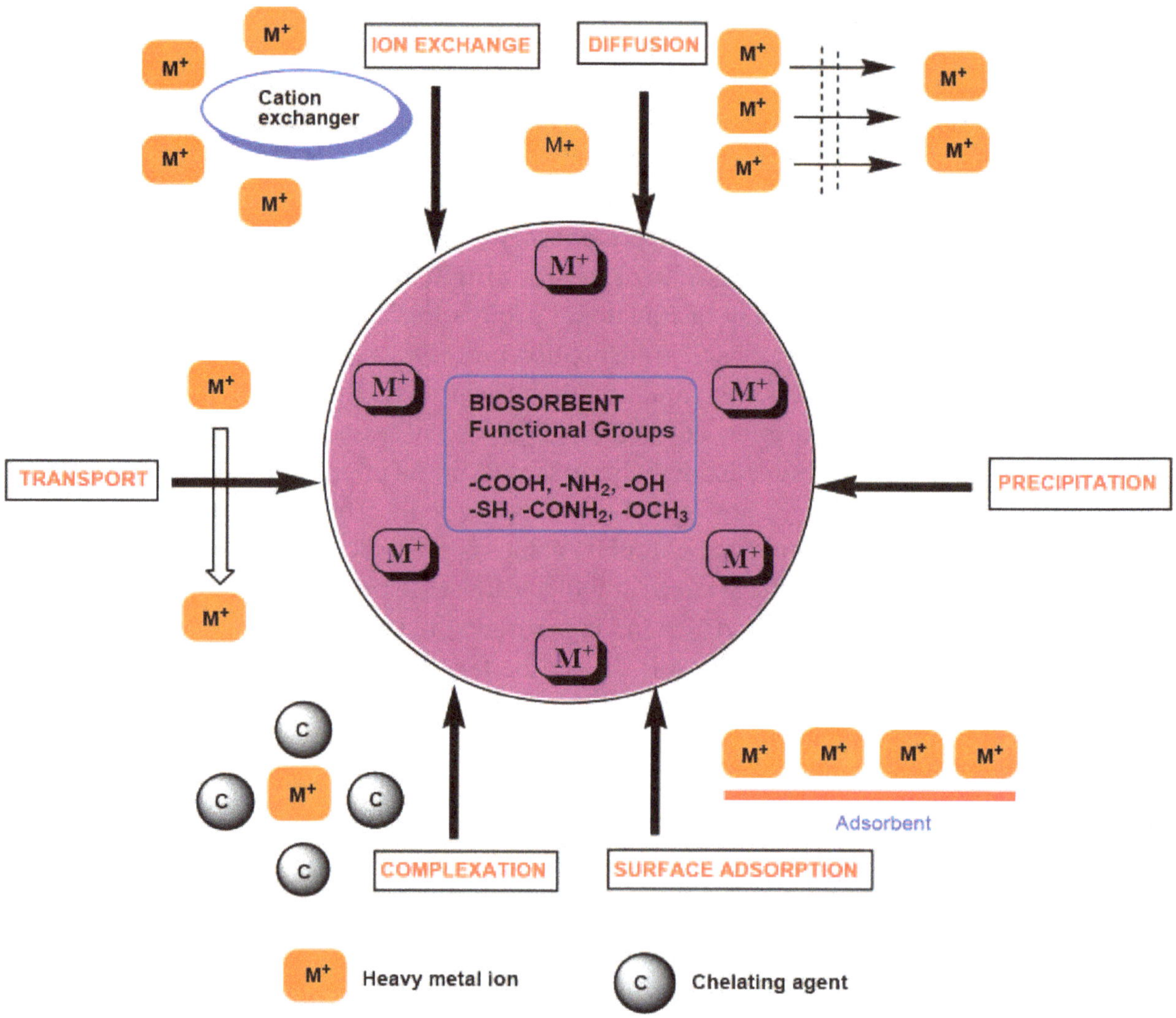

Figure 11. Plausible adsorption mechanism of heavy metal ions onto the adsorbent.

As chitosan acts as a chelating agent signifies nitrogen atom as the prominent adsorption site for Pb (II) ions adsorption. The chitosan-graphite complex binds with Pb (II) cation via the formation of coordination bond, and it is represented as follows:

$$(RNH_2 - \text{Graphite}) + Pb^{2+} \longrightarrow Pb - (RNH_2 - \text{Graphite})^{2+} \text{ OR } (Pb - GDCC)^{2+}$$

9. Adsorption kinetics

To study the mechanism and kinetics of Pb (II) ions adsorption, characteristic adsorption constants were determined using pseudo first order, pseudo second order and intraparticle diffusion models.

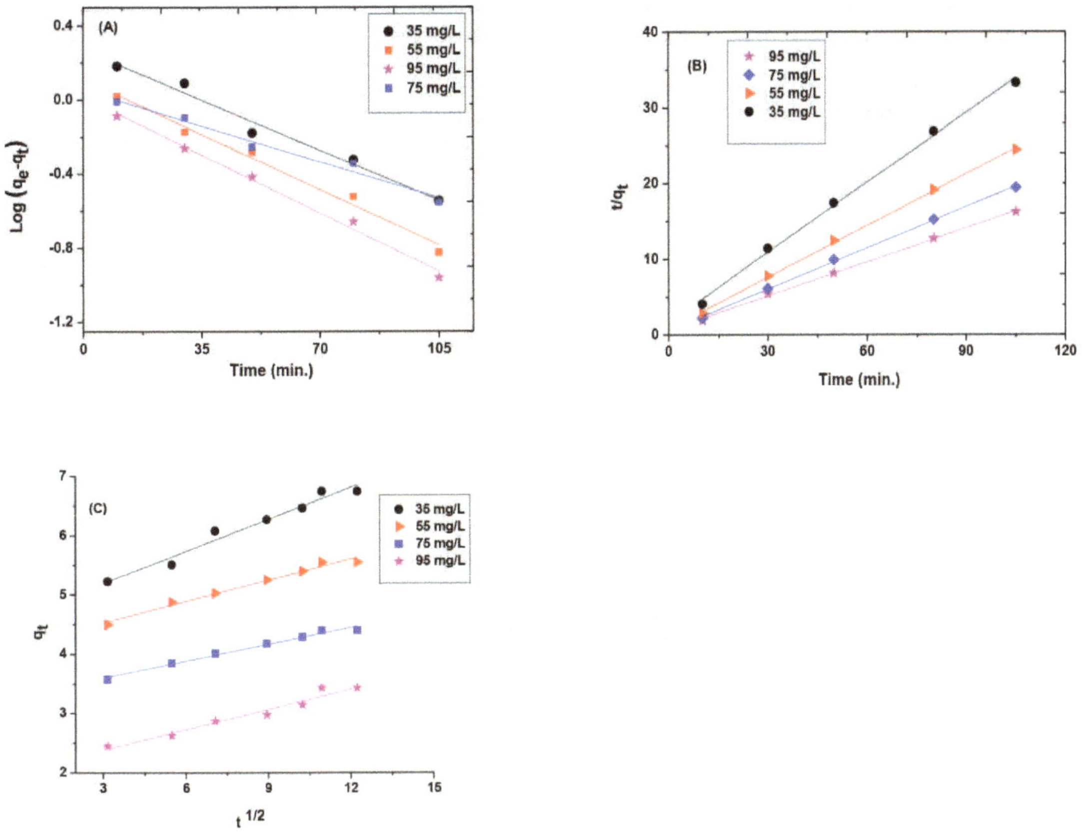

Figure 12. Linear plot of (A) pseudo first-order kinetics (B) pseudo second-order kinetics and (C) Weber-Morris Intraparticle diffusion model.

The Lagergren pseudo first-order plot, Ho presented Pseudo second-order kinetics plot and Weber Morris Intraparticle diffusion plot are shown in **Figure 12(A-C)**. Similarly, the kinetic parameters viz. rate constant (k_1 and k_2), equilibrium adsorption capacity (q_e) and correlation coefficient (R^2) are indicated in **Table 4**. Equilibrium adsorption capacity (q_e) and the pseudo first-order rate constant (k_1) can be obtained from the intercept and slope of plot between Log (q_e–q_t) against time t. Similarly, the equilibrium adsorption capacity (qe) and the pseudo-second-order rate constant K_2 were obtained from the slope and intercept of the plots of t/qt against t.

The linear correlation coefficient values of pseudo first order are comparatively lower than pseudo second-order kinetics mechanism. The calculated q_e values are much higher than the experimental q_e values for 35 to 95 mg/L Pb (II) ions concentration and thus does not represent the good fit of pseudo first order with the experimental adsorption data. Consequently, it can be concluded that the adsorption of Pb (II) ions onto GDCC is not better explained by the pseudo first-order kinetics mechanism. Results presented in the table clearly show that the correlation coefficient for pseudo second-order equation is higher than pseudo first order for all 35 to 95 mg/L concentrations of Pb (II) ions. Similarly, high k_2 of pseudo second order suggested that the metal could be rapidly sequestered by carbon functional groups, resulting in the system quickly reaching equilibrium. The calculated q_e values from pseudo second order are much closer and in good agreement with the experimental q_e values, which indicate that

Pb (II) ions (mg/L)	Pseudo-first-order			q_e (cal.)	Pseudo-second-order			Intraparticle diffusion model		
	K_1 (min^{-1})	q_e (mg/g)	R^2		K_2 (g/ mg^{-1} min^{-1})	q_e (mg/g)	R^2	$K_i t$ (mg/g $min^{0.5}$)	C (mg/g)	R^2
35	0.0115	1.1350	0.97	3.43	0.0545	3.2573	0.99	0.114	2.042	0.95
55	0.0207	1.0423	0.99	4.4	0.0628	4.4052	0.99	0.094	3.319	0.98
75	0.0184	1.3001	0.98	5.55	0.0505	5.5248	0.99	0.119	4.176	0.98
95	0.0161	1.8663	0.97	6.745	0.0324	6.7114	0.99	0.179	4.658	0.96

Table 4. Adsorption kinetics parameters for Pb(II) ions onto GDCC.

the adsorption of Pb(II) ions by GDCC follows pseudo second-order kinetics. The confirmation of pseudo second-order kinetics indicates that during the adsorption process, concentration of both adsorbent and adsorbate is involved in rate-determining step, which may be chemical adsorption or chemisorptions [26].

Intraparticle diffusion parameters are shown in **Table 4**. The intraparticle diffusion k_i values were obtained from the slope of a plot of q_t versus $t^{1/2}$. From figure, it follows that the correlation coefficient values are lower for varying Pb (II) ions concentration (35, 55, 75 and 95 mg/L) than pseudo second-order kinetics. Similarly, intraparticle diffusion plot is not linear, and the straight line does not pass through the origin, indicating that intraparticle diffusion was involved in adsorption but was not the only rate-controlling step.

10. Conclusion

The quality of water is an ever growing concern throughout the developing countries. The natural and manmade activities have a large impact on drinking water contamination that ultimately affects the human health, ecological balance and social and economic progress of countries. The chemical contamination due to heavy metal ions makes water unsuitable for drinking. Among the prominent chemical pollutants, arsenic, mercury and lead threatens health of billions of world population. It is very important for a rapidly developing country like India to be vigilant of these heavy metal problems and to ascertain preventive and remedial measures for their management. Sometimes modern and expensive remedial measures are problematic for a country like India, and hence the attempts have to be focused on the prevention and mitigation of the environmental pollutants. This chapter dealt with the mitigation of heavy metal Pb (II) ions from contaminated water using graphite doped chitosan composite (GDCC). Maximum Pb (II) ions adsorption capacity was 6.711 mg/g (from Langmuir) at optimum pH 6 with dosage of 1 g/L in 120 min. The choice of these materials was concerned with its good adsorption efficiency, safe and simple to use, easy to maintain, minimal production of residual mass, low capital cost and nontoxicity.

Author details

Asha H. Gedam[1]*, Prashil K. Narnaware[2] and Vrushali Kinhikar[3]

*Address all correspondence to: agedam.ccoew@gmail.com

1 Department of Chemistry, Cummins College of Engineering for Women, Nagpur, India

2 Visvesvaraya National Institute of Technology, Nagpur, India

3 G H Raisoni Institute of Engineering and Technology, Nagpur, India

References

[1] Pastor A. Quitina y Quitosano: Obtención, caracterización y aplicaciones. Programa CYTED, CIAD, A.C., Fondo Editorial de la Pontificia Universidad Católica del Perú, Lima, Perú; 2004

[2] Arrascue ML, Garcia HM, Horna O, Guibal E. Gold sorption on chitosan derivatives. Hydrometallurgy. 2003;**71**:191-200

[3] Rinaudo M. Chitin and chitosan: Properties and applications. Progress in Polymer Science. 2006;**31**:603-632

[4] Aranaz I, Mengíbar M, Harris R, Pãnos I, Miralles B, Acosta N, Galed G, Heras Á. Functional characterization of chitin and chitosan. Current Chemical Biology. 2009;**3**: 203-230

[5] Piccin JS, Vieira MLG, Goncalves JO, Dotto GL, Pinto LAA. Adsorption of FD&C Red No. 40 by chitosan: Isotherms analysis. Journal of Food Engineering. 2009;**95**:16-20

[6] Tajik H, Moradi M, Rohani SMR, Erfani AM, Jalali FSS. Preparation of chitosan from brine shrimp (Artemia urmiana) cyst shells and effects of different chemical processing sequences on the physicochemical and functional properties of the product. Molecules. 2008;**13**:1263-1274

[7] Lavorgna M, Piscitelli F, Mangiacapra P, Buonocore G. Study of the combined effect of both clay and glycerol plasticizer on the properties of chitosan films. Carbohydrate Polymers. 2010;**82**:291-298

[8] Muzzarelli RA, Boudrant J, Meyer D, Manno N, DeMarchis M, Paoletti M. Current views on fungal chitin/chitosan, human chitinases, food preservation, glucans, pectins and inulin: A tribute to Henri Braconnot, precursor of the carbohydrate polymers science, on the chitin bicentennial. Carbohydrate Polymers. 2012;**87**:995-1012

[9] Krajewska B. Membrane-based processes performed with use of chitin/chitosan materials. Separation & Purification Technology. 2005;**41**:305-312

[10] Guibal E. Interactions of metal ions with chitosan-based sorbents: A review. Separation & Purification Technology. 2004;**38**:43-74

[11] Yu L, Dean K, Li L. Polymer blends and composites from renewable resources. Progress in Polymer Science. 2006;**31**:576-602

[12] Payne GF, Chaubal MV, Barbari TA. Enzyme-catalysed polymer modification: Reaction of phenolic compounds with chitosan films. Polymer. 1996;**37**:4643-4648

[13] Grizzuti N, Buonocore G, Lorio G. Viscous behavior and mixing rules for an immiscible model polymer blend. Journal of Rheology. 2000;**44**:149-164

[14] Chiou MS, Ho PY, Li HY. Adsorption of anionic dyes in acid solutions using chemically cross-linked chitosan beads. Dyes and Pigments. 2004;**60**:69-84

[15] Chen T, Kumar G, Haris MT, Smith PJ, Payne GF. Enzymatic grafting of hexyloxyphenol onto chitosan to alter surface and rheological properties. Biotechnology and Bioengineering. 2000;**70**:564-573

[16] Dutta PK, Dutta J, Tripathi VS. Chitin and chitosan: Chemistry, properties and applications. Journal of Scientific and Industrial Research. 2004;**63**:20

[17] Xu HY, Yang L, Wang P, Liu Y, Peng M. Kinetic research on the sorption of aqueous lead by synthetic carbonate hydroxyapatite. Journal of Environmental Management. 2008;**86**:319-328

[18] Ground water pollution scenario in India. (Source: CPCB). http://nihroorkee.gov.in/rbis/india_information/ground%20water%20pollution.htm

[19] Current Drinking Water Standards, EPA. Office of Water, 2002. http://www.epa.gov/safewater/mcl

[20] Gedam AH, Dongre RS, Bansiwal AK. Synthesis and characterization of graphite doped chitosan composite for batch adsorption of lead (II) ions from aqueous solution. Advanced Materials Letters. 2015;**6**:59-67

[21] Gedam AH, Dongre RS. Adsorption characterization of Pb (II) ions onto iodate doped chitosan composite: Equilibrium and kinetic studies. RSC Advances. 2015;**5**:54188-54201

[22] Gedam AH, Dongre RS. Activated carbon from Luffa cylindrica doped chitosan for mitigation of lead(II) from an aqueous solution. RSC Advances. 2016;**6**:22639-22652

[23] Cardenas G, Miranda SP. FTIR and TGA studies of chitosan composite films. Journal of the Chilean Chemical Society. 2004;**49**:291-295

[24] Krishna Rao KSV, Naidu BV, Subha MCS, Sairam M, Aminabhavi TM. Novel chitosan-based pH-sensitive interpenetrating network microgels for the controlled release of cefadroxil. Carbohydrate Polymer. 2006;**66**:333-344

[25] Nomanbhay SM, Palanisamy K. Removal of heavy metal from industrial wastewater using chitosan coated oil palm shell charcoal. Electronic Journal of Biotechnology. 2005;**8**:43-53

[26] Ho YS, McKay G. Sorption of dye from aqueous solution by peat. Chemical Engineering Journal. 1998;**70**:115-124

11

Chitosan-Clay Based (CS-NaBNT) Biodegradable Nanocomposite Films for Potential Utility in Food and Environment

Asmae Laaraibi, Fatiha Moughaoui, Fouad Damiri, Amine Ouakit, Imane Charhouf, Souad Hamdouch, Abdelhafid Jaafari, Abdelmjid Abourriche, Noureddine Knouzi, Ahmed Bennamara and Mohammed Berrada

Abstract

The aim of this work is to design newer material for food packaging applications and to valorize the Moroccan marine wastes using chitosan (CS) prepared from exoskeletons of shrimps. Biodegradable and uniform nanocomposite films developed from sodium bentonite nanoparticles dispersed in chitosan matrix were carefully studied. The montmorillonite is used as nanofiller, and aqueous acetic acid solution is employed as a medium for dissolving and dispersing chitosan and montmorillonite. The existence of dialdehyde chitosan as cross-linking agent was examined. Morphology, thermal behavior, and mechanical properties of the nanocomposite films have been studied using FTIR, TGA, FEGSEM, TEM, XRD, and a tensile test. The XRD results indicate the formation of an intercalated and exfoliated nanostructure at low bentonite content and an intercalated and flocculated nanostructure at high bentonite content. Plastic deformation of the chitosan film is carried out using a thermomechanical treatment in the presence of a solvent and a plasticizer. The nanocomposite films obtained show a good tensile strength due to the reinforcement of chitosan intercalation in the silicate, which is an interesting mechanical property needed for food packaging applications. These nanocomposite films made from naturally occurring materials might play an important role in advanced research in food and environmental science.

Keywords: chitosan, food, environment, montmorillonite, sodium bentonite, films, nanocomposite, biodegradable

1. Introduction

The increase in living standards, changing consumer habits, and industrial development led to a high consumption of biodegradable plastic materials [1]. One of the best options to reduce current packaging waste is the use of biodegradable films that allow the final replacement of plastic packaging bags which are not recycled and are thus a pollution source. The present research is directed toward the development of biodegradable ecofriendly materials with enhanced properties [2]. Chitosan (CS) is a deacetylated derivative of chitin, which is the most abundant polysaccharide in nature after cellulose; it is a natural polysaccharide, biocompatible, and biodegradable in addition to the antibacterial properties that can be useful in many areas as the food packaging industry. Chitosan which consists of a linear (1–4) linked 2-amino-2-deoxy-D glucan as shown in **Figure 1** is a relatively inexpensive material, next to cellulose.

Chitin and chitosan are biopolymers having immense structural possibilities for chemical and mechanical modifications to generate novel properties, functions, and applications [3, 4], as biomedicine [5, 6], pharmaceuticals [7–9], metal chelation [10, 11], and food additives [12] and in the fabrication of sensors or biosensors [13].

Chitosan is highly soluble in acid aqueous solution. The positive charge and molecular arrangement confer to chitosan's interesting properties [14]. **Figures 2** and **3** illustrate the protonation of chitosan which leads to soluble material. **Figure 4** shows the simulation of protonated chitosan backbone (positive charges of NH^{3+} onto chitosan polymer) in acid aqueous solution. These positive charges cause a mutual repulsion and thus a swelling behavior and good solubility of chitosan. The protonation constants p*Ka* of chitosan decrease slightly, from 6.3 when the molecular weight reduces. The degree of deacetylation effects on p*Ka* values. The decrease in degree of deacetylation increases the p*Ka*. The degree of deacetylation influenced the balance of hydrophobic interactions and hydrogen bondings on chitosan [15].

In the same way, bentonite clays are also abundant and low-cost natural materials. Sodium bentonite is the name for the ore whose major constituent is the mineral, sodium montmorillonite. Montmorillonite is a three-layer mineral consisting of two tetrahedral layers sandwiched around a central octahedral layer. Bentonite is rich in montmorillonite (usually more

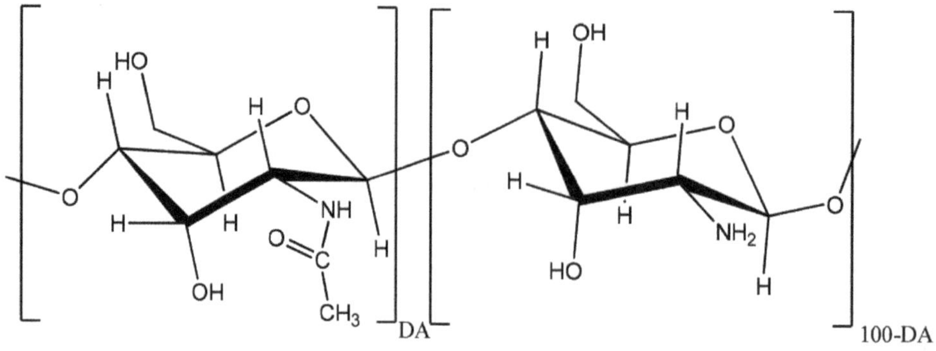

Figure 1. Chemical structure of Chitosan.

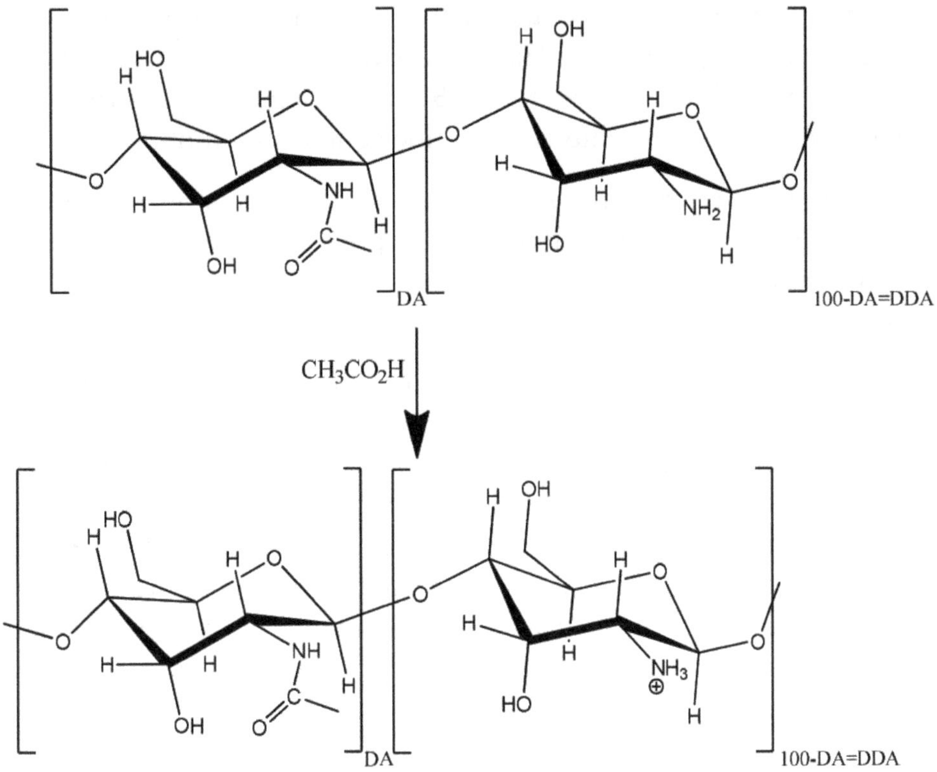

Figure 2. Protonation of chitosan in acetic acid aqueous solution.

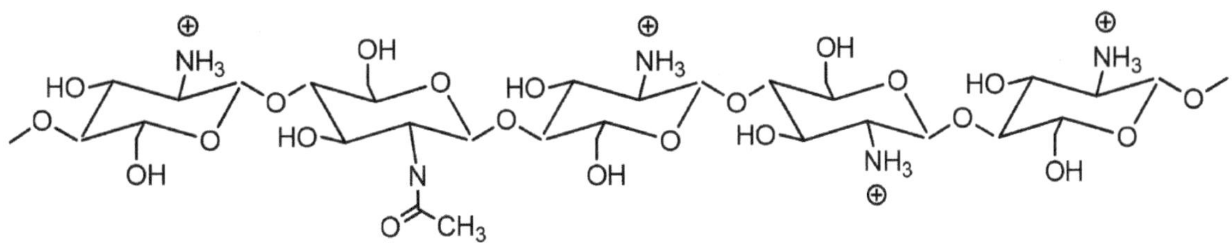

Figure 3. Polymer backbone of protonated chitosan in acid aqueous solution.

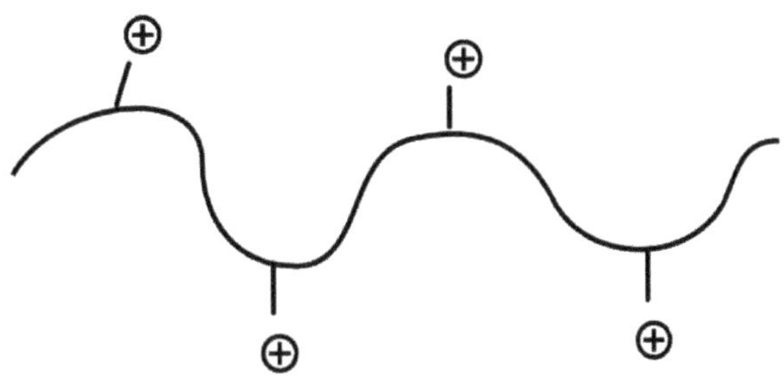

Figure 4. Simulation of protonated chitosan backbone in acid aqueous solution.

than 80%) [16–19]. Bentonite and montmorillonite names are often used interchangeably. However, the terms represent materials with different degrees of purity. Bentonite is the ore that comprises montmorillonite, inessential minerals, and other impurities.

When sodium bentonite comes into contact with water, the atoms and molecules dissolve, and ions with negative charges develop. These negative charges cause a mutual repulsion and thus a swelling within the clay structure. **Figure 5** shows the three-layer structure of a particle of bentonite and its exfoliation in sodium hydroxide aqueous solution.

The cations residing between the layers of sodium bentonite (Na-BNT or simply BNT) are exchangeable with ammonium ions of chitosan. This process converts the hydrophilic surface of the layer into a hydrophobic one, thereby improving the compatibility of nanoclay into polymer matrix.

The three processes that may occur in chitosan polymer and clay mixture, as shown in **Figure 6**, are intercalation, exfoliation, and conventional distribution. Intercalation is a physical process by which a macromolecule like a polymer is inserted in the clay sheets. Such a molecule is flanked by two clay layers and is immobilized and shielded. Exfoliation is a delaminating process wherein the gallery is expanded from its normal size of 1 nm to about 20 nm or higher. Thus there is a clear disruption of the layers which get spatially separated apart bringing about nanoscale dispersion in the polymer matrix. Thus exfoliated clays represent true nanomaterials. Intercalation and exfoliation of the clays can be accomplished using polymer from its solution or a melt [20].

Chitosan/bentonite composites are economically interesting because they are easy to prepare and involve inexpensive chemical reagents. Nanocomposites prepared from chitosan/bentonite shape natural films with a great potential and provide physical protection, they are biocompatible and biologically active toward microbial growth while being nontoxic and biodegradable. These nanobiocomposites obtained by adding nanofillers to biopolymers like chitosan result in very promising materials since they show improved properties with preservation of the material biodegradability without eco-toxicity [21].

Although chitosan/clay nanocomposites are very interesting materials, they were not extensively investigated as potential film packaging for food application. Thus, the aim of this work is to analyze the role of the chitosan/bentonite ratio, the DDA of chitosan, and a plasticizer

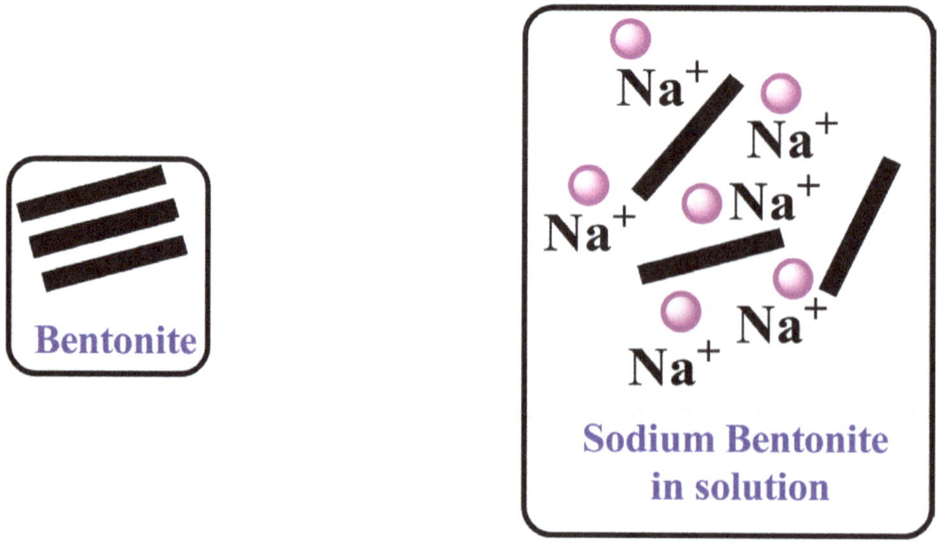

Figure 5. The three sheet structure of a particle of bentonite.

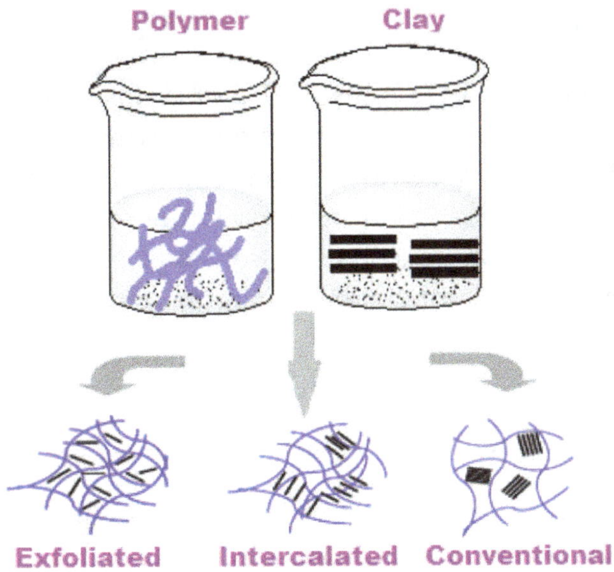

Figure 6. The three processes that may occur in polymer and clay mixture.

specially glycerol in the solution casting process for the achievement of chitosan/clay nanocomposite films. In order to investigate the combined effect of glycerol and unmodified clay on the properties of chitosan-based nanocomposites, films containing different amounts of clay and glycerol were prepared and characterized with particular regard to structural, thermal, and mechanical properties. Finally, new nanocomposite active films were proposed for safe packaging of edibles [22].

2. Material and methods

2.1. Materials

Chitosan (CS) with different degree of deacetylation (DDA) (from 70 to 100%) in powder form was prepared in our laboratory from exoskeletons of shrimp waste and purified [23, 24]. The chitosan chemical structure is schematically shown in **Figure 1**. The degree of deacetylation (DDA %) was determined by conductimetric titration [20].

Analytical grade sodium periodate ACS reagent, 99.8–100% dry basis, and bentonite were purchased from Sigma-Aldrich. Sodium bentonite used in this study was purified according to the method reported by H. Sedighi et al. [25]. The bentonite ore was beneficiated to improve its montmorillonite content by removing the impurities, generally albite, calcite, dolomite, orthoclase cristobalite, and quartz. The crude sample was primarily crushed at the size of 2 mm, and 5 g of bentonite powder was added to 100 mL of hot deionized water and stirred for 2 h. The separation of the impurities is obtained by a sedimentation process of the solution. The settled precipitate is mostly impurities, and the solid collected from the supernatant is generally pure montmorillonite. The slurry and solid phases were separated by filtration, and the remaining solid was dried at 150°C. All the other chemicals used are of analytical grade and used as received.

2.2. Preparation of chitosan film

Chitosan solution was prepared by dissolving 1 g of chitosan powder in 100 ml of aqueous acetic acid solution (1%, v/v), under continuous stirring at room temperature for 2 h followed by vacuum filtering to remove the insoluble residue. This solution was cast into Petri dishes and dried at 50°C for 20 h to evaporate the solvent and form the films. The dried films were soaked with an aqueous solution of 0.05 M NaOH to remove residual acetic acid, followed by rinsing with distilled water to neutralize, and then dried at room temperature.

2.3. Preparation of chitosan/bentonite (CS/Na-BNT) films

Chitosan/Na-BNT (also described as CSBNT) films were prepared using the casting/solvent evaporation technique. Firstly, 1% chitosan solutions were prepared by dissolving 1 g of chitosan powder in 100 ml of aqueous acetic acid solution (1%, v/v), under continuous stirring at room temperature for 2 h followed by vacuum filtering to remove the insoluble residue. Nanocomposite samples were obtained by dispersing selected amounts of bentonite in aqueous solution and stirred at 50°C until swelling was completed. After, the dispersion was slowly added to the CS solution to reach a final clay concentration of 1, 2, 3, and 5 wt% followed by stirring at room temperature for 5 h and then for 30 min at 25°C in ultrasonic bath. The amounts of chitosan, clay, and plasticizer used for each sample are listed in **Table 1**. For example, the composite film CSBNT1% is 1% BNT and 99% CS prepared from 1 g chitosan and 0.0101 g bentonite.

The nanocomposite solutions were then poured into Petri dishes and dried at 50°C for 20 h to evaporate the solvent and form the films. Free chitosan and nanocomposite films plasticized with glycerol were obtained by adding glycerol (30% (wt/wt) on solid CS) to the CS solution while stirring for 20 min at room temperature.

Following the same procedure used for chitosan films, the dried films were soaked with an aqueous solution of 0.05 M NaOH to remove residual acetic acid, followed by rinsing with distilled water to neutralize, and then dried at room temperature.

Chitosan/Na-BNT/cross-linker films were prepared using same method of manufacturing of chitosan/Na-BNT films. The dialdehyde chitosan was used as cross-linker and prepared according to I. Charhouf et al. [26] method and added after dispersing BNT in CS solution.

2.4. Oxidation of chitosan (cross-linker)

2.4.1. Preparation of dialdehyde chitosan

Mix 1 g chitosan ([CS] = 5.34 mM) in suspension with 50 ml HCl (10^{-3} M) (pH ranging from 4 to 5) with magnetic stirring. Mix with 1 ml aqueous solution of sodium periodate 0.534 mM, P0 = 0.1 (P0 = moles of $NaIO_4$ x moles of CS). The reaction was carried out at 4°C in the dark

Sample	Chitosan (g)-wt%	BNT (g)-wt%	Glycerol (g)-wt%
CS	1–100%	–	–
CSBNT1%	1–99%	1%	–
CSBNT2%	1–98%	2%	–
CSBNT3%	1–97%	3%	–
CSBNT5%	1–95%	5%	–
CSG	1–70%	–	30%
CSGBNT1%	1–69%	1%	30%
CSGBNT2%	1–68%	2%	30%
CSGBNT3%	1–67%	3%	30%
CSGBNT5%	1–65%	5%	30%

CS: Chitosan, BNT: bentonite, CSBNT1%: Film chitosan/bentonite 1%, CSBNT2%: Film chitosan/bentonite 2%, CSBNT3%: Film chitosan/bentonite 3%, CSBNT5%: Film chitosan/bentonite 5%, CSG: Film Chitosan/Glycerol, CSGBNT1%: Film Chitosan/Glycerol/bentonite 1%, CSGBNT2%: Film Chitosan/Glycerol/bentonite 2%, CSGBNT3%: Film Chitosan/Glycerol/bentonite 3%, CSGBNT5%: Film Chitosan/Glycerol/bentonite 5%.

Table 1. Amounts (g and wt %) of chitosan (CS), glycerol (G), bentonite (BNT), used for the preparation of chitosan, chitosan/glycerol, chitosan/BNT, and chitosan/glycerol/BNT.

for 30 minutes. After reaction, to eliminate the unreacted periodate, add 1 ml ethylene glycol. The oxidized chitosan was washed by distilled water for 4 h and freeze-dried.

The oxidation of chitosan using NaIO4 was well characterized as reported by I. Charhouf et al. [26]. In this work, we partially oxidized chitosan with a very few amount of sodium periodate. It is clearly seen from **Figure 7** that the oxidation reaction leads to opened structure of chitosan with dialdehyde functions.

2.5. Characterization and measurements

2.5.1. *Infrared spectroscopy (FTIR)*

Fourier transforms infrared (FTIR) spectra of the chitosan films and the chitosan/clay films were collected using a Tensor 37 FT-IR spectrophotometer (Spectrum 400 Perkin Elmer) operating in the range of 400–4000 cm^{-1} at a resolution of 4 cm^{-1}.

2.5.2. *Mechanical properties: tensile measurement*

Mechanical properties of chitosan/clay nanocomposite were measured with a Universal Testing Machine Ludwig mpK, tensile strength (TS), and percentage elongation at break (EL) of the films at 25°C according to ASTM D882 standard procedures [27]. The films were cut to a dog bone shape with a rectangular midsection (100 mm long x15 mm wide) flaring to 25 mm x

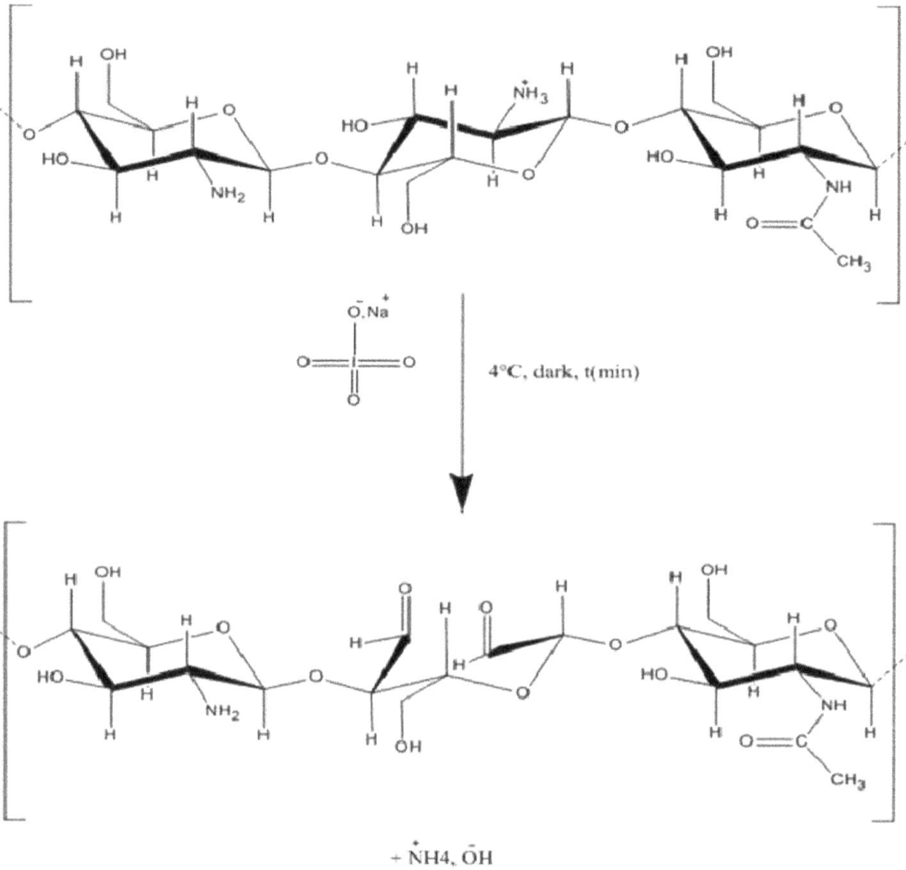

Figure 7. Oxidation reaction of chitosan by sodium periodate.

35 mm sections on each end. The thickness of each sample was measured with micrometer at three different locations and averaged. The films were conditioned at 50% RH for 72 h before each test. A 100 N load cell was used and the extension rate was set at 5 mm/min [28].

The tensile strength (σ) and percentage of elongation at break (E) were calculated using the following equations:

$$\sigma = F_{max}/A. \tag{1}$$

$$E = \Delta l/L_0 \times 100\ \%. \tag{2}$$

where F maxis the maximum load (N), A is the initial cross-sectional area (m^2), Δl is the extension of film strips (m), and L_0 is the initial length (m).

2.5.3. Thermal stability analysis

The thermal properties of nanocomposites and pure chitosan were investigated by thermogravimetric analysis (TGA). Samples were placed in the balance system and heated from 25 to 600°C at a heating rate of 10°C/min under a nitrogen atmosphere. Three replicates were tested for each sample.

2.5.4. XRD analysis

The X-ray diffraction analysis of the obtained films was performed by diffractometer with Cu Kα radiation (λ = 1.5418 A°) at room temperature. XRD scans were performed on sodium bentonite, chitosan films, and chitosan/bentonite films with a 2θ range between 5°-30°, at a scanning rate of 1°/min and scanning step of 0.01°.

2.5.5. Morphological analysis

Conventional high-vacuum scanning electron microscopy (SEM) images were also taken to visualize the structure of chitosan and oxidized chitosan. Chitosan and oxidized chitosan were freeze and dried for 24 h and were sprayed on silicon wafer substrate, then sputter-coated with gold (Agar Manual Sputter Coater; Marivac Inc., Montreal, QC, Canada), and imaged using a Quanta 200 FEG Environmental Scanning Electron Microscope (FEI Inc., Hillsboro, OR). Observations were performed at 20 kV using the high-vacuum mode.

The microstructural characterization of nanocomposites was carried out on the following samples: CSBNT3%, chitosan/bentonite 3%; CSBNT5%, chitosan/bentonite 5%; CSBNT10%, chitosan/bentonite 10%; CSBNT20%, chitosan/bentonite 20%; and CSG, chitosan/glycerol. Samples in powder form were coated in epoxy and ultramicrotomed with a diamond knife at -100°C. The recovered thin sections were observed at transmission electron microscope (TEM) (JEOL 2011) operating at 200 kV and the remaining blocks at field emission gun scanning electron microscope (FEGSEM) (Hitachi S 4700) at 1 kV after a slight platinum metal deposition.

3. Results and discussion

3.1. Preparation of bionanocomposite films

The intercalation of the cationic biopolymer chitosan into layered silicate clay (bentonite) through a cation exchange process results in nanocomposites with interesting structural and functional properties. Chitosan/Na-BNT films were prepared using the same method of manufacturing of chitosan films. However, Na-BNT was exfoliated in sodium hydroxide aqueous solution, purified, and washed prior to be added to chitosan solution. Organic matter is present in bentonite as intrinsic impurities composed predominantly of humid substances. Since competitive reactions can take place between the organic matter present in the bentonite and the chitosan, the extent of intercalation and polymer/clay interactions can be affected. Purification capable of removing of organic matter from bentonite before intercalation is fundamental.

The plasticization action of water molecules on hydrocolloid-based films has been widely reported in the literature [29, 31, 32]. In addition to water, the most commonly used plasticizer was glycerol (G), thus nearly systematically incorporated in most of biopolymer films [30]. Glycerol is indeed a highly hygroscopic molecule generally added to film-forming solutions to prevent film brittleness [31, 32]. The interest in use of the glycerol is that it acts as plasticizer and reduces the intermolecular forces by increasing the mobility of the biopolymer chains. The glycerol reduces the extent interactions between Na-BNT stacks making it possible to achieve a better dispersion of nano-sized filler and can modify the ability of water to swell BNT in the aqueous solution, due to the ability to reduce the surface energy of aqueous solution.

In this study as shown in **Figure 8**, chitosan (CS)/(BNT) nanobiocomposite and chitosan (CS)/(BNT)/cross-linker films were prepared by the intercalation of chitosan in bentonite to form miscible, biodegradable nanocomposite material used as packaging films for food preservation.

Periodate oxidation of chitosan have been relatively little explored, with only a few studies on the periodate oxidation reaction and products formed. Recently, Charhouf et al. [33] studied the periodate oxidation and the physical characterization of oxidized chitosan more in detail. The periodate oxidation of chitosan obviously leads to changes in the chemical structure. The cleavage of C2–C3 in chitosan (CS) units leads to the formation of a dialdehyde. Reaction of cross-linking chitosan and dialdehyde chitosan takes place when dialdehyde group reacts with amine moiety of unmodified chitosan as shown in **Figure 9** giving a cross-linked material.

The nanocomposite films prepared by casting technique using two inexpensive resources available and biocompatible (chitosan and Na-bentonite) were obtained as shown in **Figure 10**.

The presence of a group like hydroxyl, on the surface of chitosan, facilitates encapsulation of essential oils (EOs) or bioactive compound. The nanoemulsions were used to stabilize the EOs in the chitosan matrix, without altering its film-forming properties [34]. We investigated different emulsion formulations to encapsulate essential oils and to study their effects on the in vitro antimicrobial activity against various microorganisms. **Figure 11** shows images of antimicrobial films casted from solutions containing modified chitosan (2% w/w), dyes, and essential oils (0.05% w/w).

Rosemary essential oil, with its warm and penetrating aroma, is one of the most stimulating oils used in aromatherapy. Rosemary was one of the earliest plants to be used in medicine, as well as for cooking. It has a very strong antiseptic action so it is terrific to use in aromatherapy recipes for cleaning. Incorporation of essential oils (EOs) in chitosan films was studied in order to prepare antimicrobial barriers to be applied to food surfaces. Essential oils

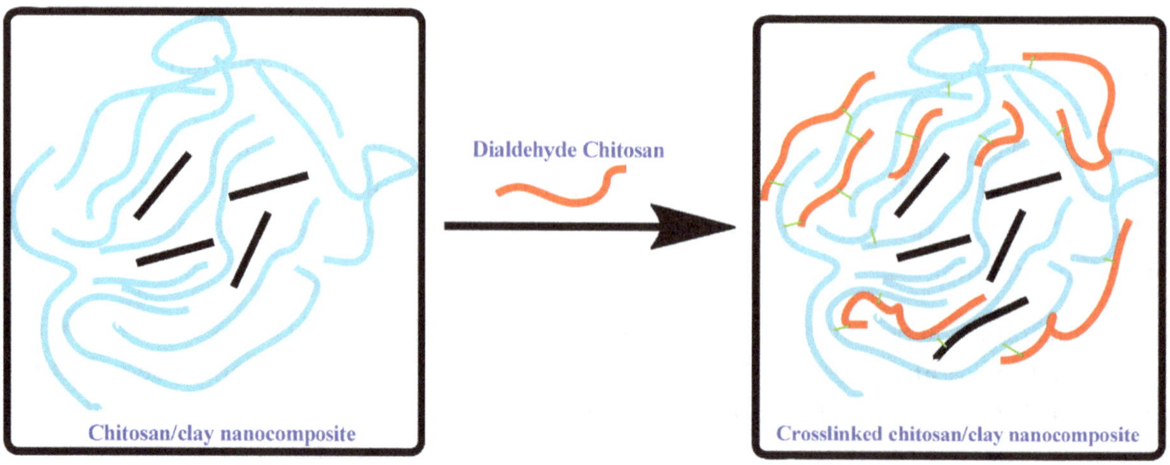

Figure 8. The sheets structure of exfoliated bentonite and dispersed in crosslinked chitosan matrix.

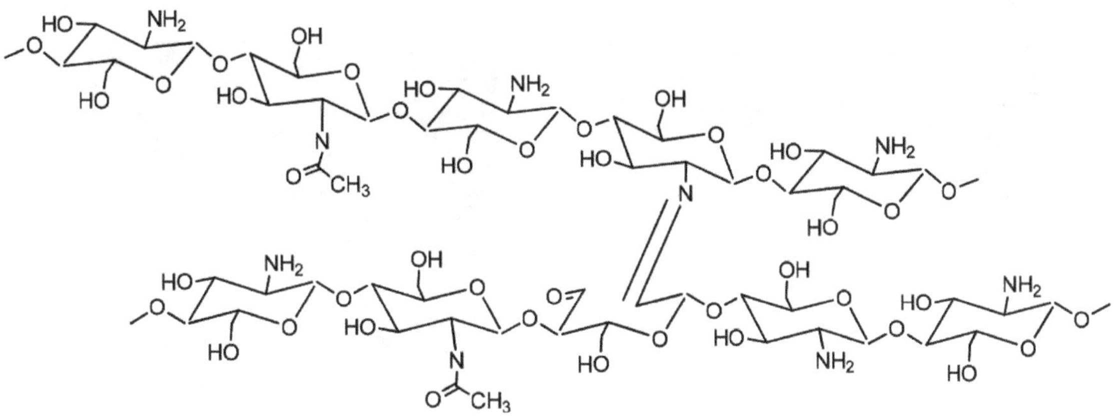

Figure 9. Crosslinking reaction between chitosan and dialdehyde chitosan.

were directly incorporated in the chitosan nanocomposite matrix. The EOs were selected by their ability to develop antimicrobial synergies against *Listeria* bacteria (monocytogenes or innocua) with chitosan, which is characterized by intrinsic antimicrobial properties.

3.2. Infrared spectroscopy (FTIR)

FTIR spectra of chitosan (CS), bentonite (BNT), and chitosan/bentonite nanocomposite (CSBNT) films are displayed in **Figure 12**. The spectrum of chitosan shows a broad peak at 3475.80 cm^{-1} corresponding to amine N–H symmetrical vibration and H bonded O–H group; the peaks at 2924.44 cm^{-1} were assigned to the symmetric and asymmetric $-CH_2$ vibrations of carbohydrate ring. The absorption peak observed at 1618.79 cm^{-1} was assigned to (C = O in amide group, amide I vibration), 1545 cm^{-1} was attributed to ($-NH_2$ bending of amide II), and

FCSBNT1% FCSBNT2% FCSBNT3%

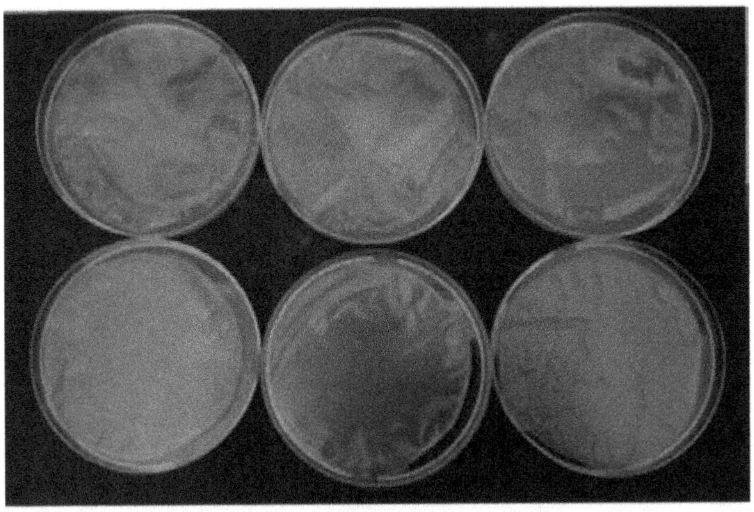

FCSBNT5% FCSBNT7% FCSBNT10%

Figure 10. Images of chitosan/bentonite nanocomposite films casted from solutions containing chitosan (2% w/w) and bentonite at various amount of pure clay.

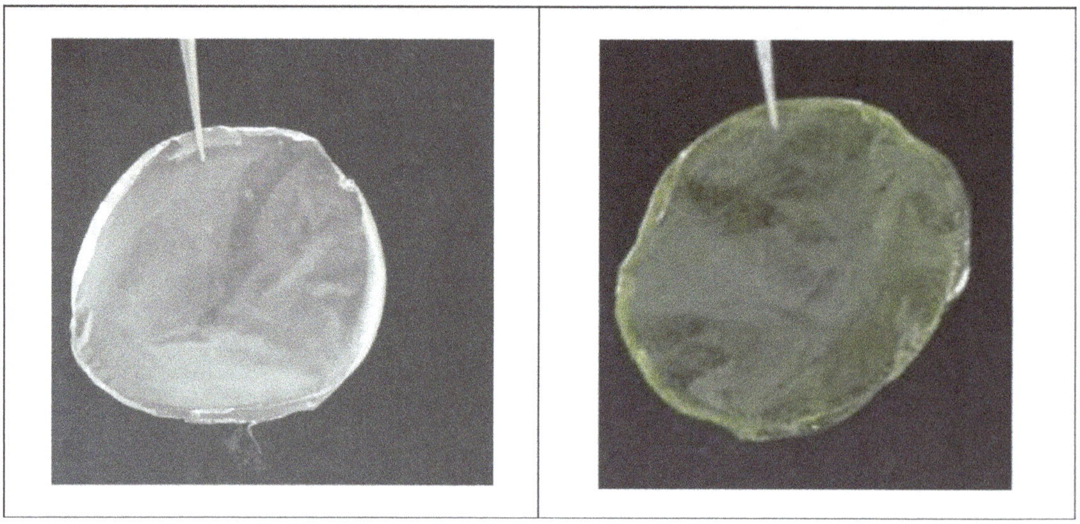

Figure 11. Images of antimicrobial films casted from solutions containing modified chitosan (2% w/w) and essential oils (0.05% w/w) added as pure rosemary essential oil.

1390 cm^{-1} was given to (N–H stretching or C–N bond stretching vibrations, amide III vibration). The peak at 1116.93 cm^{-1} corresponds to the symmetric stretching of C–O–C groups. The absorption peaks in the range 900–1200 cm^{-1} are due to the antisymmetric C–O stretching of saccharide structure of chitosan.

As can be seen in **Figure 12**, the FTIR spectrum of BNT shows a peak at 1010 cm^{-1} that belongs to Si-O-Si linkage. In addition, the characteristic absorption peaks are found at around 3670 cm^{-1} (stretching vibration of Al-OH and OH), at 3465 cm^{-1} (stretching vibration of O-H and H-O-H groups), at 1638 cm^{-1} (H-O-H bending vibration), at 933 cm^{-1} (Al-Al-OH bending frequency), and at 509 cm^{-1} (bending vibration of Si-O).

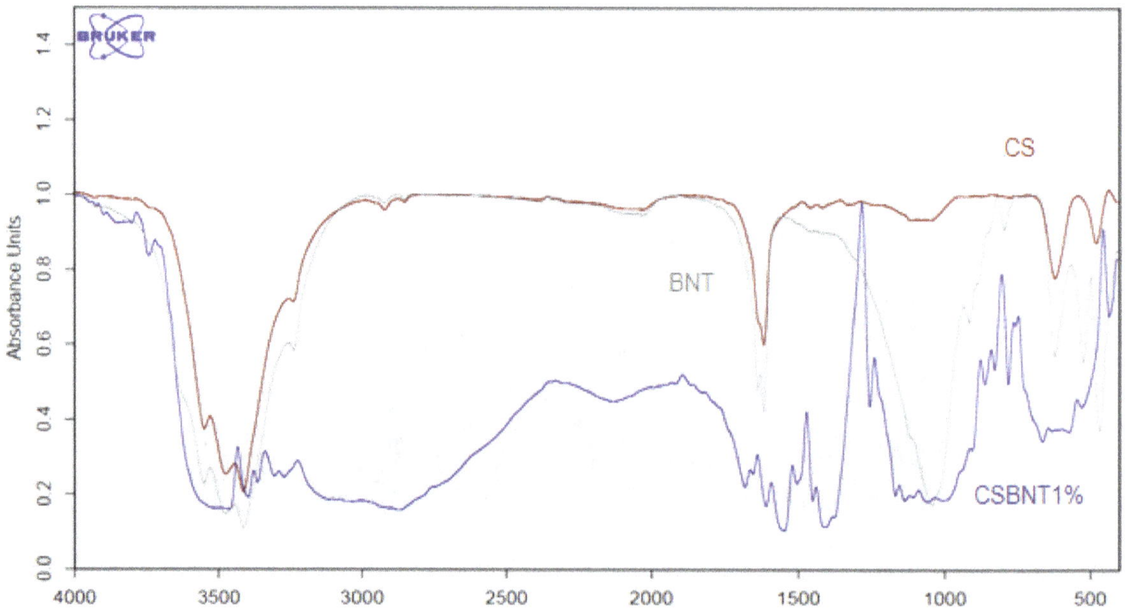

Figure 12. FTIR spectrum of: Chitosan film (FCSBNT0%), Chitosan/BNT films respectively (FCSBNT3%) and (FCSBNT5%).

The FTIR was also used to study the polymer/clay interaction, since a shift in the NH_3^+ vibration may be expected when – NH_3^+ groups interact electrostatically with the negatively charged sites of the clay [35]. Nevertheless, this shift is higher for CSBNT nanocomposite film with the lowest amounts of CS, while the chitosan/clay films with the highest amounts of biopolymer show a frequency value that trends to that observed in the films of pure chitosan (CS). This fact may be related to the $-NH_3^+$ groups that do not interact electrostatically with the clay substrate. The spectrum of CBNT nanocomposite film (**Figure 12**) shows a characteristic band at 3462.78 cm^{-1} attributed to hydrogen bonding formation due to the functional groups of CS (O-H and N-H groups) and BN (O-H groups) [36, 37]. The intensity of the NH_3^+ band also increases for higher amounts of intercalated chitosan. The secondary amide band at 1645 cm^{-1} of chitosan is overlapped with the HOH bending vibration band at 1628 cm^{-1} of the water molecules associated to the chitosan/clay films, which are present as in the starting clay, as expected for a biopolymer with high water retention capability [38, 39].

3.3. Tensile measurements of chitosan (CS), chitosan/bentonite (CSBNT), chitosan/glycerol (CSG), and chitosan/glycerol/bentonite (CSGBNT) films

The stress-strain curves of the tested specimens are being presented in **Figure 13**, while the average values along with the standard deviation of Young's modulus, tensile strength and elongation at break of the films on the stress-strain behavior of the chitosan and chitosan/glycerol films, respectively [40], are shown in **Figure 14**. The higher strength obtained in the case of the CS films can be attributed to more efficient stress transfer between the adjacent chains due to the strong electrostatic interactions between the NH_2 and NH_3^+ groups. The CSG specimen presents almost double strength (at yield and break) and elongation at break. Due to the lower acidity of the diluted films, a weaker hydrogen bond network was established between the amino groups and the glycerol chains. On the other hand, the extensive deformation strengthening in undiluted systems (CSG) suggests the creation of a long-range order and the formation of hydrogen bonding after the addition of glycerol.

The effect of BNT addition on the tensile response of the chitosan and chitosan/glycerol films is being depicted. The stress–strain curves of BNT composite films prepared from the 1 w/v% chitosan solution is being presented. The addition of BNT results in a pronounced enhancement of the stiffness and a dramatic decrease in the elongation at break of all clay-added systems. Further addition of BNT leads in intercalated structures which limited the polymer-clay interactions and thus their reinforcing ability.

The results on mechanical properties showed the increase in the tensile strength (TS) and elastic modulus (EM) of such nanocomposite films can be attributed to the high rigidity and aspect ratio of the nanoclay as well as the high affinity between the chitosan and the bentonite. On the other hand, the CS/BNT nanocomposites have shown significant decrease in elongation at break (EB). This reduction can be attributed to the restricted mobility of macromolecular chains.

In **Figure 14**, effect of BNT addition is being illustrated for the diluted systems (CS nanocomposites). The ductile response of the CS films is maintained after the addition of BNT with strength and a relative lower decrease in the elongation at break. The systems with 3 wt% BNT presented the lowest enhancement in all mechanical properties.

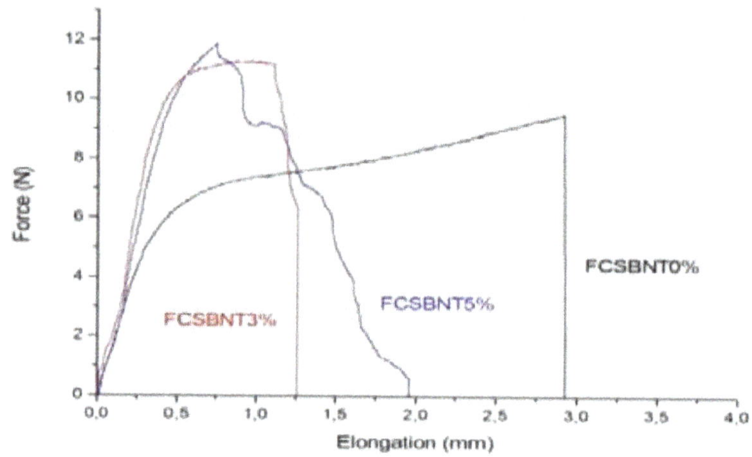

Figure 13. Stress-strain curves of chitosan film (FCSBNT0%), chitosan/BNT films (FCSBNT3%), (FCSBNT5%), respectively.

Figures 13 and **14** present the combined effect of glycerol and BNT on the tensile response of the CS-based nanocomposites. The first observation is that the addition of BNT results in a direct reduction of the strength of the chitosan/glycerol. A completely different stress–strain behavior is being obtained after the addition of BNT in diluted chitosan/glycerol systems (**Figure 14**). The CS/glycerol-based nanocomposites behave like hyperelastic materials rather than like ductile polymers. It is assumed that more water and glycerol are distributed in the chitosan network, inducing a very obvious plasticization effect. The extent of hydrated chitosan crystals was confirmed from the intensities of the XRD patterns. It is very interesting to note that although the mechanical properties of the unreinforced chitosan are comparable before and after the application of the reflux processing, reflux resulted in a fourfold increase of the stiffness and strength of the nanocomposite films.

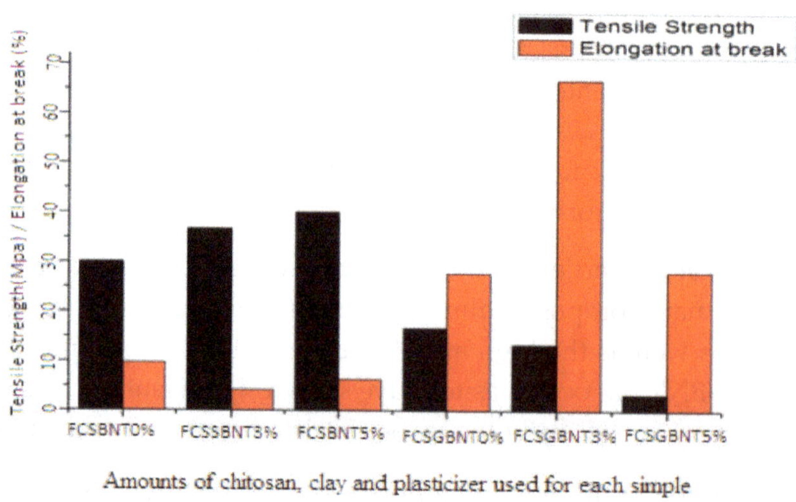

Figure 14. Mechanical properties of chitosan/BNT particles films. Chitosan film (FCSBNT0%), chitosan/BNT films (FCSBNT3%), (FCSBNT5%), respectively, and chitosan/glycerol film (FCSGBNT0%), chitosan/glycerol/BNT films (FCSGBNT3%), (FCSGBNT5%), respectively.

3.4. Thermal stability analysis

The thermal stability of the chitosan (CS) and its nanocomposites has been investigated by TGA under nitrogen (**Figure 15**). There are two steps of degradation. The first range (50–200°C) is associated with the loss of water, whereas the second range at 270°C corresponds to the deacetylation and degradation of chitosan, and the third step, in the temperature range 450–550°C, can be associated with the oxidative degradation of the carbonaceous residue formed during the second step.

The nano-dispersed clay in the chitosan matrix exhibits a significant delay in weight loss. The nanocomposite forms char with a multilayered carbonaceous-silicate structure, which may keep its multilayered structure in the polymer matrix. This high-performance carbonaceous-silicate char builds up on the surface during burning, thus insulating the underlying material and slowing the escape of the volatile products generated during decomposition. The decomposition temperature CS/BNT nanocomposites show higher thermal stability than those of the pure CS. The thermal stability of the nanocomposites increases systematically with increasing clay.

For nanocomposites containing glycerol, a further degradation step at T≈250°C is observed, related to the loss of unbound glycerol, as indicated in **Figure 16**. Furthermore, it can also be observed that the presence of glycerol plasticizer increases of about 20°C the degradation temperature for the third step, irrespective of the presence or not of the clay.

3.5. XRD analysis

The XRD patterns of chitosan and chitosan-based nanocomposite films in the range of 5–30° are shown in (**Figure 17**). The basal plane of BNT shows a reflection peak at about 2θ = 8.8°. After incorporating BNT within CS, with CS/BNT, the basal plane of BNT at 2θ = 8.8° disappears, substituted by a new weakened broad peak at around 2θ = 12.8°–13.0° (CSBNT3%, CSBNT5%). It is suggested that the BNT form intercalated and flocculated structures.

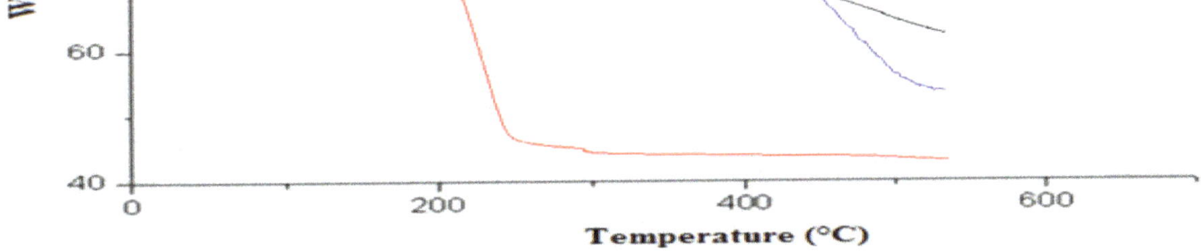

Figure 15. Thermal properties of: Chitosan (FCSBNT0%) and Chitosan/Bentonite films (FCSBNT3%), (FCSBNT5%).

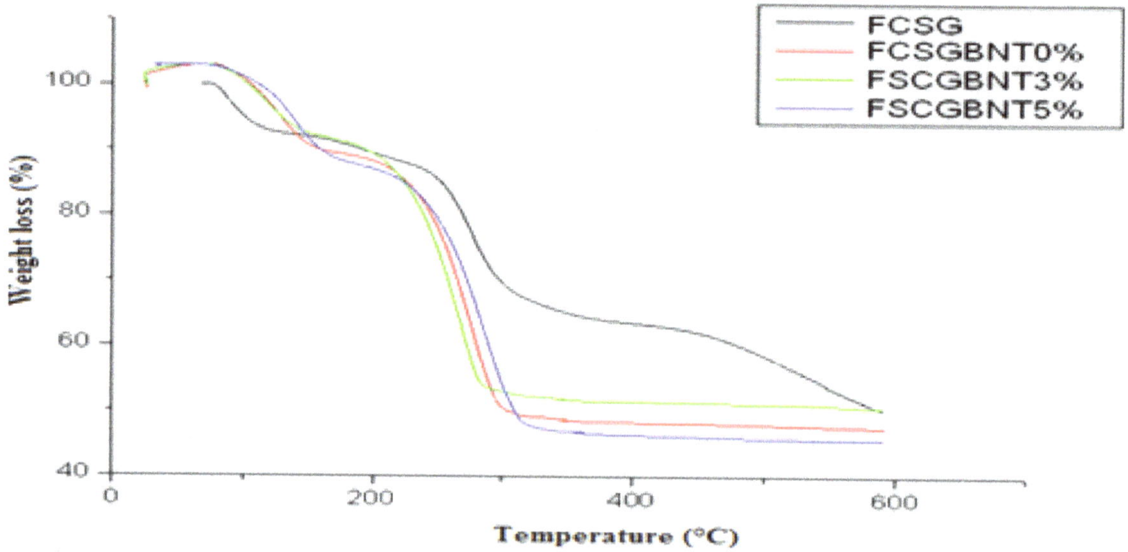

Figure 16. Thermal properties of: Chitosan/Glycerol (FCSG), Chitosan/Glycerol (FCSGBNT0%) and Chitosan/Glycerol/Bentonite (FCSGBNT3%) and (FCSGBNT5%) films.

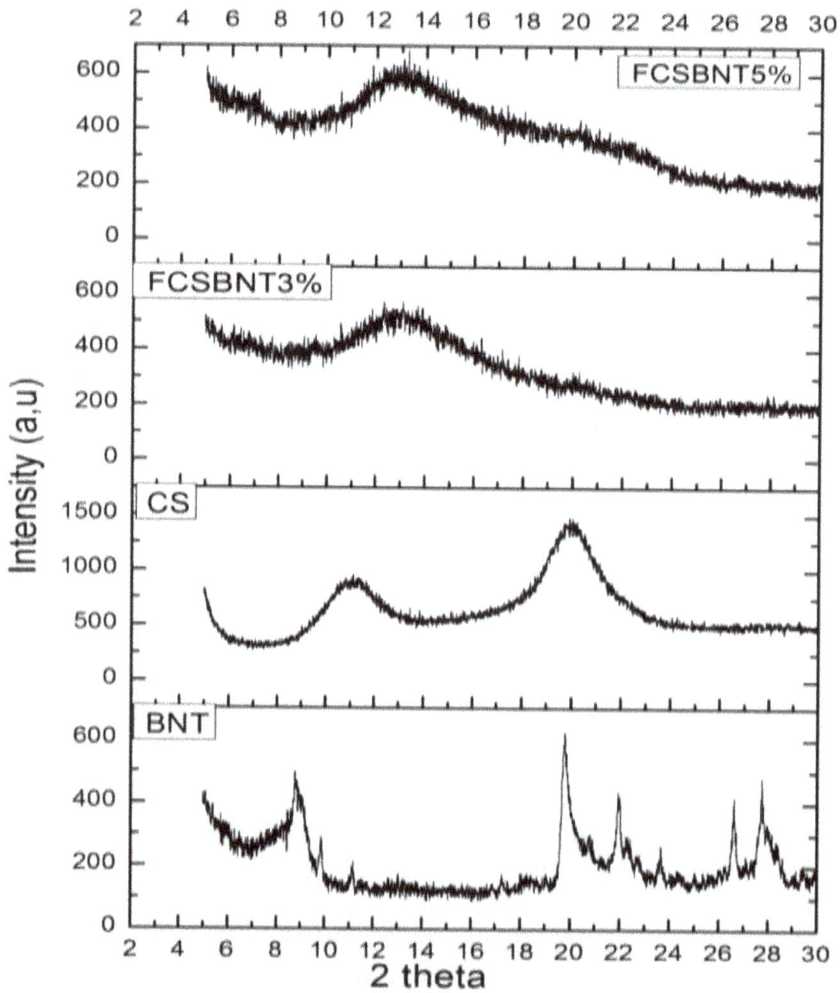

Figure 17. XRD patterns of: Chitosan (CS), Bentonite (BNT) and Chitosan/Bentonite films (FCSBNT3%) and (FCSBNT5%).

On the base of XRD patterns, it is suggested that the BNT forms intercalated and exfoliated structures at higher CS content (CSBNT5%), while decreasing the CS content (CSBNT3%), clay layers (BNT) form intercalated and flocculated structures. According to [23], the formation of flocculated structure in CS/clay nanocomposites can be due to the hydroxylated edge-edge interactions of the clay layers. Since one chitosan unit possesses one amino and two hydroxyl functional groups, these groups can form hydrogen bonds with the clay hydroxyl edge groups. This strong interaction is believed to be the main driving force for the assembly of BNT in the CS matrix to form flocculated structures.

The XRD patterns of chitosan/glycerol films obtained from 30 w/v% solutions are shown in **Figure 18**. The addition of glycerol results in a pronounced peak at 12.5°. Because of the hydrophilic and polycationic nature of chitosan in acidic media, this biopolymer has good miscibility which is attributed to the interaction of glycerol molecules with chitosan

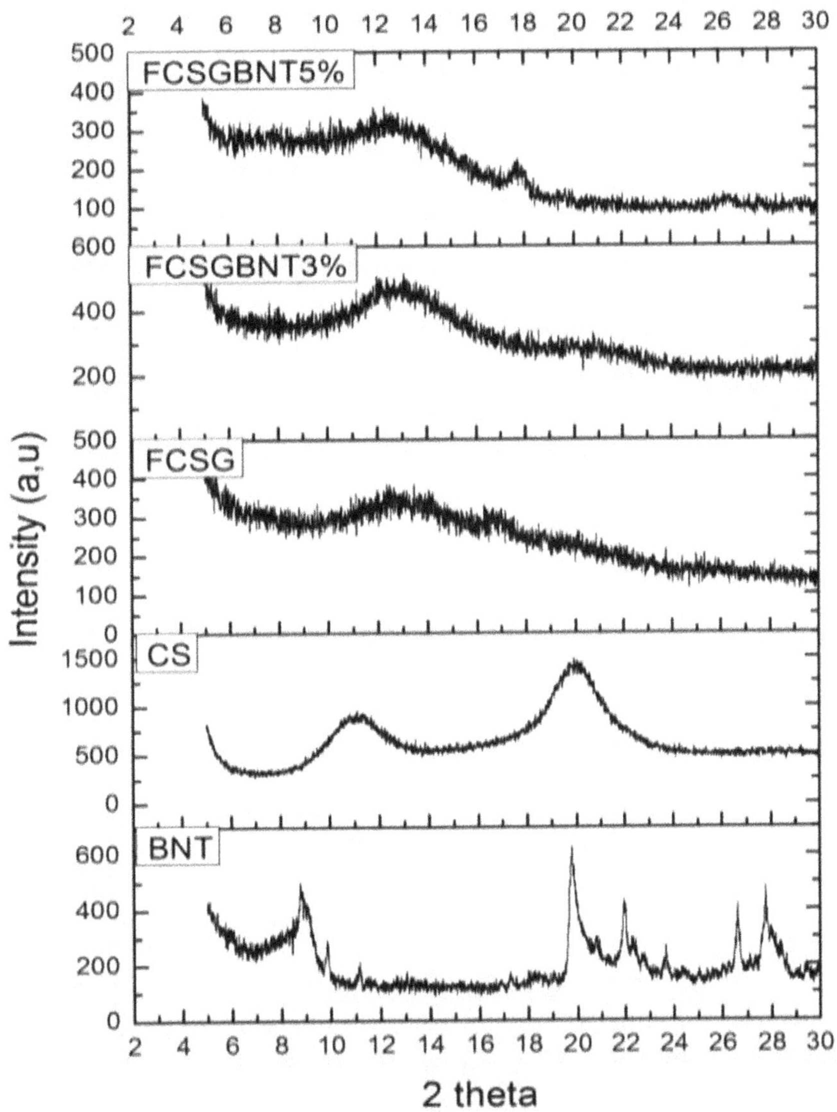

Figure 18. XRD patterns of: Chitosan (CS), Bentonite (BNT), Chitosan/Glycerol (FCSG) and Chitosan/Glycerol/Bentonite films (FCSGBNT3%) and (FCSGBNT5%).

macromolecules. Glycerol favors the chains mobility and thus the chitosan crystallization process in the early stage of the post-processing aging the effect of glycerol addition. The XRD patterns of chitosan/glycerol/BNT films obtained from chitosan solution are shown in **Figure 18**. The combined addition of glycerol and clays resulted in great enhancement of the chitosan crystallinity of the nanocomposite films prepared with 1 w/v% chitosan solution. This indicates that the presence of clay facilitates the distribution of glycerol within the chitosan matrix and the interaction of glycerol molecules with chitosan macromolecules. The combined addition of glycerol and clay had an opposite effect in films obtained from low content chitosan solution leading to decrease of the XRD peaks intensities. In addition a new peak at 18.2° appeared in XRD patterns of all obtained films. This diffraction peak is characteristic for chitosan films prepared using acetic acid solution as solvent.

The addition of glycerol favors the opening of the clay galleries resulting in intercalated nanocomposites in comparison to samples without glycerol.

Chitosan/Na-BNT nanocomposites exhibit an intercalated or intercalated/orientated structure of clays. In particular, the X-ray diffraction results show that in film without glycerol, the BNT stacks lay with their platelet surface parallel to the casting surface. The presence of glycerol, on the other hand, enhances the chitosan intercalation in the silicate galleries and hinders the flocculation process, leaving the BNT stacks randomly orientated in the space.

3.6. SEM and MET images of chitosan, oxidized chitosan, and bionanocomposite materials

The SEM images of chitosan and oxidized chitosan at high vacuum and at different magnifications are shown in **Figure 19**, showing that there is no change of elongated and fibrous network of chitosan, but on the surface of oxidized chitosan, we can see a slight degradation of some leaves.

3.6.1. FEGSEM analysis

We present in this study the microstructural characterization results obtained on the chitosan/Na-BNT nanocomposites too. The dispersion and the exfoliation of the clay were observed at field emission gun scanning electron microscope (FEGSEM) and transmission electron microscopy (TEM).

The results of the observation of the microtome block of chitosan/bentonite (CSBNT3%) and (CSBNT5%) samples are presented in **Figure 20**.

At low magnification, the CSBNT 3% powders were observed, and clusters were found grouped into in the epoxy more matte and dark appearance (**Figure 20a**). Decohesion between the epoxy and the sample powders is visible in greater or lesser proportion, probably due to preparation and cutting. At higher magnification, small clay particles of a few hundred nanometers are observed, which are relatively well dispersed in the polysaccharide (**Figure 20b**). However, a large clay aggregate of about 10 microns was also observed (**Figure 20c** and **d**).

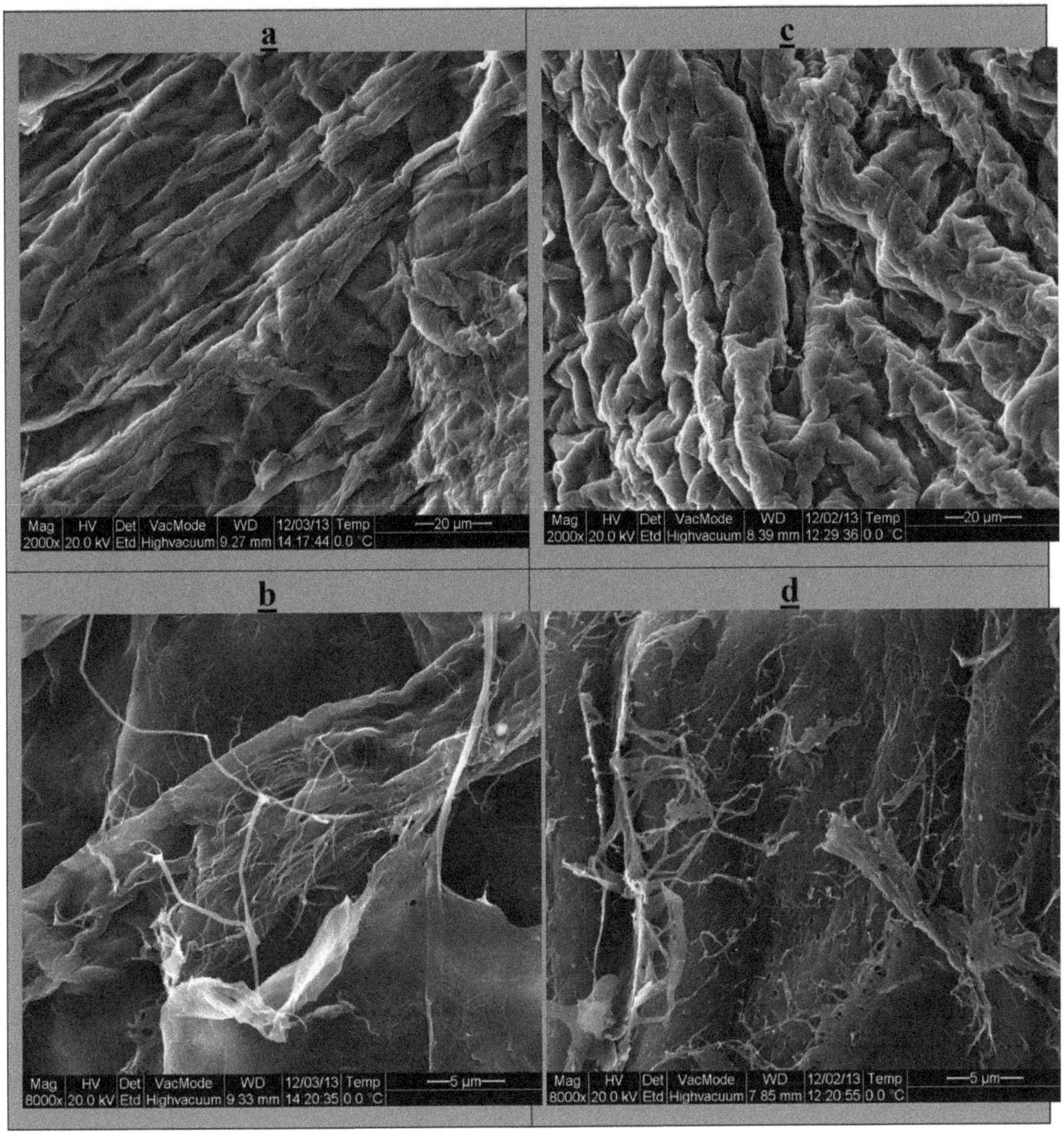

Figure 19. SEM images of (a & b) chitosan and (c & d) oxidized chitosan at different magnifications.

At higher magnification, small clay particles similar in size to those observed in the CSBNT 3% sample are regularly observed with chitosan/BNT 5% as shown in **Figure 21**.

3.6.2. TEM analysis

Transmission electron microscopy (TEM) images indicated that the silicate layers were dispersed in the chitosan matrix. The results of the TEM observation of the chitosan/BNT 3% sample show that small particles of clay from less than 100 nanometers to a few hundred nanometers are observed at low magnification which is consistent with the FEGSEM observations. Larger aggregates of clay are also observed but more rarely.

Depending on the level of the clay particles, the leaflets are sometimes well aligned (**Figure 22**) and sometimes of more unstructured appearance (**Figure 23**). This unstructured aspect of the

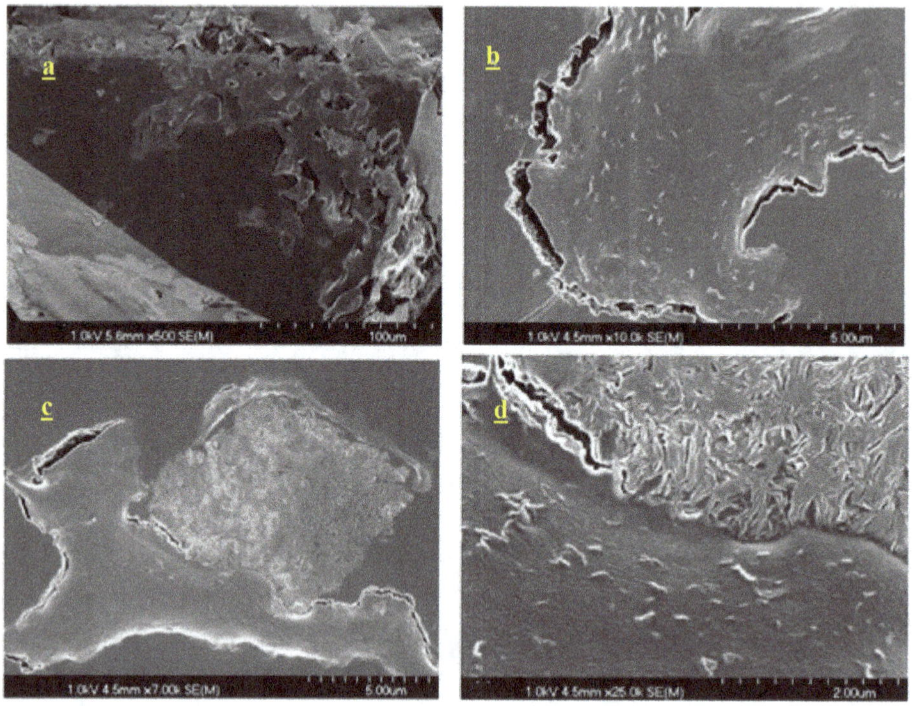

Figure 20. Images of FEGSEM of chitosan/BNT 3%.

clay sheets may be a sign of a more advanced level of intercalation. There are also some isolated single or double leaflets around other clay particles (**Figure 23**), which is a clear sign of exfoliation.

As expected, the concentration of clay in the polysaccharide affects the dispersion of the clay. The higher the concentration, the poorer the dispersion is obtained which is the effect of the greater aggregation of the clay. A good but not always uniform dispersion is observed in chitosan/BNT 3% and 5%. Similarly, the concentration of clay also appears to affect the level

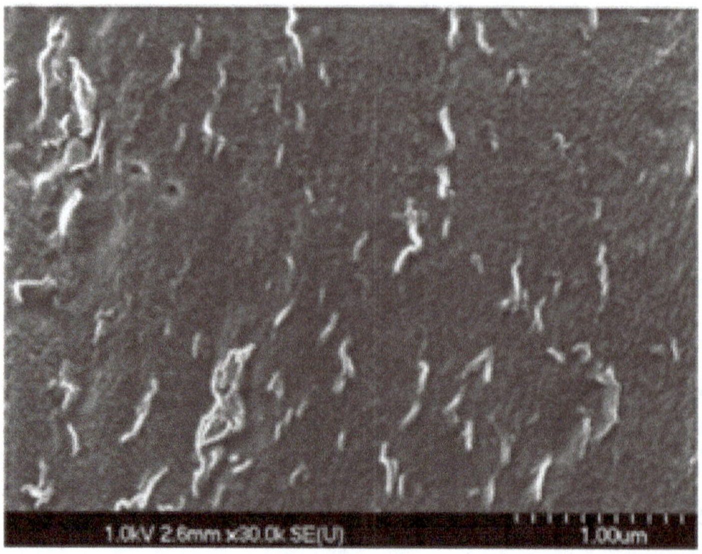

Figure 21. Image of FEGSEM of chitosan/BNT 5%.

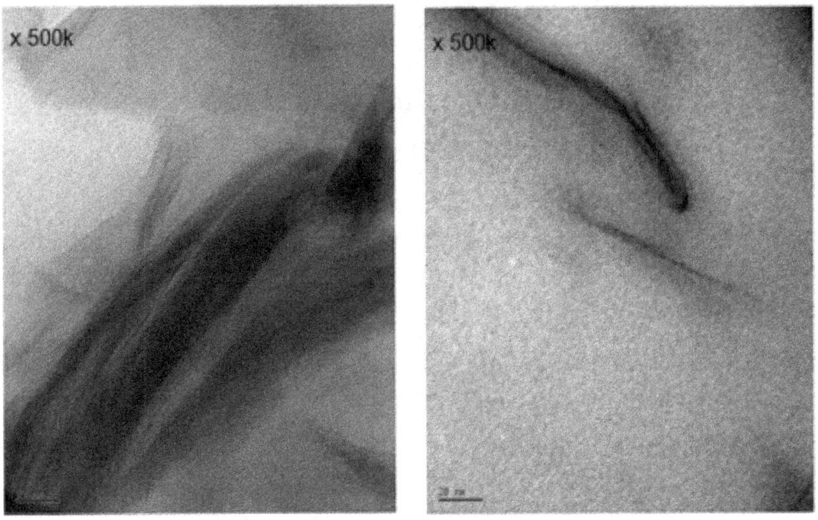

Figure 22. Images TEM of chitosan/BNT 3% (x 500K).

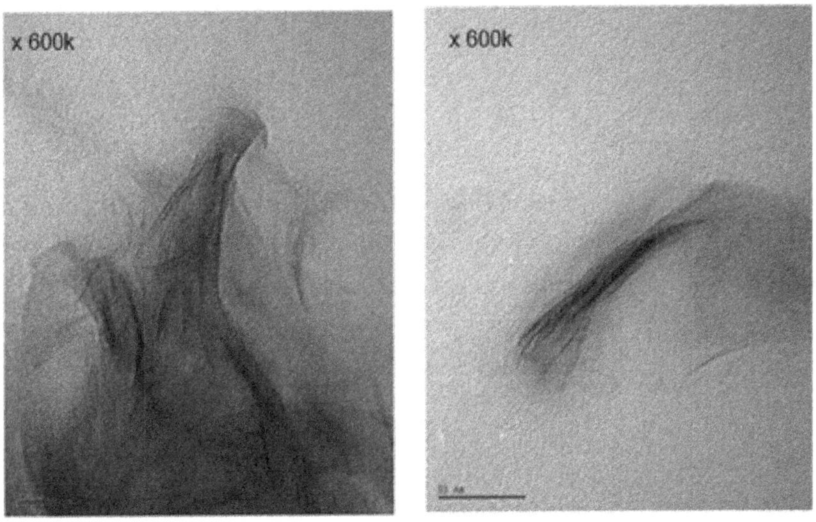

Figure 23. Images TEM of chitosan/BNT 3% (x 600K).

of intercalation/exfoliation. Based on MET observations, signs of exfoliation (the presence of isolated single or double clay leaflets and more unstructured appearance of the leaflets in the particles) are visible in samples of lower clay concentration less than 10%.

4. Conclusion

Natural biopolymer-based biodegradable packaging materials are a new generation of polymers emerging on the packaging market, and driven by the perception that biodegradable plastics are "environmentally friendly," their use is predicted to increase chitosan, a natural material that has interesting antimicrobial and film-forming activities. Its application in films can contribute to food preservation and shelf-life extension. In this study, various films were successfully prepared by the solution casting technique and characterized with particular regard to structural, thermal, and mechanical properties. Films of chitosan/bentonite, chitosan/glycerol/bentonite, and chitosan/glycerol/bentonite/essential oil nanocomposites were prepared with purified bentonite (BNT), and according to the process used, they might be less expensive than other packaging materials.

Exfoliated chitosan/clay nanocomposites of varying clay contents have been successfully prepared with or without the presence of glycerol (plasticizer) and oxidized chitosan (cross-linker). This approach represents a new route to prepare high-performance nanocomposite materials. The oxidized chitosan could partially react with the amine groups on chitosan; as a result, high mechanical properties can be obtained. The Na-BNT layers are exfoliated by chitosan chains and disorderly dispersed in the chitosan matrix, as confirmed by XRD and TEM characterization. The incorporation of a small amount of clay into the chitosan matrix results in obvious enhancement in the thermal properties of chitosan. The chitosan/clay nanocomposites retain good mechanical properties. Once the clay is exfoliated and efficiently dispersed into the chitosan matrix, the storage modulus and tensile property of the chitosan/clay nanocomposites are significantly improved with respect to that of neat chitosan.

Acknowledgements

This work would not have been possible without the great support of Julian Zhu, a polymer chemist professor at Montreal University, Department of chemistry, Quebec, Canada. This work was supported by the Francophone University Association (AUF).

Author details

Asmae Laaraibi[1]*, Fatiha Moughaoui[1], Fouad Damiri[1], Amine Ouakit[1], Imane Charhouf[1], Souad Hamdouch[1], Abdelhafid Jaafari[2], Abdelmjid Abourriche[1], Noureddine Knouzi[1], Ahmed Bennamara[1] and Mohammed Berrada[1]

*Address all correspondence to: berrada_moh@hotmail.com

1 Department of Chemistry, Laboratory of Biomolecules and Organic Synthesis (BIOSYNTHO), Faculty of Sciences Ben M'Sik, University Hassan II of Casablanca, Morocco

2 Department of Chemistry, Laboratory of Applied Chemistry and environment, Faculty of Sciences and Technologies, University Hassan I of Settat, Morocco

References

[1] Martinho G, Balaia N, Pires A. The portuguese plastic carrier bag tax: the effects on consumers' behavior. Waste Management. 2017;**61**:3-12

[2] Tan W, Zhang Y, Szeto Y, Liao L. A novel method to prepare chitosan/montmorillonite nanocomposites in the presence of hydroxy-aluminum oligomeric cations. Journal of Computer Science and Technology. 2008;**68**:2917-2921

[3] Christensen B, Vold I, Vårum KM. Vårum, chain stiffness and extension of chitosans and periodate oxidized chitosans studied by size-exclusion chromatography combined with light scattering and viscosity detectors. Journal of Carbohydrate Polymers. 2008;**74**:559-565

[4] Mourya VK, Inamdar NN. Chitosan-modifications and applications: opportunities galore. Reactive & Functional Polymers. 2008;**68**:1013-1051

[5] Luo H, Li J, Chen X. Antitumor effect of N-succinyl-chitosan nanoparticles on K562 cells. Biomedicine & Pharmacotherapy. 2010;**64**:521-526

[6] Drumright R, Siegwart DJ, Matyjaszewski K. The development of microgels/nanogels for drug delivery applications. Progress in Polymer Science. 2008;**33**:448-477

[7] Baldrick P. The safety of chitosan as a pharmaceutical excipient. Regulatory Toxicology and Pharmacology. 2010;**56**:290-299

[8] Muzzarelli RA. Genipin-crosslinked chitosan hydrogels as biomedical and pharmaceutical aids. Carbohydrate Polymers. 2009;**77**:1-9

[9] Sinha V, Singla A, Wadhawan S, Kaushik R, Kumria R, Bansal K, Dhawan S. Chitosan microspheres as a potential carrier for drugs. International Journal of Pharmaceutics. 2004;**274**:1-33

[10] Feng Y, Yang L, Li F. A novel sensing platform based on periodate-oxidized chitosan. Analytical Methods. 2010;**2**

[11] Kofuji K, Qian C, Nishimura M, Sugiyama I, Murata Y, Kawashima S. Relationship between physicochemical characteristics and functional properties of chitosan. European Polymer Journal. 2005;**41**:2784-2791

[12] Ferraro V, Cruz I, Jorge RF, Malcata FX, Pintado ME, Castro PML. Valorisation of natural extracts from marine source focused on marine by-products: a review. Food Research International. 2010;**43**:2221-2233

[13] Jerónimo P, Araújo AN, Montenegro M. Development of a sol–gel optical sensor for analysis of zinc in pharmaceuticals. Sensors and Actuators B: Chemical. 2004;**103**:169-177

[14] Bhuvaneshwari S, Sruthi D, Sivasubramanian V, Niranjana K, Sugunabai J. Development and characterization of chitosan film. Journal of Engineering Research and Applications. 2011;**1**:292-299

[15] Wang QZ, Chen XG, Liu N, Wang SX, Liu CS, Meng XH, Liu CG. Protonation constants of chitosan with different molecular weight and degree of deacetylation. Carbohydrate Polymers. 2006;**65**:194-201

[16] Holzer L, Münch B, Rizzi M, Wep R, Marschall P, Graule T. 3D-microstructure analysis of hydrated bentonite with cryo-stabilized pore water. Journal Applied Clay Science. 2010;**47**:330-342

[17] Li Q, Yue QY, Sun H, Su Y, Gao Y. Comparative study on the properties, mechanisms and process designs for the adsorption of non-ionic or anionic dyes onto cationic-polymer/bentonite. Journal of Environmental Management. 2010;**91**:1601-1611

[18] Utracki L. Clay-containing polymeric nanocomposites. Shrewsbury: Rapra Technology Ltd; Polymer International. May 2005;**54**:860

[19] Wei J, Zhu R, Zhu J. Simultaneous sorption of crystal violet and 2-naphthol to bentonite with different CECs. Journal of Hazardous Materials. 2013;**166**:195-199

[20] Nguyen QT, Baird DG. An improved technique for exfoliating and dispersing nanoclay particles into polymer matrices using supercritical carbon dioxide. Polymer. 2007;**48**:6923-6933

[21] Pillai SK, Ray S. Chitosan-based nanocomposites. Journal of Natural Polymers. 2012:33-68

[22] DeLeo C, Augusto C, do Carmo Goncalves PM, Velankar S. Preparation and characterization of clay nanocomposites of plasticized starch and polypropylene polymer blends. Journal of Polymers and the Environment. 2011;**19**:689-697

[23] Charhouf I. Thesis titled Synthesis and characterization of hydrogel of Chitosan for medical applications. Hassan II University of Casablanca, Faculty of Science Ben M'Sik; 10 January 2015

[24] Lavertu M, Xia Z, Serreqi AN, Berrada M, Rodrigues A, Wang D, Buschmann MD, Ajay Gupta. A validated 1H NMR method for the determination of the degree of deacetylation of chitosan. Journal of Pharmaceutical and Biomedical Analysis. 2003;**32**:1149-1158

[25] Sedighi H, Irannajad M, Gharabaghi M, Amirkabir. Silica impurities removal on bentonite sample for nanoclay production. Journal of Science & Research (Civil & Environmental Engineering) (AJSR – CEE), Summer. 2013;**45**(1):11-13

[26] Charhouf I, Bennamara A, Abourriche A, Berrada M. Characterization of chitosan and fabrication of chitosan hydrogels matrices for biomedical applications. MATEC Web of Conferences 5. 2013;**04030**. DOI: 10.1051/matecconf/20130504030

[27] Standard Test Method for Tensile Properties of Thin Plastic Sheeting, Annual Book of ASTM Standards. Current edition approved April 10, 2002

[28] ASTM D882 published on June 2002, originally published as D 882-46 T. and Last previous edition is D 882-01

[29] Cheng LH, Karim AA, Seow CC. Effects of water-glycerol and water-sorbitol interactions on the physical properties of konjac glucomannan films. Journal of Food Science. March 2006;**71**(2). DOI: 10.1111/j.1365-2621.2006.tb08898.x

[30] Cuq B, Gontard N, Cuq J, Guilbert S. Selected functional properties of fish myofibrillar protein-based films as affected by hydrophilic plasticizers. Journal of Agricultural and Food Chemistry. 1997;**45**(3):622-626

[31] Karbowiak T, Hervet H, Léger L, Champion D, Debeaufort F, Voiley A. Effect of plasticizers (water and glycerol) on the diffusion of a small molecule in iota-carrageenan biopolymer films for edible coating application. Biomacromolecules. 2006;**7**(6):2011-2019

[32] Kristo E, Biliaderis CG. Water sorption and thermo-mechanical properties of water/sorbitol-plasticized composite biopolymer films: caseinate-pullulan bilayers and blends. Food Hydrocolloids. 2006;**20**(7):1057-1071

[33] Charhouf I, Bennamara A, Abourriche A, Chenite A, Zhu J, Berradae M. Characterization of a dialdehyde chitosan generated by periodate oxidation. International Journal of Sciences: Basic and Applied Research (IJSBAR). 2014;**16**(2):336-348

[34] Bilia AR, Guccione C, Isacchi B, Righeschi C, Firenzuoli F, Bergonzi MC. Essential oils loaded in nanosystems: a developing strategy for a successful therapeutic approach. Evidence-based Complementary and Alternative Medicine. 2014;**2014**:651593. DOI: 10.1155/2014/651593

[35] Wang L, Wang A. Adsorption characteristics of congo red onto the chitosan/montmorillonite nanocomposite. Journal of Hazardous Materials. 2007;**147**:979-985

[36] Paluszkiewicz C, Stodolak E, Hasik M, Blazewicz M. FT-IR study of montmorillonite–chitosan nanocomposite materials. Spectrochimica Acta Part A: Molecular and Biomolecular Spectroscopy. 2004;**79**:784-788

[37] Tan W, Zhang Y, Szeto Y, Liao L. A novel method to prepare chitosan/montmorillonite nanocomposites in the presence of hydroxy-aluminum oligomeric cations. Journal of Computer Science and Technology. 2008;**68**:2917-2921

[38] Darder M, Colilla M, Ruiz-Hitzky E. Chitosan-clay nanocomposites: application as. 500 electrochemical sensors. Journal Appl Clay Science. 2005;**28**:199-208

[39] Han Y, Lee SH, Choi KH, Park I. Preparation and characterization of chitosan–clay nanocomposites with antimicrobial activity. Journal of Physics and Chemistry of Solids. 2010;**71**: 464-467

[40] Laaraibi A, Charhouf I, Bennamara A, Abourriche A, Berrada M. Berrada. Valorization of marine wastes in a preserving film based on chitosan for food applications. Journal of Materials and Environmental Science. 2015;**6**(12):3511-3516

Permissions

All chapters in this book were first published in CCMFST, by InTech Open; hereby published with permission under the Creative Commons Attribution License or equivalent. Every chapter published in this book has been scrutinized by our experts. Their significance has been extensively debated. The topics covered herein carry significant findings which will fuel the growth of the discipline. They may even be implemented as practical applications or may be referred to as a beginning point for another development.

The contributors of this book come from diverse backgrounds, making this book a truly international effort. This book will bring forth new frontiers with its revolutionizing research information and detailed analysis of the nascent developments around the world.

We would like to thank all the contributing authors for lending their expertise to make the book truly unique. They have played a crucial role in the development of this book. Without their invaluable contributions this book wouldn't have been possible. They have made vital efforts to compile up to date information on the varied aspects of this subject to make this book a valuable addition to the collection of many professionals and students.

This book was conceptualized with the vision of imparting up-to-date information and advanced data in this field. To ensure the same, a matchless editorial board was set up. Every individual on the board went through rigorous rounds of assessment to prove their worth. After which they invested a large part of their time researching and compiling the most relevant data for our readers.

The editorial board has been involved in producing this book since its inception. They have spent rigorous hours researching and exploring the diverse topics which have resulted in the successful publishing of this book. They have passed on their knowledge of decades through this book. To expedite this challenging task, the publisher supported the team at every step. A small team of assistant editors was also appointed to further simplify the editing procedure and attain best results for the readers.

Apart from the editorial board, the designing team has also invested a significant amount of their time in understanding the subject and creating the most relevant covers. They scrutinized every image to scout for the most suitable representation of the subject and create an appropriate cover for the book.

The publishing team has been an ardent support to the editorial, designing and production team. Their endless efforts to recruit the best for this project, has resulted in the accomplishment of this book. They are a veteran in the field of academics and their pool of knowledge is as vast as their experience in printing. Their expertise and guidance has proved useful at every step. Their uncompromising quality standards have made this book an exceptional effort. Their encouragement from time to time has been an inspiration for everyone.

The publisher and the editorial board hope that this book will prove to be a valuable piece of knowledge for researchers, students, practitioners and scholars across the globe.

List of Contributors

Carlos Escudero-Oñate and Elena Martínez-Francés
Norwegian Institute for Water Research (NIVA), Oslo, Norway

Xue Luo and Li Li
Institute of Textiles and Clothing, China
The Hong Kong Polytechnic University, China

Thomas Hahn and Susanne Zibek
Fraunhofer-Institute of Interfacial Engineering and Biotechnology, Stuttgart, Germany

Rajendra Sukhadeorao Dongre
Department of Chemistry, RTM Nagpur University, Nagpur, India

Claudia A. Vega-Cázarez, Dalia I. Sánchez-Machado and Jaime López-Cervantes
Instituto Tecnológico de Sonora, Ciudad Obregón, Sonora, México

Porfirio Gutierrez-Martinez, Aide Ledezma-Morales, Luz del Carmen Romero-Islas, Anelsy Ramos-Guerrero, Jovita Romero-Islas, Carolina Rodríguez-Pereida, Paloma Casas-Junco, Leonardo Coronado-Partida and Ramsés González-Estrada
Tecnológico Nacional de México, Mexico City, Mexico
Laboratorio Integral de Investigación en Alimentos, Biochemical Engineering Department, Instituto Tecnológico de Tepic, Tepic, Mexico

Daniel Hernandez-Patlan and Bruno Solis-Cruz
Laboratorio 5: LEDEFAR, Unidad de Investigacion Multidisciplinaria, Mexico
Facultad de Estudios Superiores (FES) Cuautitlan, Mexico
Universidad Nacional Autonoma de Mexico (UNAM), Mexico
Cuautitlan Izcalli, Estado de Mexico, Mexico

Billy M. Hargis and Guillermo Tellez
Department of Poultry Science, University of Arkansas, Fayetteville, Arkansas, USA

Marilyn Porras-Gómez, Jose Vega-Baudrit and Santiago Núñez-Corrales
National Nanotechnology Laboratory, National Center for Advanced Technology Studies, San José, Costa Rica

Suha M. Dadou, Milan D. Antonijevic, and Babur Z. Chowdhry
Department of Pharmaceutical, Chemical and Environmental Science, UK
Faculty of Engineering and Science, University of Greenwich, Kent, UK

Adnan A. Badwan
Research and Innovation Centre, The Jordanian Pharmaceutical Manufacturing Co. (Plc), Naor, Jordan

Boukhlifi Fatima
Equip "Materials and Applied Catalysis", Faculty of Science, University Moulay Ismail, Meknes, Morocco

Mamouni Fatima Zahrae
Equip "Materials and Applied Catalysis", Faculty of Science, University Moulay Ismail, Meknes, Morocco
Equip "Science of Water and Environmental Engineering", Faculty of Science, University Moulay Ismail, Meknes, Morocco

R. Razouk
Equip Research Agronomy and Plant Physiology, Regional Center for Agricultural Research, Meknes, Morocco

Asha H. Gedam
Department of Chemistry, Cummins College of Engineering for Women, Nagpur, India

Prashil K. Narnaware
Visvesvaraya National Institute of Technology, Nagpur, India

Vrushali Kinhikar
G H Raisoni Institute of Engineering and Technology, Nagpur, India

Asmae Laaraibi, Fatiha Moughaoui, Fouad Damiri, Amine Ouakit, Imane Charhouf, Souad Hamdouch, Abdelmjid Abourriche, Noureddine Knouzi, Ahmed Bennamara and Mohammed Berrada
Department of Chemistry, Laboratory of Biomolecules and Organic Synthesis (BIOSYNTHO), Faculty of Sciences Ben M'Sik, University Hassan II of Casablanca, Morocco

Abdelhafid Jaafari
Department of Chemistry, Laboratory of Applied Chemistry and environment, Faculty of Sciences and Technologies, University Hassan I of Settat, Morocco

Index

www.ingramcontent.com/pod-product-compliance
Lightning Source LLC
LaVergne TN
LVHW060557210926
848485LV00005B/1326
* 9 7 8 1 6 3 9 8 9 2 5 3 2 *